UNIX Unbounded

A Beginning Approach

FOURTH EDITION

Amir Afzal, D. Sc.

Upper Saddle River, New Jersey
Columbus, Ohio

Editor in Chief: Stephen Helba
Assistant Vice President and Publisher: Charles E. Stewart, Jr.
Production Editor: Alexandrina Benedicto Wolf
Production Coordination: Custom Editorial Productions Inc.
Design Coordinator: Diane Ernsberger
Cover Designer: Jeff Vanik
Cover Art: Digital Vision
Production Manager: Matthew Ottenweller
Marketing Manager: Adam Kloza

This book was set in Times Roman by Custom Editorial Productions Inc. It was printed and bound by The Banta Company. The cover was printed by Phoenix Color Corp.

Notice to the Reader: All product names listed herein are trademarks and/or registered trademarks of their respective manufacturer.
The publisher and the author do not warrant or guarantee any of the products and/or equipment described herein, nor has the publisher or the author made any independent analysis in connection with any of the products, equipment, or information used herein. The reader is directed to the manufacturer for any warranty or guarantee for any claim, loss, damages, costs, or expense arising out of or incurred by the reader in connection with the use or operation of the products or equipment.
The reader is expressly advised to adopt all safety precautions that might be indicated by the activities and experiments described herein. The reader assumes all risks in connection with such instructions.

Pearson Education Ltd.
Pearson Education Australia Pty. Limited
Pearson Education Singapore, Pte. Ltd.
Pearson Education North Asia Ltd.
Pearson Education Canada, Ltd.
Pearson Educación de Mexico, S.A. de C.V.
Pearson Education—Japan
Pearson Education Malaysia Pte. Ltd.
Pearson Education, *Upper Saddle River, New Jersey*

10 9 8 7 6 5 4 3 2 1
ISBN 0-13-092736-8

To my wife, Emma

Preface

The price break on the UNIX operating system for microcomputers and recent hardware advances have boosted the acceptance and popularity of UNIX for microcomputers. Consequently, there are students and novice UNIX users with computer skills but no experience with any operating system.

This book is for this group of users and students. It is neither an operating system book per se nor a UNIX reference book. It is a textbook written in a tutorial manner, intended as a teaching/learning tool in a classroom/lab environment. It is a book for introductory operating system courses. It discusses operating systems concepts in general, leading into a presentation of UNIX and the UNIX environment. It covers the topics necessary for the UNIX user to function independently and do most of the everyday, routine jobs. It also gives readers a good knowledge base so they can move on to more advanced courses or books.

I wrote this book relying mostly on my experience as a UNIX teacher. The organization of the chapters is what I follow, and the examples are what I usually use in my UNIX classes. This book is an introductory book, but not a simple book. I did not try to change the technical aspect of the material through storytelling, nor did I use irrelevant stories to make the material lighter or more interesting.

The chapters are short, and in cases where a topic requires more discussion, the material is presented in two chapters. The format of the chapters is kept the same as much as possible. However, consistency is sacrificed when the format is not appropriate for presenting the material.

Each chapter starts with a general explanation of concepts and topics. Simple concrete examples clarify the explanations or show how to use the commands and are followed by more detailed and complex commands and examples as the chapter progresses. Each chapter ends with questions, a review section, and, when appropriate or necessary, practical exercises for using a terminal are added.

Chapter 1 briefly describes the fundamentals of computer hardware and software and explains basic computer terms and concepts. It discusses the types of software and moves the emphasis to the system software. It explains the importance of the operating system and explores its primary functions.

Chapter 2 presents a brief history of the UNIX operating system. It explores the historical development of UNIX, discusses the major UNIX versions, and explains some of the system's important features.

Chapter 3 explains how to start and end a UNIX session. Simple UNIX commands are introduced, and their applications are explained. The process of establishing contact with UNIX is explored, and some internal UNIX operations are discussed.

Chapters 4 and 6 cover the UNIX operating system vi editor. After a brief discussion of the editors that are supported by UNIX, Chapter 4 introduces the vi editor, and presents the basic commands necessary for a simple editing job. Chapter 6 shows more of the vi editing power and flexibility by covering the more advanced vi commands and explains various ways the vi editor can be customized.

Chapter 5 is the first of two chapters that discuss the file structure of the UNIX system. It covers the basic concepts of files and directories and their arrangement in a hierarchical tree structure. It presents commands that facilitate the manipulation of the file system. Chapter 7 is the second chapter about the UNIX file system and its associated commands. It presents more file manipulation commands, explains the shell input/output redirection operators, and introduces file substitution metacharacters.

Chapter 8 covers the shell and its role in the UNIX system. It explains the shell features and capabilities, the shell variables, and the shell metacharacters. Startup files and process management under UNIX are also covered.

Chapter 9 concentrates on the UNIX communication utilities. It explains the UNIX e-mail facilities and shows the commands and options available. It discusses the shell and other variables that affect the e-mail environment. It shows how to make a startup file that customizes use of the e-mail utilities.

Chapter 10 discusses the essentials of program development. It explains the steps in the process of creating a program. It gives an example of a simple C program and walks through the process of writing the source code and creating an executable program.

Chapter 11 concentrates on shell programming. It explains the capabilities of the shell as an interpretive high-level language. It covers shell programming constructs and particulars. It shows the creation, debugging, and running of shell programs.

Chapter 12 builds on the commands and concepts of the previous chapter and covers more of the shell programming commands and techniques. It also presents a simple application program and shows the process of developing programs using the shell language.

Chapter 13 presents a few additional important UNIX commands. Disk commands, file manipulation commands, and security are the major topics of this chapter.

New to This Edition

I have received suggestions from my colleagues and from professors who have adopted this textbook to add other topics, UNIX capabilities, and features in this edition. On the other hand, I also have received suggestions to keep this textbook as is. You realize my dilemma. However, I made the decision to keep the book's structure and level as in previous editions and not to create a thousand-page cover-it-all reference book.

This fourth edition of *UNIX Unbounded: A Beginning Approach* includes commands from the Linux operating system and its Bourne Again shell *(bash)*. Some inconsistencies have been removed. All known typos and errors have been corrected. All programs have been tested using different shells. The end-of-the-chapter exercises and terminal sessions have been reviewed and new questions and exercises have been added.

Resources

An instructor's manual (ISBN 0-13-092737-6) is available for this book. It includes lecture notes on each chapter, answers to review questions at the end of each chapter, and test questions for assistance in preparing examinations on the material in this book. The following teaching resources are also available: Blackboard, ISBN 0-13-049828-9; Course Compass, ISBN 0-13-049829-7; Test Manager, ISBN 0-13-049826-2; and PowerPoint masters of figures in the text, ISBN 0-13-049820-3.

Acknowledgments

This fourth edition of *UNIX Unbounded* would not have been possible without the help of my colleagues in academia and industry. I am grateful to all of them.

Thanks to students in my C/C++ and UNIX classes for their suggestions and feedback.

Thanks to Dean Farzan Soroushi at Strayer University for testing the new material in his classes.

Thanks to Tammy Dingler, Georgia Southwestern State University, and Lawrence Osborne, Lamar University, TX, for their many corrections and suggestions.

Thanks to my colleagues at the General Dynamics–Chantilly Office, particularly Jim Schmid, for accommodating my teaching/writing schedule.

Thanks to Tom Swanson, my co-author of the upcoming book *UNIX Administration Unbounded,* for being the guru that he is, and being so generous with his time.

Thanks to Charles Stewart at Prentice Hall for his patience and continued support of my writing projects.

How to Read This Book

If this is the first time you are learning about the UNIX operating system, I suggest that you start with the first chapter and continue working through the chapters in the sequence in which they are presented. If you already know some aspects of the UNIX operating system, I suggest that you browse through topics you know and review the main points to help you understand the other chapters. Most of the chapters are interrelated in the sense that your skills from the previous chapter help you—and sometimes are necessary for you—to go to the next one.

A number of concrete examples clarify concepts or show different ways you can use a command. I encourage you to try them on your system. UNIX comes in many dialects and is also easily modifiable. This means you may find discrepancies between the manual and your system, and some of the screen displays or command sequences in this book may not exactly match those on your system.

Typographical Notes

Throughout this book, certain words are emphasized by using different typefaces. In the running text, **bold** words are UNIX commands or specific characters that you type on the keyboard as part of an example; `sans serif` words are directory names, pathnames, or filenames; and *italic* words are keywords or terms being introduced for the first time.

The following shows an example of a terminal screen. This is what you expect to see on your system when you practice the commands:

```
UNIX System V release 4.0
login: david
password:
```

The following is an example of the command sequences. Characters you type on the keyboard are indicated in **bold** type. The information on the right is a commentary on the action being performed on the left. This format is used when a line-by-line explanation of the commands or outputs is necessary.

$ **pwd [Return]** . . . . Check your current directory.
`/usr/david` . . . . . . You are in `david`.
$ **cd source [Return]** . . . . . Change to `source` directory.

Icons

Icons are used throughout the text to draw your attention, list some features, or present action to be taken. Four icons are used throughout the text.

Note

- Lists the important points
- Draws your attention to a particular aspect of a command or a screen display

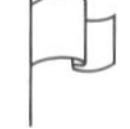

Flag

- Draws your attention (flags you) to common user mistakes
- Warns you of the consequences of your action

Computer

- Shows how the commands work on the system
- Lets you try the commands on your system

Box

- Shows a sequence of keys that you must press to perform a specified task

Keyboard Conventions

[**Return**]: This represents the Return key, sometimes called CR (for carriage return) or the Enter key. You usually press this key at the end of your command or input line.

[**Ctrl-d**]: This means you should simultaneously hold down the key labeled Ctrl (for Control) and press letter d key. Other control characters that consist of the Ctrl key and a letter are shown similarly.

Contents

8 Exploring the Shell, 215

CHAPTER 1

First Things First

This chapter briefly describes the fundamentals of computer hardware and software and explains basic computer terms and concepts. It discusses the types of software, explains the importance of the operating system, and explores its primary functions.

In This Chapter

1.1 INTRODUCTION

Most people acquire fundamental knowledge about computers by taking introductory computer courses or by using computers in a work or home environment. In case you have not had any computer courses (or you have forgotten what you have learned), this chapter starts with a brief introduction to computers in general and explains some common hardware and software terms. The chapter goes on to discuss operating system software.

1.2 COMPUTERS: AN OVERVIEW

What is a computer? *Merriam Webster's Collegiate Dictionary* defines *computer* as "a programmable electronic device that can store, retrieve and process data." This chapter expands on that definition and explores each component of a computer system.

Computers can be grouped according to their sizes, capabilities, and speed into four classifications as follows:

- Supercomputers
- Mainframe computers
- Minicomputers
- Microcomputers

These are rather arbitrary classifications: the low-end systems of one category can overlap the high-end systems of the other.

Supercomputers *Supercomputers* are the fastest and the most expensive computers, 1,000 times faster than mainframe computers. Supercomputers are designed for the most demanding of computational applications, such as weather forecasting, three-dimensional modeling, and computer animation. All these tasks require an extremely large number of complex calculations and need supercomputer performance. Supercomputers usually have hundreds of processors and are used with the very latest and most expensive hardware devices.

Supercomputers are used for numerous other applications. Even Hollywood uses the advanced graphics capabilities of supercomputers to create special effects for movies.

Mainframe Computers *Mainframe computers* are large, fast systems designed to meet the information processing needs of large organizations. Mainframes can support several hundred users and execute hundreds of programs simultaneously. They have extensive *input/output* (I/O) capabilities and support a large amount of primary and secondary storage. Mainframes are used mostly in large business environments, such as banks and hospitals, and in other large institutions, such as universities. They are costly and usually require a trained support staff for operation and maintenance.

I/O devices are the means by which a computer communicates with the external world (humans or other computers). They vary in speed and communication medium.

See page x for an explanation of the icons used to highlight information in this chapter.

Minicomputers Until the late 1960s, all computers were mainframes, and only large organizations could afford them. Then *minicomputers* were developed. Their original function involved performing specialized tasks, and they were used primarily in universities and scientific environments. Minicomputers quickly became popular among small and medium-sized organizations with data processing needs. Some of today's "minis" rival mainframes in power and capability, and most are general-purpose computers. Like mainframes, minicomputers are capable of providing information processing services for multiple users and can execute many application programs concurrently. However, they are cheaper than mainframes and are easier to install and maintain.

Microcomputers Also called *personal computers* or *PCs*, *microcomputers* are the least costly and most popular computers on the market. They are small enough to fit on top of a desk or in a briefcase. *Microcomputers* vary widely in cost and power, with some models rivaling the minicomputers and older mainframes. They are capable of running many business applications and can function as stand-alone units or can be hooked up with other computers to extend their capabilities.

Table 1.1 shows each class of computers, their typical specifications (memory size, number of users, etc.), and their approximate speeds.

Table 1.1
Computer Classifications

Class	Typical Specifications	Approximate Speed
Microcomputer	64+ million main memory cells 4 billion disk storage cells single user	10+ million instructions per second
Minicomputer	128+ million main memory cells 10 billion disk storage cells 1 tape drive 128 interactive users	30+ million instructions per second
Mainframe	1+ billion main memory cells 100 billion disk storage cells multiple tape drives 100s interactive users 4+ central processing units or more	50+ million instructions per second
Supercomputer	1+ billion main memory cells 100 billion disk storage cells 64+ central processing units or more	2+ billion floating-point operations per second

1.3 COMPUTER HARDWARE

Regardless of how complex or simple a computer system is, it has four fundamental functions: *input, processing, output,* and *storage.* Also, computers consist of two distinct parts: *hardware* and *software.* These two parts complement each other, and the

Figure 1.1
Four Functional Parts of a Computer System

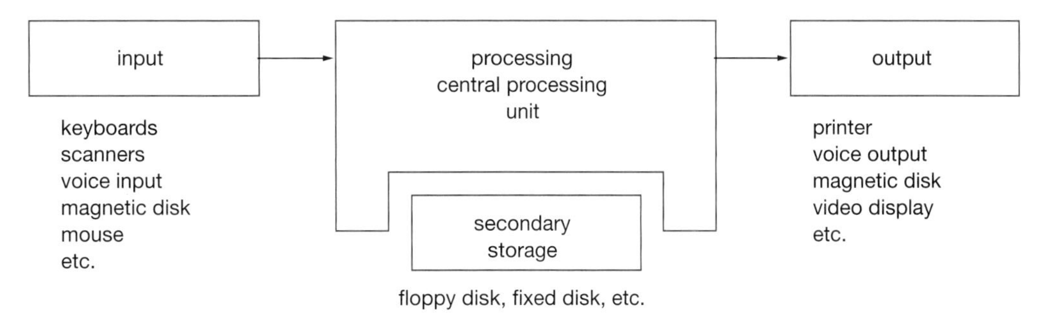

integration of hardware and software enables computers to perform their fundamental functions.

Figure 1.1 shows the four functions of a computer system and the typical hardware devices associated with each function.

Most computer systems have five basic hardware components that work together to accomplish the computer's required tasks. The number, implementation, complexity, and power of these components varies among computer systems. However, the functions performed by each component are generally quite similar. These components are as follows:

- Input devices
- Processor unit
- Internal memory
- External storage unit
- Output devices

The way these components are put together and arranged is called the *system hardware configuration.* For example, in one system the processor unit and the external storage unit may be housed in a single component; in another system, they may be separate components.

1.3.1 Input Devices

Input devices are used to enter instructions or data into the computer. Numerous input devices are available, and their number is growing. The keyboard, light pen, scanner, and mouse are the most popular input devices, with the keyboard used almost universally.

Certain devices can be used for either input or output—magnetic disks and touch screen terminals, for example.

1.3.2 Processor Unit

The *processor unit* is the intelligent part of the computer system. It is called the *central processing unit,* or *CPU,* and it directs computer activities. The CPU controls the performance of tasks such as sending keyboard data to the main memory, manipulating stored data, or sending the results of an operation to the printer. The CPU consists of three basic sections:

- Arithmetic and Logic Unit (ALU)
- Registers
- Control Unit (CU)

The CPU is also called the brain, heart, or thinking part of the computer.

Arithmetic and Logic Unit The *arithmetic and logic unit,* or *ALU,* is a section of the CPU's electronic circuitry that controls all arithmetic and logic operations. *Arithmetic operations* include addition, subtraction, multiplication, and division. More advanced computational functions such as exponentiation and logarithms also can be implemented. The *logic operations* include comparisons of letters, numbers, or special characters to recognize whether they are equal to, greater than, or less than one another.

To summarize, the ALU is responsible for the following:

- Performing arithmetic operations
- Performing logical operations

Registers A small set of temporary storage cells generally is located within the CPU. These storage cells are called *registers.* A register can typically hold a single instruction or data item. Registers are used to store data and instructions that are needed immediately, quickly, and frequently. For example, if two numbers are about to be added, each is kept in a register. The ALU reads these numbers and stores the result of the addition in another register. Because registers are located within the CPU, their contents can be accessed quickly by other CPU components.

To summarize, registers are used for the following:

- Storing instructions and data within the CPU

Control Unit The *control unit* is a section of the CPU's electronic circuitry that directs and coordinates the other parts of the system to carry out program instructions. It does not execute the directions itself; instead it sends electronic signals along the appropriate circuitry to activate other parts. It is primarily responsible for the movement of data and instructions from the main memory and for controlling the ALU. As program instructions and data are needed by the CPU, they are moved from primary storage to registers by the control unit.

To summarize, the control unit is responsible for the following:

- Activating the other components
- Transferring instructions and data from the main memory to the registers

1.3.3 Internal Memory

A computer is a two-state machine. These two states can be interpreted as 0 or 1, yes or no, up or down, and so on. Almost any device that can store either of two states can serve as storage. However, most computers use integrated circuit memory, which holds binary digits or bits. Each *bit* can be either 0 or 1, therefore representing one state or the other. A small computer might have enough memory to store millions of bits; a larger one might store billions. The difference is one of degree, not of memory functionality.

The internal memory (also called main or primary memory) is storage for the following:

- Holding current program instructions
- Holding data to be processed by a program
- Holding intermediate results created by executing program instructions

Main memory is short term and retains data only for the period that a program is running.

The main memory storage consists of storage cells that hold programs that are currently being executed and their associated data. The execution of programs requires a great deal of data movement between the main memory and the CPU.

The CPU is a fast device; thus it is desirable to implement main memory using devices capable of fast access. In the current computer hardware, main memory is implemented with silicon-based semiconductor devices called *memory chips*. Computers usually have two types of main memory:

- Random access memory (RAM)
- Read only memory (ROM)

Random Access Memory *Random access memory* (RAM) is the working memory of the computer. It provides the access speed the CPU requires and allows the CPU to read and write a specific memory location by referring to its address. While the computer is on, programs and data are stored temporarily in RAM. Data stored in RAM can be changed, modified, and erased.

RAM is a volatile memory and does not provide permanent storage. When you turn off the computer (or turn off power to RAM by any other means), its contents are lost.

Read Only Memory The second type of memory, *read only memory* (ROM), contains the permanently stored programs and data that manufacturers place in the system. The CPU can only read instructions from ROM; it cannot alter, erase, or write over them. When you turn the computer off, none of the information stored in ROM is lost. Programs in ROM are sometimes called *firmware,* something between hardware and software.

Data Representation

We are accustomed to using the decimal numbering system. The decimal system (base 10) consists of 10 digits that range from 0 to 9. On the other hand, computers work with a binary system (base 2) that consists of 2 digits, 0 and 1.

Bit (Binary Digit) Each bit can hold either a 0 or a 1. A *bit* is the smallest unit of information a computer can understand.

Byte Computer memory must be able to store letters, numbers, and symbols, and a single bit by itself is not of much use for this. Bits are combined to represent some meaningful data. A group of eight bits is called a *byte* (pronounced *bite*), and it can represent a character. A character could be an uppercase or lowercase letter, a number, a punctuation mark, or a special symbol.

ASCII When you input data to a computer, the system must change it from what you recognize (letters, numbers, and symbols) into some format that the computer understands.

The American Standard Code for Information Interchange (ASCII, pronounced *ask-ee*) is one of the coding schemes used to represent characters in 8-bit bytes.

The ASCII code can represent a maximum of 256 characters, including all uppercase and lowercase letters, numbers, punctuation marks, and special symbols.

Word Bytes are fine for storing characters but are too small to hold large numbers. In fact, 256 is the largest number that can be stored in one byte. Certainly, to be of any use, computers need space for large numbers. Most computers are able to manipulate a group of bytes called a *word.* The *word* size is system dependent and could vary from 16-bit (2 bytes) to 32-bit (4 bytes) or even 64-bit (8 bytes).

The Memory Hierarchy

As already stated, the basic unit of storage is a bit. Bits are grouped together to form a byte; in turn, bytes are grouped together to form a word. Figure 1.2 depicts the memory hierarchy and the numerical value that each storage size can store.

Memory Size The letter K is used to express the size of the main memory, or disk space. K stands for kilo, which means *thousand* in the metric system. However, when measuring computer memory, K stands for *kilobytes,* which represent 1,024 bytes of storage (2 to the power of 10). For example, 32K of memory means 32,768 bytes (32 times 1024). There are other measurements referring to the size of the computer memory:

- *Megabyte* (MB) is approximately one million bytes
- *Gigabyte* (GB) is approximately one billion bytes

Addressing Memory

You can think of *main memory* (primary storage) as a sequence of continuous or adjacent memory cells. Each physical storage unit (cell) is assigned a unique address corresponding to its location in the sequence. In byte-addressable computers, each byte in the memory has its own *electronic address,* a code name identifying its exact location in the memory.

On most computers, the bytes or word addresses are assigned sequentially, 0, 1, and so on. The CPU uses these addresses to specify which instructions or data are to be loaded (read) into the CPU and which data are to be written. For example, if you input the letter H from the keyboard, the character is read into main memory and stored in a

Figure 1.2
The Memory Hierarchy

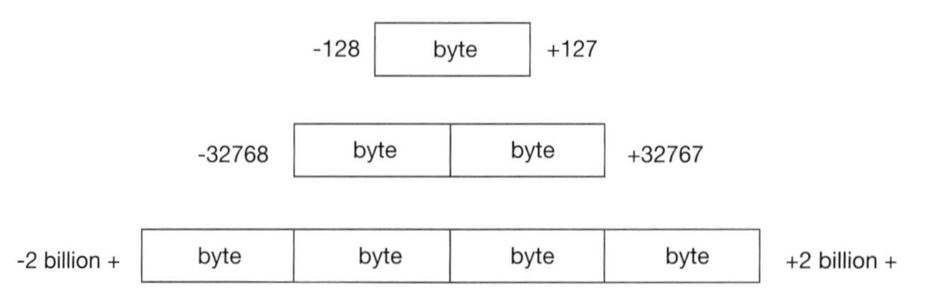

specific address, say address 1000. Address 1000 references a single byte of memory, and at any time the system can reference and process the data stored at this address.

The CPU requires instructions and data from many parts of the main memory; thus, it is necessary for the main memory to be what is referred to as a *direct access,* or a *random access device.* It must have the ability to specify a particular location to be read from or written to.

1.3.4 External Storage

External or secondary storage is a *nonvolatile* extension of the main memory. Main memory is expensive, and on most computers it is a scarce resource. Main memory is *volatile;* it loses its contents when the power is turned off. Hence, it makes sense to save your programs and data on another medium. Secondary storage media could be a floppy disk (diskette), hard disk, and/or magnetic tape.

1. *Secondary storage is an extension of main memory, not a replacement for it. A computer cannot execute a program or manipulate data stored on disk unless the data are first copied into main memory.*
2. *Main memory holds the current programs and data, whereas secondary storage is for long-term storage.*

Table 1.2 summarizes a computer's various types of storage devices and their usual contents.

Table 1.2
Summary of the Different Storage Types

Storage Type	Location	Usage
Registers	Within the CPU Very high-speed devices	Currently executing instructions; instructions; part of the related data
Primary Storage	Outside of the CPU High-speed devices (RAM)	Entire programs or part of the programs currently being executed; part of the associated data
Secondary Storage	Low-speed devices Electromagnetic or optical	Programs not currently being executed; large amount of data

1.3.5 Output Devices

A basic output device is the *display terminal.* The terms *CRT* (cathode ray tube), *VDT* (video display terminal), and *monitor* refer to the TV-style device that displays character images. The image displayed on a screen is temporary and is called *soft copy*. By routing the output to a printer, a permanent copy, called a *hard copy,* is obtained. Of course, other output devices are available, including a voice response device or a plotter to generate hard copy graphic outputs.

1.4 PROCESS OPERATION

A complex chain of events occurs when a computer executes a program. To begin, the program and its associated data are loaded into the main memory. The *control unit* (CU) reads the first instruction and its data inputs from the main memory into the CPU. If the instruction is a computational (or comparison) instruction, the control unit signals the ALU which function to perform, where the input data are located, and where to store the output data. Some instructions, such as input and output to secondary storage or I/O devices, are executed by the control unit itself. After the first instruction is executed, the next instruction is read and executed. This process continues until the last instruction of the program is read from the main memory into the registers in the CPU and is executed.

Each instruction is known to the ALU by a unique number called the instruction code or the operation code ("op code").

The steps required to process each instruction can be grouped into two phases: the *instruction cycle* and the *execution cycle.* The operation of the instruction and execution cycles is depicted in Figure 1.3, and the explanation for the sequence of events is as follows:

Instruction Cycle The sequence of events during the *instruction cycle,* also referred to as the *fetch cycle,* is as follows:

Step 1: The control unit reads an instruction from main memory into a CPU register called the *instruction register.*

Step 2: The control unit increments the *instruction pointer register* to show the location of the next instruction in the main memory.

Step 3: The control unit generates a signal to the ALU to execute the instruction.

Figure 1.3
Sequence of the Processor Operation

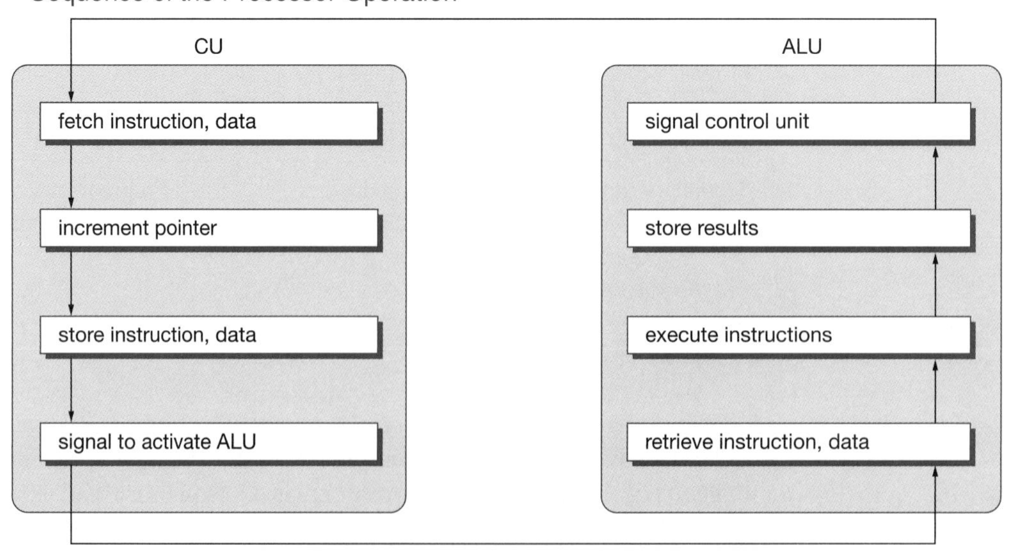

Execution Cycle The sequence of events during the *execution cycle* is as follows:

Step 1: The ALU accesses the operation code of the instruction in the instruction register to determine which function to perform and to obtain the input data for the instruction.
Step 2: The ALU executes the instruction.
Step 3: The results of the instruction are stored in registers or are returned to the control unit to be written to memory.

The instruction cycle *is performed within the control unit; the* execution cycle *is performed within the arithmetic and logic unit.*

1.4.1 Performance Measurement

Most computers are designed to be general-purpose computers; they support a wide variety of application programs. It is not possible for computer manufacturers or vendors to know the workload of your computer and to provide you with performance measurements accordingly. Instead, they provide performance measurements for various machine actions, such as executing an instruction, accessing main memory, or reading from a magnetic disk. Performance measurement generally is given as specific measures of each component of the computer, a measure of the communication capacity among the components, and a general measurement that attempts to summarize all of the performance measurements.

CPU Speed Within the CPU, the primary performance consideration is the speed at which an instruction can be executed. This is usually specified in terms of *millions of instructions per second* (MIPS). Unfortunately, not all instructions take the same amount of time to execute. Certain instructions, such as addition of whole numbers, execute quickly; others, such as division of fractional (floating-point) numbers, might execute more slowly. The measurement that is specially oriented to computation of fractional numbers is *millions of floating-point operations per second* (MFLOPS).

The MFLOPS performance measure is commonly given for workstations and supercomputers. It is expected that the applications running on these machines perform mostly floating-point computations.

Access Time The speed at which the CPU can retrieve data from storage or input/output (I/O) devices depends on the *access time.* Access time is normally measured in milliseconds (millionths of a second) or nanoseconds (billionths of a second).

Channel Capacity When access time is stated for a specific device, there is no guarantee that the communication channel between the device and the CPU is capable of supporting that speed. The ability of the communication channel to support data movement between the CPU and a device, or vice versa, is normally specified as a *data transfer rate.* This rate specifies the amount of data that can be moved over the channel in a specified interval of time—for example, one million data items per second.

Overall Performance The overall performance of a computer system is a combination of the CPU speed, the access time of the storage and I/O devices, and the capacity of the communication channels connecting them to the CPU. Different applications create different demands on the various computer components. Thus, overall performance measurement is valid only for a particular instance or class of applications.

1.5 WHAT IS SOFTWARE?

A computer off the assembly line with no software is just a machine (hardware) and is not capable of doing much. It is the software that gives a computer its diverse capabilities. Through software, a computer can take on any one of many personalities. With the right software, a computer can become a word processor, a calculator, a database manager, or a sophisticated communications device—or even all of them at the same time.

In general, computer programs are called software. *You can run numerous programs on the same machine* (hardware) *to perform different tasks.*

Program A *program* is a set of instructions that directs the activities of a computer system. It consists of instructions that are logically sequenced to perform a specific operation. Programs are written in one of the available computer programming languages. Computer programming languages are necessary to facilitate the creation of the application programs. A wide range of program development tools are available to aid programmers in developing application and system programs.

Software Categories Computer software can be categorized into two types: *system software* and *application software.* The operating system is the most important system software and the only software that is absolutely necessary for the computer to function. However, it is the application software that makes computers useful in different environments and for so many jobs. Figure 1.4 shows the general categories of software and examples of their associated software programs.

Figure 1.4
Types of Software

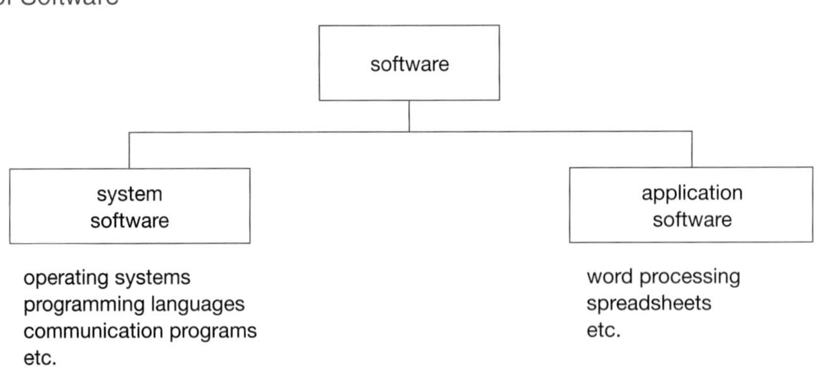

Users generally interact with application software and part of the system software. Figure 1.5 shows a layered approach to software. The advantage of this approach is that the user and application programmers do not need to comprehend and deal with the technical details of physical processing.With the layered approach, the machine's physical details and many basic processing tasks are embedded within the system software and hidden from the user.

1.5.1 System Software

The *system software* is a set of programs that mostly control the internal performance of a computer. The most important software in this category is the *operating system,* which

Figure 1.5
User Interaction with Software Layers

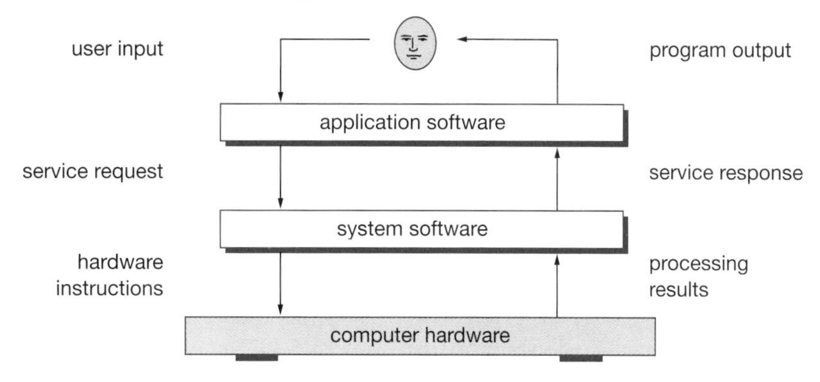

controls the basic functions of the computer and provides a platform for application programs. Other system software includes database management system (DBMS) and communication software, for example.

Who Is the Boss?

The operating system is the boss and is the most important system software component of a computer. It is a collection of programs that controls all hardware and software in a computer. The necessary parts of the operating system are loaded into the main memory when you turn the computer on and remain there until you turn it off. The operating system plays different roles as *service provider, hardware manager,* and facilitator of the *user interface.* Instead of looking for a universally accepted definition of the operating system, let's explore its roles and responsibilities. These are the primary purposes and functions of an operating system:

- To provide an interface for users and application programs to low-level hardware functions
- To allocate hardware resources to users and their application programs
- To load and accept the application programs on behalf of the users

The necessary parts of the operating system are always resident in the main memory.

Operating System As Resource Manager

The operating system is in control of computer resources: main memory, CPU time, and peripheral devices. In a typical computer system, several jobs are executing concurrently. All are competing for the available resources, and the operating system allocates resources to programs according to the availability of the resources and priority of the running programs. It allocates the CPU time. After all, there is only one CPU and many users, and only one user at a time can have the CPU's attention.

The operating system monitors and allocates the main memory among programs, which have different sizes and memory requirements, preventing your program from being mixed with someone else's program that happens to be in the main memory. It coordinates use of the peripheral devices, including things such as whose turn it is to read from the disk or write to it, and what job goes to the printer first.

The operating system continually responds to the program's resource requirements, resolves resource conflicts, and optimizes the allocation of resources.

Operating System As User Interface

The operating system provides the means for users to communicate with the computer. Every operating system provides a set of commands for directing the operation of the computer, a *command-driven user interface.* Command syntax is difficult to learn, remember, and use. One alternative is the *menu-driven user interface,* which provides a set of menus and lets users choose desired functions from the menus. Another popular alternative is the icon-driven, graphically based user interface, called a *graphic user interface* (GUI). You can execute most of the operating system commands by manipulating graphical images on the video screen. For example, the graphical image (*icon*) of a file folder may represent files, or a picture of a typewriter may represent a word processing program. This visual metaphor for commands and functions provides an easy-to-learn interface. With a GUI, users can choose an icon, usually with a pointing device such as a mouse, to activate programs.

Operating System Model

Using the software layers approach, you can view the operating system as a layered set of software. Figure 1.6 shows a model of an operating system and its program layers. Input from a user or an application program travels through the layers to reach the hardware, and results from the hardware travel back to the user through the same layers. Let's explore the general responsibilities of each layer.

Kernel Layer The *kernel* is the innermost layer of the operating system software. It is the only layer that interacts directly with the hardware. This provides a measure of machine independence within the operating system. In theory at least, an operating system can be altered to interact with a different set of computer hardware by making changes to the kernel only. It provides the most basic functions of an operating system, including loading and executing programs and allocating hardware resources (CPU time, access to disk drives, and so on) to individual programs. Localizing the interaction between hardware and software in this layer insulates the users of the application level from direct knowledge of hardware specifics.

Service Layer The *service layer* accepts service requests from the *command layer,* or the application programs, and translates them into detailed instructions to the kernel. Processing results, if any, are passed back to the program that requested

Figure 1.6
Operating System Layers

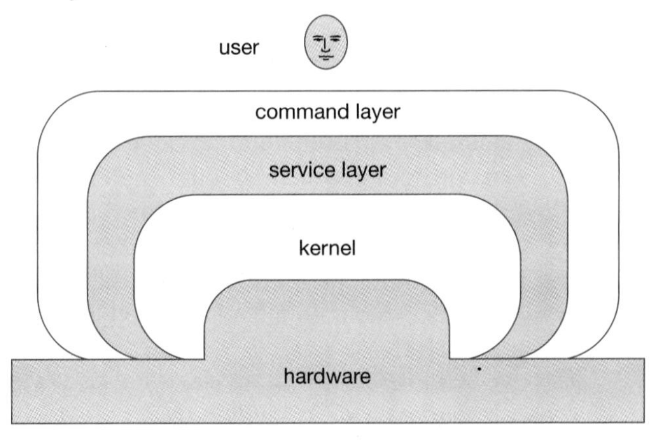

the service. The service layer consists of a set of programs that provide the following types of services:

- Access to I/O devices—for example, the movement of data from an application to a printer or terminal
- Access to storage devices—for example, the movement of data from a tape drive or a magnetic disk to an application program
- File manipulation—for example, opening and closing files, reading from a file, and writing to a file
- Other services such as window management, access to communication networks, and basic database services

Command Layer The *command layer,* also called the *shell* (because it is the outermost layer), provides the user interface and is the only part of the operating system with which users can interact directly. The command layer responds to a set of specific commands supported by each operating system. The set of the commands and their syntax requirements is referred to as a *command language.* There are other alternatives to the command language.

Operating Systems Environment

Operating system functions can be fulfilled in many ways. In single-user environments, like most microcomputers, all available resources are allocated to a single program. That program is the only program in the computer. On larger computers (and on microcomputer networks), where more than one user is sharing the computer resources, the operating system must resolve conflicts arising from the request for the same resource by different programs. Let's explore some basic concepts and terminology describing the different operating systems and their environments.

Although several programs can be in the memory of a microcomputer, only one program is active at a time.

Single-Tasking A *single-tasking (single-programming)* operating system is designed to execute only one process at a time. This is the usual arrangement with the *single-user* environment, and it is generally restricted to microcomputers and certain specialized applications.

Multitasking A *multitasking (multiprogramming)* operating system is capable of executing more than one program at a time for a user. It can run several programs in the background while you are working on another task in the foreground. For example, you can direct the operating system to sort a large file in the background while you are typing memos using a word processor in the foreground. The operating system informs you when a background task is finished. Multitasking is the capability that allows one user (terminal) to execute more than one program concurrently.

Multiuser In a multiuser environment, more than one user (terminal) can use the same host computer. The multiuser operating system is complex software that provides services for all users concurrently. The users' programs are in the main memory, and it appears that they are executed simultaneously; however, there is only one CPU, and the processor can execute only one program at a time. The multiuser operating system takes advantage of the speed disparity between the computer and its peripheral devices

Figure 1.7
A Multiuser Computer System

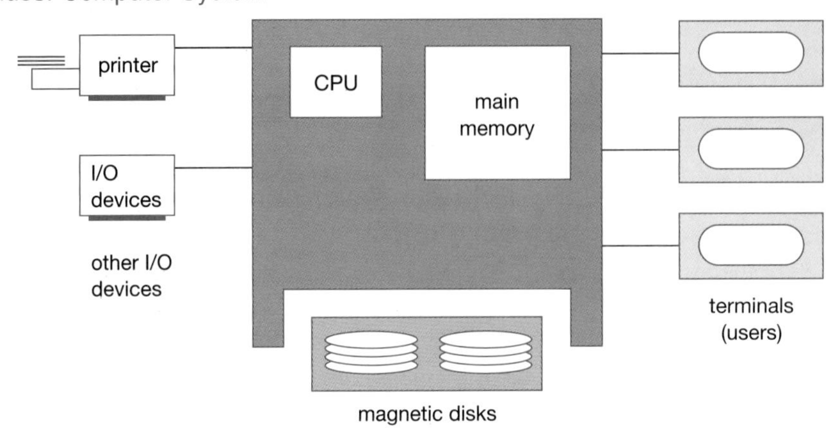

(disks, printers, and so on). In comparison with the processor speed, input/output devices are very slow. Thus, when one program is waiting for its I/O request, the processor has plenty of time and can turn its attention to another program in the memory. The process of switching programs is concealed from the user and can give the illusion that the user is the only one using the system.

Figure 1.7 shows an example of a *multiuser* system. There are four users, one printer, and some other I/O devices. The users share the same host computer and its resources.

The multiuser operating system is capable of providing services for many users (terminals), using the same host computer.

Another way to set up a *multiuser* system is to connect two or more computers together and create a network. Figure 1.8 shows a network computer system consisting of four computers. Three of the computers stand alone as single-user hosts, and one functions as a server, providing disk space for others.

Time-Sharing A *time-sharing* operating system is designed for online processing environments where active user involvement is required. It refers to multiple users sharing time on a single host computer. A time-sharing operating system allocates a slice of a time to each user task. It rapidly switches among tasks and each slice of time executes a small portion of each user task.

Batch The *batch* operating system is designed to execute programs *(batch process)* that do not require active user intervention. A *batch process* normally uses noninteractive I/O devices such as disks or document scanners for input and returns results to those same devices. Batch processing is used only in especially large transaction processing environments, such as nightly processing of checks in a bank.

Memory Capacity Limitation

How large can a computer program be? Remember, the entire program and its associated data must be in the memory throughout its execution. Thus the capacity of the computer's main memory becomes the limitation on the size of an application program. However, that is not as limiting as it might seem at first. The computer is a sequential machine; it executes one instruction at a time in each processing cycle, so it is not necessary for the entire application to be in the main memory continually.

Figure 1.8
The Multiuser in a Network Environment

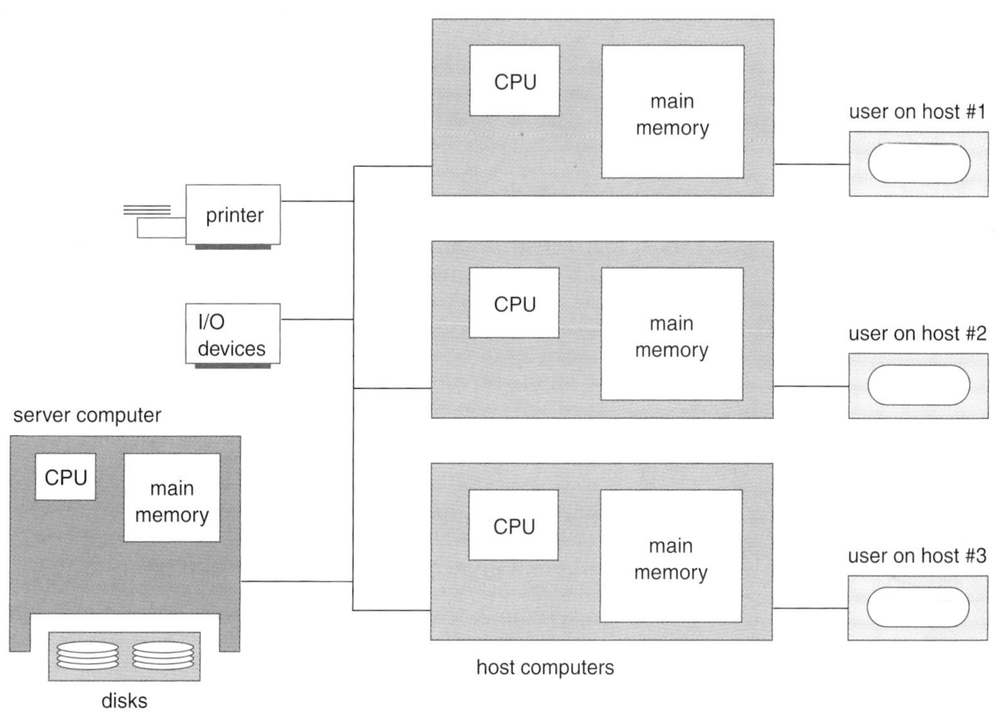

Virtual Memory To support large memory requirements when only small amounts of physical RAM are available, operating systems usually support *virtual memory*. Using *virtual memory,* programs are divided into smaller parts called *pages.* Under the control of the operating system, only necessary pages of a program are read into the main memory to support the ongoing process of running the program. A *swap* space is created on the hard disk and pages of memory are written to this reserved area of the disk. This area on the disk is treated as an extension of physical memory. By moving pages back and forth between the swap space and RAM, the operating system effectively can behave as if it had much more physical memory than it does. This approach permits the use of a relatively large program with minimal demands for the main memory. However, memory efficiency is achieved at the cost of some speed due to the hard drive's slower access.

Within a virtual memory system, secondary storage devices can be regarded as an extension of the main memory.

1.5.2 Application Software

Application software is designed and written to solve problems or provide automation and efficiency in personal, business, and scientific environments. Application programs are available for most data processing needs. You can buy off-the-shelf programs such as payroll systems, inventory systems, word processing, electronic spreadsheets, and so on. All you need to do is to select the right application program for the job. If you are not happy with off-the-shelf application programs, you can write your own program, using one of the available computer programming languages.

REVIEW EXERCISES

1. What is a CPU? What are its primary components?
2. What are registers? What are their functions?
3. What is main memory?
4. Explain the steps of the instruction cycle. Explain the steps of the execution cycle.
5. What is the instruction set of a computer system?
6. Explain single-tasking and multitasking. How do they differ?
7. Explain single-user and multiuser environments. How do they differ?
8. What is system software?
9. What is application software?
10. What are the primary components (functions) of an operating system?
11. Explain the operating system software layers.
12. What is a kernel layer? What function does it perform?
13. What is a service layer? What function does it perform?
14. What are the differences between primary storage and secondary storage?
15. Explain virtual memory. Why is it used?

CHAPTER 2

The UNIX Operating System

This chapter briefly describes the history of the UNIX operating system. It explores UNIX development over the years, discusses the major UNIX versions, and explains some of the system's important features.

In This Chapter

2.1 THE UNIX OPERATING SYSTEM: A BRIEF HISTORY

During the early 1960s, many computers were working in *batch mode,* running single jobs. Programmers had to use punch cards to input their programs and then wait for the output on the line printer. The UNIX operating system was born in 1969 as a response to the frustration of the programmers and the need for new computing tools to help them with their projects.

The UNIX operating system is the brainchild of Ken Thompson and Dennis Ritchie, two Bell Laboratories researchers. At the time, Ken Thompson was working on a program called Space Travel, a program simulating the motion of the planets in the solar system. The program was under an operating system called *Multics,* one of the first operating systems that provided a multiuser environment, and ran on a General Electric 6000 Series computer. But Multics was large, slow, and required substantial computer resources. Thompson found a smaller computer and transferred the Space Travel program to run on it. The smaller computer was a little-used PDP-7, one of the series of machines made by Digital Equipment Corporation (DEC). On that computer, Thompson created a new operating system that he called UNIX, and he adapted some of the advanced Multics concepts to that operating system. Other operating systems had more or less the same capabilities, and UNIX took advantage of the work that had gone into those operating systems by combining some of the most desirable aspects of each of them.

UNIX was transferred to the PDP-11/20 in 1970, and then to the PDP-11/40, the PDP-11/45, and eventually the PDP-11/70. Each of these machines had features that gradually added to the complexity of the hardware that UNIX could support. Dennis Ritchie and others at Bell Labs continued the development process of UNIX, adding utilities (such as a text processor).

Like most of the operating systems, UNIX was originally written in assembly language. *Assembly language* is a primitive set of instructions that depend on the computer architecture. Programs written in assembly language are machine dependent and work on only one computer (or one family of computers). Therefore, moving UNIX from one computer to another involved significant rewriting of the programs.

Thompson and Ritchie were experienced users of Multics, which was written in a high-level language called PL/1, and they were aware of the advantages of using a high-level language to write operating systems. (A high-level language is much easier to use than assembly language.) They decided to rewrite the UNIX operating system in a high-level language. The language they chose was C. The C programming language is a general-purpose language featuring the commands and structures of modern, high-level computer languages. In 1973, Ken and Dennis successfully rewrote the UNIX operating system in C.

About 95 percent of the UNIX operating system is written in C. A very small part of UNIX is still written in assembly language; that part is mostly concentrated in the kernel, the part that interacts directly with the hardware.

Universities and colleges have played an important role in the popularity of the UNIX operating system. In 1975, Bell Labs offered the UNIX operating system to

See page x for an explanation of the icons used to highlight information in this chapter.

educational institutions at minimal cost. UNIX courses were incorporated into the computer science curriculum, and students, in turn, became familiar with UNIX and its sophisticated programming environment. As those students graduated and joined the work force, they carried their UNIX skills to the commercial world and UNIX was introduced to industry.

There are two major versions of the UNIX operating system:

- AT&T UNIX version V
- Berkeley UNIX

Other UNIX varieties are based on one of these two versions.

Table 2.1 lists a brief chronological history of the UNIX operating system.

Table 2.1
UNIX History

Year	Event
1969	Ken Thompson and Dennis Ritchie started working on the used PDP-7 machine at Bell Labs to create what became UNIX.
1973	UNIX was rewritten in C. This made UNIX more portable.
1975	UNIX became available outside of bell Labs. This UNIX version was known as Six Edition. The first version of BSD was derived from this edition.
1979	UNIX Seven Edition was released. This edition was an improvement over the previous one and was ported to different machines.
1980	4BSD was released by University of California at Berkeley. Microsoft introduced Xenix.
1982	UNIX system III was introduced. This version was released by AT&T's UNIX System Group (USG) and was the first public release outside Bell Laboratories.
1983	Unix System V was released. This was the first supported release from AT&T. Computer Research Group (CRG) and UNIX System Group (USG) merged to become UNIX System Development Lab.
1984	4.2BSD was released by University of California at Berkeley.
1984	UNIX SVR2 (system V release 2) was introduced.
1986	4.3BSD was released.
1987	UNIX SVR3 (system V release 3) was introduced.
1988	POSIX.1 was published. Open Software Foundation (OSF) and UNIX International (UI) were formed.
1989	SVR4 (system V release 4) was introduced.
1991	UNIX System Laboratories (USL), an AT&T spin-off, became an independent company. Linus Torvalds started Linux development.

Table 2.1
(continued)

Year	Event
1992	SVR4.2 (System V release 4.2) was released by USL.
1993	4.4BSD was released. USL was acquired Novell company. SVR4.2MP was released. This release was by Novell/USL and was the final release of system V.
1995	UNIX 95 was introduced by X/Open. Novell sold UnixWare business to Santa Cruz Operation (SCO).
1996	The Open Group was formed.
1997	Single UNIX Specification, Version 2 was introduced by the Open Group. This version's specifications were made available on the Web.
1998	UNIX 98 was introduced by the Open Group. UNIX 98 included Base, Workstation, and Server products.
1999	The UNIX system reached its thirtieth anniversary. Linux 2.2 kernel was released.
2001	Version 3 of the Single UNIX Specification was released. Linux 2.4 kernel was released.

2.1.1 UNIX System V

In 1983, AT&T released the standard UNIX System V, which was based on the UNIX that AT&T was using internally. As the UNIX development effort continued, other features were added and some existing features were improved. Over the years, UNIX System V grew larger in size as well as in the number of its tools and utilities. The early improvements and new features were incorporated into the UNIX operating system's later releases, UNIX System V Release 3.0 in 1987 and UNIX System V Release 4.0 in 1989.

UNIX System V Release 4.0 was an effort to merge most of the popular features of the Berkeley UNIX and other UNIX systems. This unification helped simplify the UNIX product and reduced the need for manufacturers to create new UNIX variants.

2.1.2 Berkeley UNIX

The Computer Systems Research Group at the University of California at Berkeley added new features and made significant changes to the UNIX operating system. This version of UNIX is called the Berkeley Software Distribution (BSD) version of the UNIX system and has been distributed to other colleges and universities.

2.1.3 UNIX Standards

The UNIX operating system is available on all types of computers (micros, minis, mainframes, and supercomputers) and is an important operating system in the computer industry. As more UNIX-based systems were introduced into the market and more application programs became available, an effort toward UNIX standardization began. AT&T's UNIX System V Release 4.0 was an effort to standardize the UNIX system and, in turn, to facilitate the writing of the applications that run on all versions. The written UNIX standard from AT&T is called *System V Interface Definition* (SVID). Other companies that market UNIX operating systems or related UNIX products have joined together and developed a standard called *Portable Operating System Interface for Computer Environments* (POSIX). POSIX is largely based on the System V Interface Definition.

2.2 OTHER UNIX SYSTEMS

Almost all major UNIX vendors offer versions of the UNIX system that are based on UNIX System V. Most of the UNIX variants share a large number of commands and features similar to System V Release 4 (SVR4). The following subsections give brief descriptions of some of these variants.

2.2.1 Linux

UNIX is one of the most widely used operating systems in the world and has long been the standard for workstations and servers. UNIX is a commercial product and must be purchased for each platform it will run on, and the licensing fees for UNIX versions for PC machines range from a few hundred to several thousand dollars.

The Linux variant of the UNIX operating system is the brainchild of Linus Torvalds, a computer science student at the University of Helsinki in Finland. It was designed to run on Intel-based personal computers. Many programmers worked on Linux, and Linux grew as programmers adapted features and programs that were originally written as commercial UNIX products to Linux. Unlike other versions of UNIX, Linux is a freely distributable version of UNIX, and there are no licensing fees involved. Some companies have undertaken the task of assembling and testing versions of Linux, which they package on a CD-ROM for a (usually) minimal price.

Linux is not called UNIX for copyright reasons, but it is a complete UNIX implementation. It conforms to many of the same standards that UNIX does.

2.2.2 Solaris

SunOS, later called Solaris, was the operating system of Sun Microsystems that was based on UNIX System V Release 2 and BSD (Berkeley Software Distribution) 4.3. Solaris 2.0 is based on SVR4, and the current version of Solaris is Solaris 2.4, which has a large installations base. There is a wide range of applications and tools available with Solaris 2.4, including the graphic user interface.

2.2.3 UnixWare

The Novell version of UNIX was based on UNIX System V and was marketed as UnixWare. Novell sold UnixWare to the Santa Cruz Operation (SCO), and UnixWare and its related products are now offered by SCO. There are two versions of UnixWare: UnixWare Personal Edition, which is designed for desktop use, and UnixWare Application Server, which is used on servers. UnixWare is available only for Intel-based computers, and there are many applications available for it.

2.2.4 Which UNIX to Learn?

Linux is UNIX in all aspects but name, and it runs on most platforms at a fraction of the cost of a UNIX system. Linux gives you a great environment to experiment with at a very reasonable cost. Therefore, learning Linux is an easy and certainly cheaper way to learn UNIX. Anything you learn with Linux is directly transferable to other UNIX systems.

This book covers core UNIX commands that are applicable to all kinds of UNIX systems, including Linux. Where necessary, the differences are pointed out and explained.

2.3 OVERVIEW OF THE UNIX OPERATING SYSTEM

As explained in Chapter 1, a typical computer system consists of the hardware, the system software, and application software. The *operating system* is system software that controls and coordinates the activities of the computer. Like other operating systems, the UNIX operating system is a collection of programs that includes text editors, language compilers, and other system utility programs.

Figure 2.1 shows the UNIX operating system architecture. The UNIX operating system is implemented in a layered-style software model.

Figure 2.1
The Unix System Components

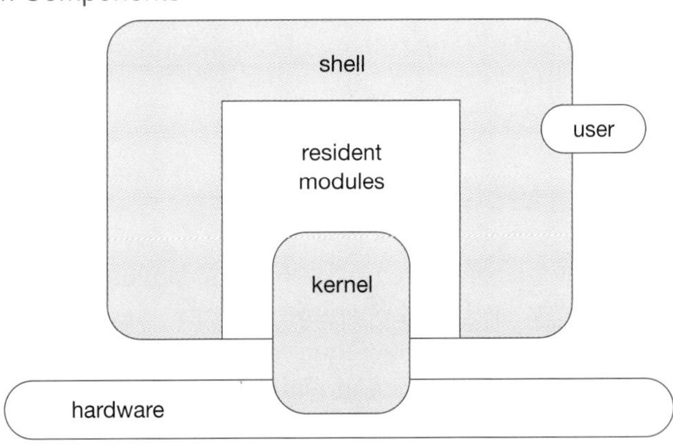

Kernel The UNIX *kernel,* also called the *base operating system,* is the layer that manages all the hardware-dependent functions. These functions are spread over a number of modules within the UNIX kernel. The kernel layer consists of modules closest to the hardware that are for the most part protected from the application programs. Users have no direct access to it.

1. *The utility programs and UNIX commands are not part of the kernel.*
2. *A user's application programs are protected from inadvertent writes by other users.*

Resident Modules Layer The *resident modules layer* provides service routines that perform user-requested services. These services include input/output control services, file/disk access services (called the *file system*), and process creation and termination services. A system call is used by application programs to access this layer.

Utility Layer The *utility layer* is the UNIX user interface, commonly referred to as the *shell.* The shell and each of the other UNIX commands and utilities are separate programs. They are part of the UNIX distribution software but are not considered part of the kernel. There are more than 100 commands and utilities in UNIX that provide various types of services to users and application programs.

Virtual Computer The UNIX operating system allocates an execution environment to each user in the system. This environment, or *virtual computer,* consists of a terminal for user interface and shared access to the other computer resources such as memory, disk drives, and most importantly, the CPU. UNIX, a multiuser operating system, is implemented as a collection of virtual computers. To the users, it appears that each of them has his or her own private pseudocomputer. The virtual computers are slower than the base computer due to sharing the CPU and other hardware resources with other virtual computers.

Processes The UNIX operating system allocates resources to users and programs by way of processes. Each process has a process identification number, and a set of resources is associated with that number. Each is executed in a virtual computer environment. What this really means is that a process runs in a virtual computer much as if it had a dedicated single-user CPU.

2.4 UNIX FEATURES

This section briefly discusses some features of the UNIX operating system. UNIX has some features that are common to most operating systems. It also has some unique features.

2.4.1 Portability

The use of the C programming language made UNIX a portable operating system. Today, the UNIX operating system works on a variety of machines ranging from microcomputers to supercomputers. The portability feature helps to decrease the user's learning time when moving from one system to another. It also provides more choices among hardware vendors.

2.4.2 Multiuser Capability

Under UNIX, a number of users can share computer resources simultaneously. Depending on the machine being used, UNIX may be able to support more than one hundred users, each one running a different program. UNIX provides security measures that permit users to access only the data and programs for which they have permission.

2.4.3 Multitasking Capability

UNIX allows the user to initiate a task and then proceed to perform other tasks while the original task is being run in the background. UNIX also allows users to switch back and forth between the tasks.

2.4.4 Hierarchical File System

UNIX provides users with the ability to group data and programs in a manner that provides easy management. Users can find data, and programs can be located with little difficulty.

2.4.5 Device-Independent Input and Output Operations

Input and output operations are device independent because UNIX treats all devices (such as printers, terminals, and disks) as files. With UNIX, you can redirect the output of your commands to any device or file. This redirection process is also possible with input data. You can redirect the input that comes from your terminal to come from a disk instead.

2.4.6 User Interface: Shell

The UNIX user interface was designed primarily for users with programming backgrounds. Experienced programmers find UNIX simple, concise, and elegant. Beginners, on the other hand, find it terse, not very friendly, and sometimes difficult to learn; it provides no feedback, warning, or hand-holding. For example, the command **rm *** deletes all the files silently and with no warning.

The user's interaction with UNIX is controlled by a program called the *shell,* which is a powerful command interpreter. The shell is the face of UNIX and the part with which most of the users interact. However, the shell is only one part of the operating system and one way of interacting with UNIX. The shell is not really part of the operating system, so it can be changed. A user might choose a technical shell (*command-driven user interface*), or prefer selecting commands from a menu (*menu-driven user interface*) or pointing at pictures (icons) in what is called *graphical user interface* (GUI). A user can even write his or her own user interface.

The UNIX shell, a complex and sophisticated user interface, provides plenty of innovative and exciting features. You can create new functions by combining the exiting commands. For example, the command **date ¦ lp** combines the two commands **date** and **lp** to print the current date on the line printer.

Shell Script Many data processing applications are run frequently—at daily, weekly, or other regular intervals. In other situations, a set of commands must be entered many times. Typing the same set of commands again and again is annoying and is a procedure prone to error. A way to remove this difficulty is to write a shell script. A *shell script* is a file that consists of a series of commands.

The UNIX shell is a highly sophisticated programming language. Chapters 11 and 12 discuss UNIX shell script capabilities and programming methods.

2.4.7 Utilities

The UNIX system includes more than 100 utility programs, also called *commands.* Utilities are part of the standard UNIX system and are designed to perform a variety of functions required by the users. These utilities include the following:

- Text editing and text formatting utilities (see Chapters 4 and 6)
- File manipulation utilities (see Chapters 5 and 7)
- Electronic mail (e-mail) utilities (see Chapter 9)
- Programmers' tools (see Chapter 10)

2.4.8 System Services

The UNIX system provides a number of services that facilitate administration and maintenance of the system. A description of these services is beyond the scope of this text. However, the following are some of these services:

- System administration service
- System reconfiguration service
- File system maintenance service
- File transfer service (called UUCP for *UNIX to UNIX Copy*)

REVIEW EXERCISES

1. What are the two major UNIX system versions?
2. What is the kernel?
3. What is a shell?
4. Briefly explain the virtual computer concept.
5. What is a process?
6. Why was UNIX rewritten? What computer language was used?
7. Is UNIX a multiuser, multitasking operating system?
8. Is UNIX portable?
9. What is a shell script?
10. Name some of the UNIX variants.
11. What is the UNIX variant on your system?

CHAPTER 3

Getting Started

This chapter explains how to start and end (log in and log out) a UNIX session. It also explains the function of passwords and how to change your password. Then after showing the command line format, the chapter introduces a few simple UNIX commands and explains their applications. It also describes how you can correct typing mistakes. Finally, the process of establishing contact with UNIX is explored in more detail, and some internal UNIX operations are discussed.

In This Chapter

3.1 ESTABLISHING CONTACT WITH UNIX

The process of establishing contact with the UNIX operating system consists of a number of prompts and user inputs that start and end a session. A *session* is a period of time during which you use the computer.

3.1.1 Logging In

UNIX is a multiuser operating system, and you are not likely to be the only person using the system. The process of identifying yourself and letting UNIX know that you want to use the system is called the *logging-in process.* The first thing you must do to use a UNIX system is log in.

To get the attention of the UNIX operating system, you need a means of communicating with UNIX. In most cases, a keyboard (input) and a screen display (output) provide the means of communication. After turning on the screen display on your computer, you begin the logging-in process by pressing [Return]. When UNIX is ready for you to log in, it displays some messages (these vary from system to system) and then the **login:** prompt (see Figure 3.1).

Figure 3.1
The Login Prompt

```
UNIX System V release 4.0
login:
```

Login Name Most UNIX systems require that you set up an account before you can use the computer. When your account is created, your *login name* (also called *User ID* or *User Name*) and password also are established. A User ID (user identification) is usually issued by the system administrator (or your professor if you are working on a college or university computer system). Your User ID is unique and identifies you to the system.

Respond to the **login:** prompt by typing your User ID and then pressing [Return]. After you type your User ID (for this example, we will use **david**), the UNIX system displays the **password:** prompt (see Figure 3.2).

Figure 3.2
The Password Prompt

```
UNIX System V release 4.0
login: david
password:
```

If no password has been assigned to your account, UNIX does not display the **password:** prompt, and the logging-in process ends.

See page x for an explanation of the icons used to highlight information in this chapter.

Password Like the User ID, a password is supplied by the system administrator. A password is a sequence of letters and digits used by UNIX to verify that the user is allowed to use the User ID.

Type your password and then press [Return]. In order to protect your password from others, UNIX does not echo the letters you type; that is, you do not see them on the screen. UNIX verifies your User ID and password; if they are correct, a pause occurs while the system sets things up for you.

After verifying the User ID and password, the UNIX system displays some messages. Often the date is displayed, along with a "message of the day" (which contains messages from the system administrator). The UNIX system indicates that it is ready to accept your commands by displaying the system prompt. The *standard prompt* is usually a dollar sign (**$**) or a percent sign (**%**). Let's assume the prompt is a dollar sign (**$**). It will be displayed as the first character on the line (see Figure 3.3).

Figure 3.3
The Logging-in Process

```
UNIX System V release 4.0
login: david
password:

Welcome to super duper UNIX system
Sat Nov 29 15:40:30 EDT2001
* This system will be down from 11:00 to 13:00
* This message is from your friendly system administrator!

$_
```

You must enter your ID and password exactly as they are known to the system. That is, lowercase or uppercase characters must be entered as lowercase and uppercase respectively. At this point, UNIX is not forgiving, and you cannot correct your entries using the [Backspace] or [Del] keys. If you make a mistake and enter the wrong user ID or password, UNIX displays a message and the **login:** prompt will appear again, ready for you to enter your User ID (see Figure 3.4).

Figure 3.4
Incorrect Login

```
UNIX System V release 4.0
login: David
password:
Login incorrect
login:
```

3.1.2 Changing Your Password: The passwd Command

The **passwd** command changes your current password. If you do not have a password, it creates one.

Type **passwd** and press [Return]. UNIX displays the **Old password:** prompt (see Figure 3.5).

Figure 3.5
The Old Password Prompt

```
$ passwd
Changing password for david
Old password:
```

For security reasons, UNIX never displays your password on the screen.

UNIX verifies your password to ensure that an unauthorized user is not changing your password. (UNIX does not display the **Old password:** prompt if you do not have a password yet.) Enter your current password and press [Return]. Next UNIX displays the **New password:** prompt (see Figure 3.6).

Figure 3.6
New Password Prompt

```
$ passwd
Changing password for david
Old password:
New password:
```

After you enter your new password, UNIX shows the **Re-enter new password:** prompt. Retype the new password. UNIX is verifying that you did not make a mistake the first time you entered the new password (see Figure 3.7).

Figure 3.7
Re-Enter New Password Prompt

```
$ passwd
Changing password for david
Old password:
New password:
Re-enter new password:
```

If you typed the new password the same way both times, UNIX will change your password. There might be a confirmation message that indicates your password was successfully changed, or else the **$** prompt will appear with no feedback about changing your password. In any case, your password is changed and the new password will be required for your future logins.

Make sure you remember your new password. You will need it the next time you want to log in.

Password Format The **passwd** command does not accept just any sequence of alphanumeric characters as a password. Your password must comply with the following criteria:

- The new password must differ from the old one by at least three characters.
- The password must be at least six characters long and must contain at least two characters and one number.
- The password must differ from your User ID.

If UNIX detects anything wrong with your password, it displays an error message and shows the **New password:** prompt again (see Figure 3.8).

Figure 3.8
Sample of Error Messages

```
Password is too short - must be at least 6 digits
New password:

Password must differ by at least 3 positions
New password:
```

Usually after a few errors, if a correct password is not provided, UINX will give up and terminate the **passwd** command. There might be a confirmation message that indicates that the command is terminated and your password was not changed, or else the **$** prompt will appear with no feedback. Of course, you can start all over and try to change your password by issuing the **passwd** command again.

The wordings of the prompts are slightly different for the Linux system, but the command and sequence of prompts are the same. Figure 3.9 shows an example of changing password in Linux environment.

Figure 3.9
Changing a Password in the Linux Environment

```
$ passwd
Changing password for david
(current) UNIX password:
New UNIX password:
Retype new UNIX password:
passwd: all authentication tokens updated successfully
$_
```

Figure 3.10 shows a few examples of error messages indicating that Linux detected an incorrect password.

Figure 3.10
Sample of Error Messages

```
$ passwd
Changing password for david
(current) UNIX password:
New UNIX password:
BAD PASSWORD: it's WAY too short
New UNIX password:
BAD PASSWORD: Dictionary word
New UNIX password:
BAD PASSWORD: it is too short
New UNIX password:
Retype new UNIX password:
Sorry, passwords do not match
passwd: Authentication information cannot be recovered
$_
```

3.1.3 General Rules for Choosing Passwords

The **passwd** command will try to prevent you from choosing a really bad password. However, it is up to you to create your password wisely. In particular, observe the following general rules:

- Do not use a dictionary word (in any language).
- Do not use the name of a person, pet, location, or a character from a book.
- Do not use any variation of your name and ID/account name.
- Do not use known information about you, such as your phone number, birthday, and so on.
- Do not use a simple pattern or easy sequence of keyboard keys.

Tips for Creating and Safeguarding Your Password

Passwords are important security risks if they are not protected. This might not be as important in a classroom environment, but make a habit of protecting your password by observing the following rules:

- Do not write down your password—try to memorize it.
- Do not place your password in a file.
- Do not use the same password for all of your password-protected systems and activities.
- Do not reveal your password to anyone.
- Do not let anyone watch you entering your password
- Choose a "hard-to-guess" password.

3.1.4 Logging Off

The process of signing off when you have finished with the system is called *logging off* or *logging out.* When you want to log off, the prompt sign must be displayed on the screen. You cannot log off in the middle of a process. To log off, at the prompt sign press [Ctrl-d], which means to simultaneously hold down the key labeled *Ctrl* (for Control) and press the key for the letter *d.* When this key sequence is pressed, nothing is displayed on the screen; the command, however, is recognized by the system. The UNIX system responds first by displaying logging off messages that your system administrator has set up. Then the screen shows your system's standard welcome message and **login:** prompt. This lets you know that you have logged off properly and that the terminal is ready for the next user.

The terminal session shown in Figure 3.11 shows the logging-in and logging-out processes.

Figure 3.11
The Logging-In and Logging-Out Process

```
Unix System V release 4.0
login: david
password:
Welcome to super duper UNIX system
Sat Nov 29 15:40:30 EDT 2005
*This system will be down from 11:00 to 13:00
*This message is from your friendly system administrator!

$[Ctrl-d]

Unix system V release 4.0
login:
```

Turning off the terminal without logging off does not terminate your session with UNIX.

3.2 USING SOME SIMPLE COMMANDS

When you are using a UNIX system, you have hundreds of UNIX commands and utilities at your disposal. Some basic commands you will use frequently, some you will use occasionally, and some you may never use! The difficult part of systems such as UNIX that have a rich set of commands is learning to master the details of each command. Fortunately, most UNIX commands share the same basic structure, and online help is available on most of the UNIX systems to assist your memory.

3.2.1 The Command Line

Every operating system has commands that facilitate use of the system. By typing the commands, you tell the UNIX system to do something. For example, the command

date tells the system to display the date and time. To see this, type **date** and press [Return]. This line of instruction is known as the *command line.* UNIX interprets the press of [Return] as the end of the command line and responds by showing the current date and time on the screen:

```
Sat  Nov  29  14:00:52 EDT 2005
$_
```

Each time you give UNIX a command, it carries out the command and then displays a new prompt, indicating that it is ready for the next command.

3.2.2 Basic Command Line Structure

Each command line consists of three fields:

- Command name
- Options
- Arguments

Figure 3.12 shows the general format of a UNIX command.

Figure 3.12
The Command Line Format

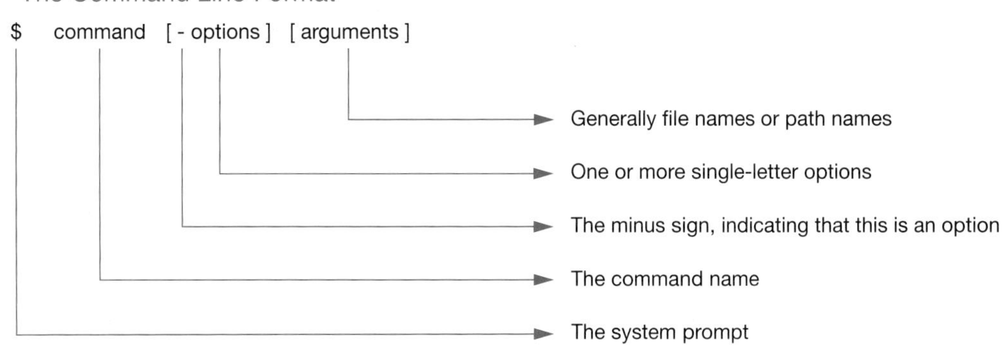

1. *Fields are separated by one or more spaces.*
2. *Fields enclosed in brackets [...] are optional for most of the commands.*

You must signal that you have completed entry of a command by pressing [Return] at the end of every command.

Command Name Any valid UNIX command or utility program functions as a *command name.* Under UNIX, commands and utilities are different. In this text, however, the word *command* refers to both of them.

UNIX is a case-sensitive system and only accepts command names in lowercase letters.

Options *Options* are variations on the command. If they are included in the command line, they are usually preceded by a minus sign (-). Most options are designated by a single lowercase letter, and more than one option can be specified on one command line. This text does not describe all possible options for every command.

Arguments *Arguments* are needed for commands that perform some sort of operation, such as printing a file or displaying information. Commands often need to operate on something, which means that you must provide additional information. The additional information the command needs to operate on is called an *argument.* For example, if the command is **print**, then you have to tell UNIX what to print (the name of a file) and where to find the file you want to print (somewhere on the disk). The name of the file plus where the file is located on the disk is an example of the *argument field.*

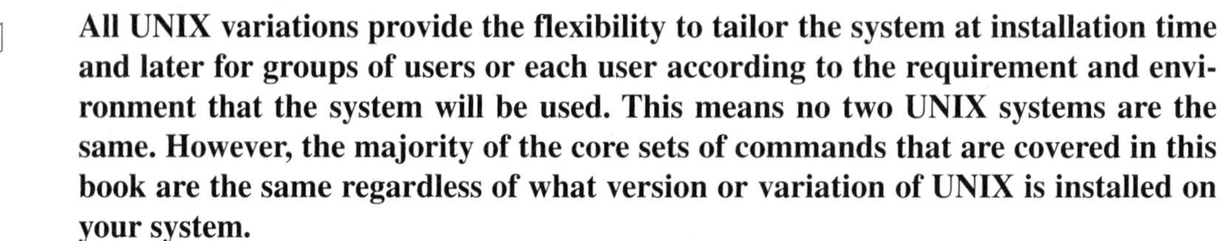

All UNIX variations provide the flexibility to tailor the system at installation time and later for groups of users or each user according to the requirement and environment that the system will be used. This means no two UNIX systems are the same. However, the majority of the core sets of commands that are covered in this book are the same regardless of what version or variation of UNIX is installed on your system.

The man **command (explained later in this chapter) is the best way to get detailed help for the commands that are available on your system. This is especially useful to check the options available for each command. In this book, only a few options for each command are used to demonstrate how a specific command is used. Depending on the variations of your UNIX system, some of these options might be different than the ones that are used in examples presented here.**

3.2.3 Date and Time Display: The date Command

The **date** command displays the current date and time on the screen. The date and time are set by the system administrator. Users cannot change them.

Show the current date and time (see Figure 3.13).

Figure 3.13
The **date** Command

```
$ date
Sat Nov 27 14:00:52 EDT 2005
$_
```

The **date** *command displays the day of the week, month, day, and time (in this case, Eastern Daylight Time), followed by the year. UNIX uses a 24-hour clock.*

3.2.4 Information on Users: The who Command

The **who** command lists the login names, terminal lines, and login times of the users who are currently logged on the system. You can use the **who** command to check the level of activity in the system or to find out whether a particular person is on the system.

Who is logged on to the system (see Figure 3.14)?

Figure 3.14
The **who** Command

```
$ who
david          tty04  Nov 28  08:27
daniel         tty10  Nov 28  08:30
$_
```

1. *The first column shows the login name of the user.*
2. *The second column identifies the terminal being used.*
3. *The tty number gives you some indication about the location of the terminal.*
4. *The third and fourth columns show the date and time that each user logged in.*

If you type **who am I** or **who am i**, UNIX displays who the system thinks you are (see Figure 3.15).

Figure 3.15
The **who** Command with **am i** Argument

```
$ who am i
david  tty04  No 28  08:48
$_
```

who Options

Table 3.1 lists some of the **who** command options. Under Linux, some of these options are not available, or some of the options outputs are slightly different. Linux also provides some alternative and new options. For example, to display a line of columns headings we can use **-H** or **--heading**. The results will be the same. These new and alternative options are listed under the Linux heading in Table 3.1.

1. *Linux new and alternative options are listed under the Linux heading in Table 3.1.*
2. *Linux new and alternative options are preceded by two minus signs (--).*

Table 3.1
The **who** Command Options

Option	Linux	Operation
-q	**--count**	The quick **who**; just displays the name and number of users.
-H	**--heading**	Displays heading above each column.
-b		Displays the time and date of the last reboot.
	--help	Displays a usage message.

1. **There must be a space between the command field and the option field.**
2. **Options are preceded by a minus sign. In Linux, some options are preceded by two minus signs.**
3. **There is no space between the minus sign(s) and the option letter.**
4. **Option letters must be typed exactly as they are indicated, uppercase or lowercase.**

Examples Using Options

The following examples show how the use of options in the command line can change the output format and level of details displayed.

Display columns headers using the **-H** Option (see Figure 3.16A).

Figure 3.16A
The **who** Command with **-H** Option

```
$ who -H
NAME            LINE         TIME
david          tty04        Nov 28   08:27
daniel         tty10        Nov 28   08:30
$_
```

Display column headers using the Linux alternative option **--heading** (see Figure 3.16B).

Figure 3.16B
The **who** Command Using Linux **--heading** Option

```
$ who --heading
USER      LINE      LOGIN-TIME
david     tty04     Nov  28  08:27
daniel    tty10     Nov  28  08:30
$_
```

Get a quick list and user count, the **-q** option (see Figure 3.17A).

Figure 3.17A
The **who** Command with **-q** Option

```
$ who -q
david daniel
# users=2
$_
```

You can use the alternative Linux option **--count** to display user count (see Figure 3.17B).

Figure 3.17B
The **who** Command Using Linux **--count** Option

```
$ who --count
david  daniel
#  users=2
$_
```

Show time and date of the last boot, the **-b** option (see Figure 3.18).

Figure 3.18
The **who** Command with **-b** Option

```
$ who -b
system boot Nov 27  08:37
$_
```

Linux help Option

Linux provides a short explanation of the command usage for most of the commands. This feature is invoked by using the option **--help** or sometimes **-?**. These help options are not available for all commands, and Linux displays appropriate error messages if the wrong option is used.

Show a short description for the **who** command usage with the **--help** option (see Figure 13.19).

Figure 3.19
The **who** Command Using Linux **--help** Option

```
$ who --help
Usage: who [OPTION]... [ FILE ¦ ARG1 ARG2 ]

-H, --heading   print line of column headings
-i, -u, --idle  add user idle time as HOURS:MINUTES, . or old
-l, --lookup    attempt to canonicalize hostnames via DNS
-m              only hostname and user associated with stdin
-q, --count     all login names and number of users logged on
-s              (ignored)
-T, -w, --mesg  add user's message status as +, - or ?
  --message     same as -T
  --writable    same as -T
  --help        display this help and exit
  --version     output version information and exit
$ _
```

3.2.5 Display a Calendar: The cal Command

The **cal** command displays the calendar for the specified year. If the year and the month are both specified, then the calendar for just that month is displayed. Month and year are examples of command line arguments. The default argument for the **cal** command is the current month.

Display the calendar for the current month (see Figure 3.20). Assuming current month, year is November, 2005.

Figure 3.20
The **cal** Command with No Option

```
$ cal
 November 2005
 S    M   Tu    W   Th    F    S
            1   2    3    4    5
 6    7    8    9   10   11   12
13   14   15   16   17   18   19
20   21   22   23   24   25   26
27   28   29   30
$_
```

Display the calendar for November, 2010 (see Figure 3.21).

Figure 3.21
The **cal** Command Output

```
$ cal 11 2010
November 2010
 S    M   Tu    W   Th    F    S
      1    2    3    4    5    6
 7    8    9   10   11   12   13
14   15   16   17   18   19   20
21   22   23   24   25   26   27
28   29   30
$_
```

1. *Type the specified year in full year value. For example, type **cal 2005** (not **cal 05**).*
2. *Use the month number (1 to 12) and not the month name.*
3. *The **cal** command without arguments displays calendar of the current month.*
4. *The **cal** command with year argument but without month argument displays a calendar for the specified year.*

3.3 GETTING HELP

UNIX does not leave novices or forgetful experienced users in the cold! The **learn** and **help** commands are two programs that provide assistance for using the UNIX operating system. These programs are designed to be user friendly and are intended to be used by novices, without assistance. However, these commands vary from one system to another, and they may or may not be installed on your system.

3.3.1 Using the learn Command

The **learn** command brings up a computer-aided instruction program that is arranged in a series of courses and lessons. It displays a menu of the courses and guides the user to select courses, presents descriptions of the courses, and so on.

To use the **learn** command, type **learn** on the command line and press [Return]. If the **learn** program is installed on your system, the **learn** main menu is displayed (see Figure 3.22). Otherwise, you see an error message similar to the following:

```
learn: not found
```

Figure 3.22
Learn Utility Main Menu

```
$ learn
These are the available courses
   files
   editor
   vi
   more files
   macros
   eqn
   C
If you want more information about the courses, or if you have
never used 'learn' before, press RETURN; otherwise type the name
of the course you want followed by RETURN
```

3.3.2 Using the help Command

The **help** command is more popular than the **learn** command and is available on many UNIX systems. The **help** command presents you with a hierarchy of the menus, and a series of menu selections and questions lead you to descriptions of the most commonly used UNIX commands. To use the **help** command, type **help** on the command line and press [Return]. UNIX displays the **help** main menu (see Figure 3.23). If **help** is not installed on your system, you see the following error message:

```
help: not found
```

Figure 3.23
Help Utility Main Menu

```
$ help

help:UNIX System On-line Help

          Choices        description



                    s       starter: general information
                    l       locate:find a command with keyword
                    u       usage: information about command
                    g       glossary: definition of terms
                    r       redirect to a file or a command
                    q       Quit
                    Enter choice
```

3.3.3 Getting More Information: The UNIX Manual

You can find a detailed description of the UNIX system in a large document called the *User's Manual.* Your installation may have a printed copy of this manual. The electronic version of the manual is stored on disk and is called the online manual. If the online manual is installed on your system, you can easily display pages from system documentation on your terminal. However, the UNIX user's manual is tersely written and difficult to read. It is more like a reference guide than a true user's manual and is most helpful to experienced users who know the basic purpose of a command but have forgotten exactly how to use it.

3.3.4 Using the Electronic Manual: The man Command

The **man** (manual) command shows pages from the online system documentation. To get information about a command, type **man** followed by the name of the command. As you begin to learn new commands, you may want to use **man** to get the details about the commands. For example, to find more information about the **cal** command, type **man cal** and then press [Return]. UNIX responds by showing a page similar to Figure 3.24. Be patient: Sometimes the system takes a while to find the information about the selected command.

The UNIX user's manual is organized into sections, and the number in parentheses after the command name refers to the section number in the manual that contains the description. For example "CAL(1)" refers to section 1, in the User Command section. The other sections include System Administration Commands and Games, for example.

*Nearly all UNIX installations provide the **man** pages, and the **man** command is the most popular way to obtain detailed information about command usage and options.*

Figure 3.24
man Utility Display of **cal**

```
cal(1)          User Environment Utilities cal(1)


NAME
   cal - print calendar
SYNOPSIS
   cal [ [ month ] year ]
DESCRIPTION


   cal prints a calendar for the specified year. If a month
   is also specified, a calendar just for that month is
   printed. If neither is specified, a calendar for the present
   month is printed. Year can be between 1 and 9999. The year is
   always considered to start in January even though this is
   historically naive.  Beware that "cal 83" refers to the
   early Christian era, not the 20th century.
   The month is a number between 1 and 12.
   The calendar produced is that for England and the United
   States.
EXAMPLES


   An unusual calendar is printed for September 1752. That is
   the month 11 days were skipped to make up for lack of leap
   year adjustments. To see this calendar, type: cal 9 1752
```

3.4 CORRECTING TYPING MISTAKES

Even the most talented typists make mistakes or change their minds when typing commands. The shell program interprets the command line after you press [Return]. As long as you have not ended your command line (by pressing [Return]), you have the opportunity to correct your typing mistakes or to cancel the whole command line.

If you mistype a command name and end the command line by pressing [Return], UNIX displays a generic error message. It responds with the same error message if you type a command that is not installed on your system.

The date command is mistyped (see Figure 3.25).

Figure 3.25
UNIX Error Message

```
$ daye
daye: not found
$_
```

Erasing Characters Use the [Backspace] key for erasing characters. When you press [Backspace], the cursor moves to the left, and the character it moves over is erased (An alternative way to erase characters is to press [Ctrl-h] once for each character you intend to erase.) For example, if you type **calendar** and then press [Backspace] five times, the cursor moves over the last five characters and **cal** remains on the screen. Then you can press [Return] to execute the **cal** command. The character that erases the command line a character at a time is called the *erase character.*

Erasing an Entire Line You can erase an entire line any time before pressing [Return] to end the command line. When you press the [Ctrl-u] key, the entire command line is removed and the cursor moves to a blank line. For example, suppose you type **passwd** and then you decide not to change your password. You press [Ctrl-u] and the **passwd** command is erased. The character that erases the entire line is called the *kill character.*

[Ctrl-h] and [Ctrl-u] are each considered a single character although each one involves two keys.

Terminating Program Execution If you have little time, and the program you are running takes a long time to perform its task, then you may want to terminate the execution of the program. The character that terminates your running program is called the *interrupt character.* On most systems, [Del] or [Ctrl-c] is assigned as the interrupt character. The interrupt character stops the running program and causes the shell prompt **$** to be displayed.

Try the kill and interrupt characters on your system; if they do not work, ask the system administrator for assistance.

3.5 USING SHELLS AND UTILITIES

Much of UNIX's strength and flexibility comes from its shells. The *shell* is a program that handles user interaction with the UNIX system. The UNIX commands are processed by a shell that lies between the user and the other parts of the operating system. Each time you type a command and press [Return], you send a command to the shell. The shell analyzes the line you typed and then proceeds to carry out your request; that is, it is a command interpreter. For example, if you type the **date** command, the shell locates where the program `date` is kept and runs it. So you, the user, are actually addressing the shell and not UNIX, technically speaking.

Shell Commands Some commands are part of the shell program; these *built-in commands* are recognized by the shell and are executed internally.

Utility Programs Most UNIX commands are executable programs (utilities) that the shell locates and executes.

In this book, the word *command* is used to refer to both shell commands and utility programs.

3.5.1 Kinds of Shells

The shell is just a program. Like other programs, the shell has no special privileges on the system. This is one reason various flavors of the shell program exist. If you are a professional programmer, you can create your own shell program, with some effort and devotion. UNIX System V Release 4 (SVR4) provides three different shells for you to choose from: the Bourne shell (*sh*), the Korn shell (*ksh*), and the C shell (*csh*).

Bourne Shell The Bourne shell (*sh*) is the standard shell that is part of most UNIX operating systems. It is usually the default shell on your system. The prompt for the Bourne shell is the dollar sign (**$**).

Korn Shell The Korn shell (*ksh*) is a superset of the Bourne shell. It has the basic syntax and features of the Bourne shell (*sh*) as well as many other features. You can generally run the shell program written for *sh* under *ksh* without modifications. The prompt for the Korn shell is the dollar sign (**$**).

C Shell The C shell (*csh*) was developed at the University of California at Berkeley and is part of the BSD (Berkeley Software Distribution) version of UNIX. The syntax of shell programming under the C shell is different from *sh* and *ksh* since it is a C language–style syntax. The prompt sign for C shell is the percent sign (%).

Bourne Again Shell As was mentioned before, most UNIX systems provide more than one shell, and you can set up your system to use any of the available shells. However, Linux comes with its own standard shell called the Bourne Again shell. The Bourne Again shell (bash) is based on the Bourne shell and is usually the default shell on Linux systems. The prompt for the Bourne Again shell is the dollar sign ($).

3.5.2 Changing Your Shell

Most of the UNIX systems provide more than one shell. Whatever shell you are using when you login to a UNIX system is assigned to your account by the system administrator. You can easily change your current shell by typing the name of the shell you want to use. For example, the following sequence of commands changes your current shell from *ksh* to *bash* and back to *ksh* again:

```
$_ . . . . . . . . . . .  Assuming you have Linux and your current shell
                          is (ksh).
$ bash [Return] . . .     Change the current shell to bash.
bash$_ . . . . . . . .    Prompt indicating the shell in use is bash.
bash$ exit [Return] .     Exit bash and return to ksh.
$_ . . . . . . . . . . .  Prompt shows the shell in use is ksh.
```

Of course, at this point, changing your current shell to another shell is a temporary change and it will remain in effect for your current session only. The next time you log in, your current shell will be the shell assigned to your account, regardless of the changes you have made in the previous session. This topic will be revisited in Chapter 8, in which the UNIX shell is fully explored.

Usually, when you change your current shell to another shell, the shell name is displayed as part of the prompt. In the above example, the prompt is ***bash$****.*

3.5.3 The Shells in This Book

As far as the basic commands and features are concerned, all shells are very similar. This book assumes you have either Bourne shell *(sh)* or Korn shell *(ksh)* and if under the Linux environment, the Bourne Again shell *(bash)*. Commands and utilities that are drastically different are explained separately for each shell.

1. *Commands applicable to the Korn shell only are identified by the ksh designation.*
2. *Commands applicable to the Bourne Again shell only are identified by the bash designation.*

3.6 MORE ABOUT THE LOGGING-IN PROCESS

When you boot (bootstrap) the UNIX system, the resident part of the operating system (kernel) is loaded into the main memory. The rest of the operating system programs (utilities) remain on the system disk and are brought into memory only when you request that the command be executed. The shell program also is loaded into memory for execution whenever you log in. Learning about the sequence of events that occur when you log in will help you gain a better understanding of the UNIX system's internal operation.

After UNIX completes the boot procedures, a program called *init* activates a program called `getty` for each terminal port on the system. The `getty` program displays the **login:** prompt at the assigned terminal, and then waits for you to type your User ID (see Figure 3.26).

Figure 3.26
`getty` Displays the **login:** Prompt

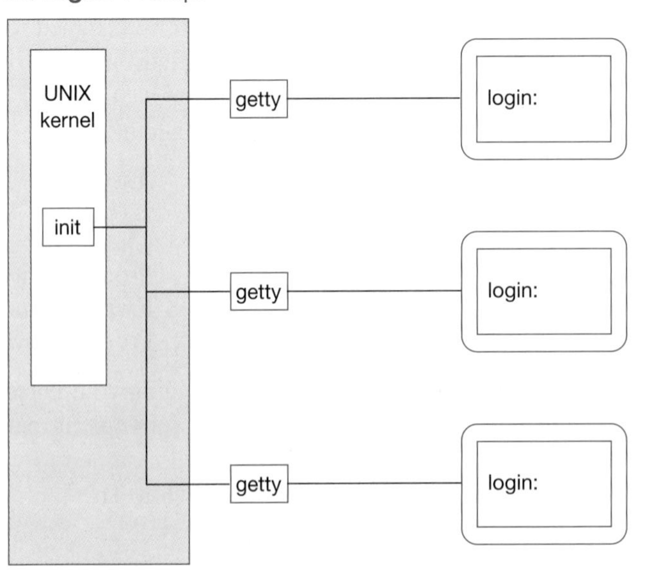

When you enter your User ID, getty reads it and starts another program, called `login`, to complete the logging in process. It also gives the `login` program the characters you typed at the terminal, which are presumed to be your User ID (login

Figure 3.27
`login` Displays the **password:** Prompt

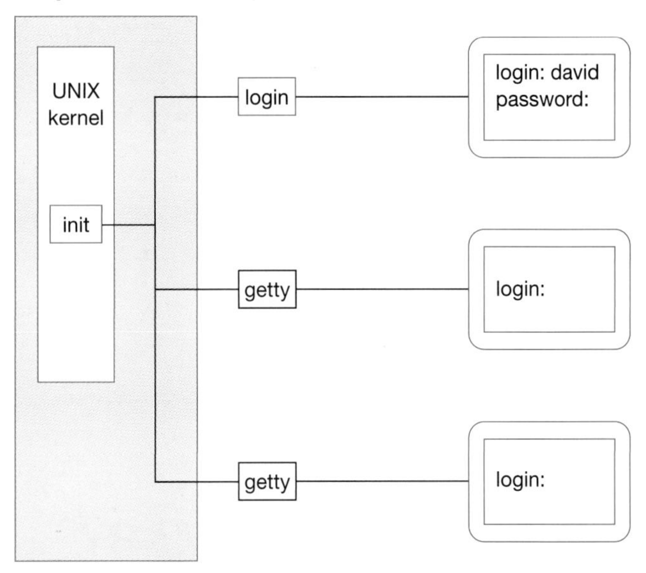

name). Next, the `login` program begins execution and displays the **password:** prompt at the terminal. The `login` program waits for you to enter your password (see Figure 3.27).

After you have typed in your password, the `login` program verifies your User ID and password. Next it checks for the name of a program to execute. In most cases, this will be the shell program. The shell program, if it is the Bourne shell, displays a **$** prompt. Now the shell is ready to accept your commands (Figure 3.28).

Figure 3.28
The Shell Program Displays the **$** Prompt

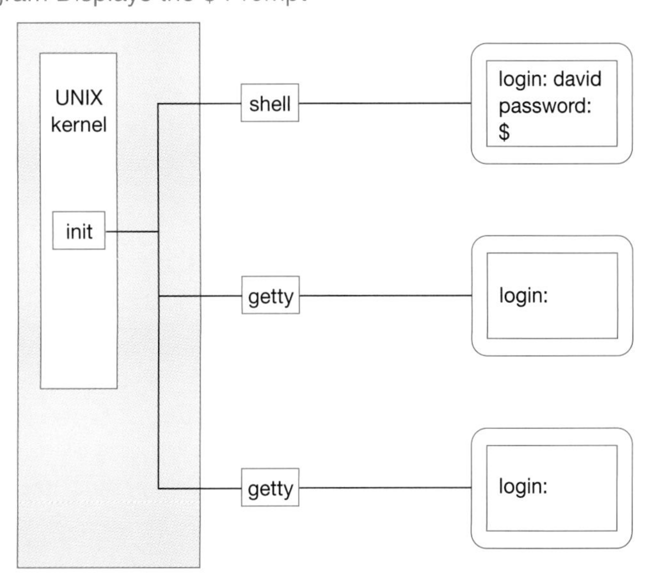

When you log off the system, the shell program is terminated and the UNIX system starts up a new `getty` program at the terminal and waits for someone to log in. This cycle continues as long as the system is up and running (Figure 3.29).

Figure 3.29
Logging-In and Logging-Off Cycle

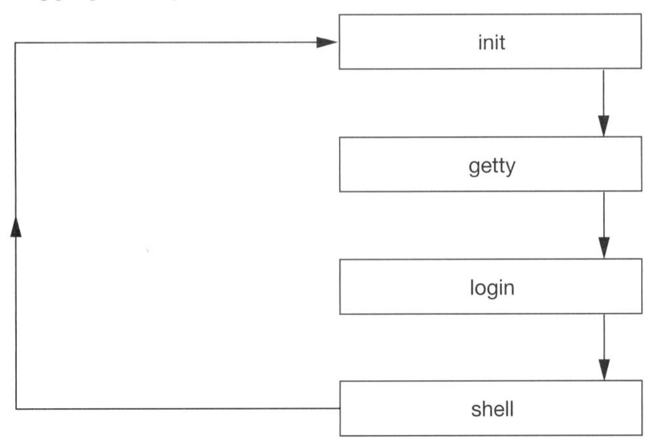

COMMAND SUMMARY

The following UNIX commands (utilities) have been discussed in this chapter. To refresh your memory, the command line format is repeated in Figure 3.30.

Figure 3.30
The Command Line Format

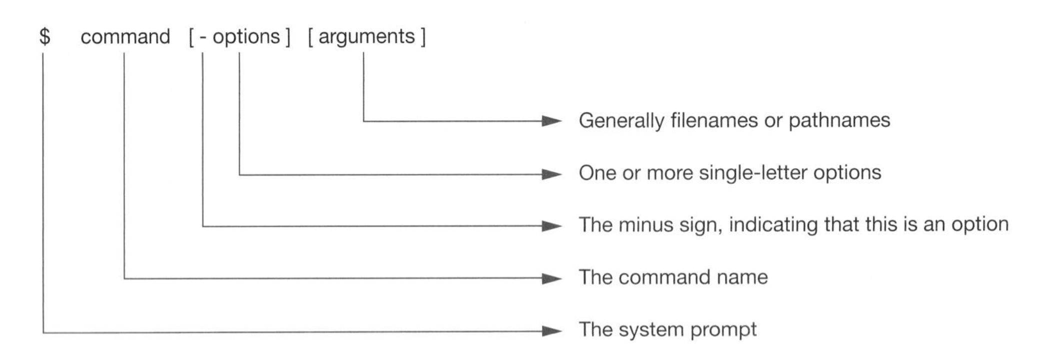

cal
Displays the calendar for a specified year or month of a year.

date
Displays the day of the week, month, date, and time.

help
Brings to the display a series of menus and questions that lead you to the descriptions of the most commonly used UNIX commands.

learn
A computer-aided instruction program that is arranged in a series of courses and lessons. It displays the menu of courses and lets you select your desired course and lesson.

man
This command shows pages from the online system documentation.

passwd
This command changes your login password.

who
Lists the login name, terminal lines, and login times of the users who are on the system.

Option	Linux	Operation
-q	**--count**	The quick **who**; just displays the name and number of users.
-H	**--heading**	Displays heading above each column.
-b		Displays the time and date of the last reboot.
	--help	Displays a usage message.

REVIEW EXERCISES

1. What is the logging-in process?
2. What is the logging-out process?
3. Why are you assigned a login name?
4. What is the sequence of events that comprise the UNIX internal operation at startup (boot) time?
5. What is the shell program, and what is its role in the UNIX environment?
6. What are the general rules for choosing a password?
7. What sort of information does the **man** command provide?
8. What is the general format of a UNIX command?
9. How do you end your command line?
10. What is the *command name*?
11. What is the *options* field in the command line?
12. What is the *arguments* field in the command line?
13. Name the differents types of UNIX shells. What are the prompt signs for UNIX shells?

14. What are the shell commands?
15. What are the utility programs?

Terminal Session

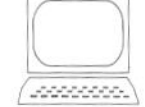

Before starting your first terminal session, spend a few minutes to determine the following information about your system:

Find out your login name (User ID).
If your system requires a password to log in, then find out your password.

Now turn on your terminal and wait for the **login:** prompt.

1. Using your User ID and password, log on to the system. Take notice of the messages appearing on the screen.
2. Check the prompt sign and find out what shell variety you are using.
3. Use the **who** command to find out who is currently logged on to the system.
4. Use the **who** command options to find out the number of users in the system and the last time the system was booted.
5. Find out what help utilities are available in your system.
6. Use the **date** command to see the current date and time.
7. Use the **cal** command to find the day of your birth date.
8. Look at the calendar for the year 2005.
9. Use the **passwd** command to change your password.
10. Try new passwords that do not meet the password format, so that you become familiar with the types of error messages that UNIX displays.
11. After you have successfully changed your password, log out and log in again using your new password.
12. Find out which keys are assigned as the following:
    - The erase key
    - The kill key
    - The interrupt key
13. Try to correct your typing mistakes using the erase key.
14. Try to terminate a line of command using the kill key.
15. Use the **who** command options. If you have Linux, use the --help and other Linux alternative options.
16. Use the **man** command to get detailed information about the commands in this chapter, such as **passwd** and **date** commands.
17. Change your current shell to another shell. For example, if your current shell is the Bourne shell *(sh),* change it to the Korn shell *(ksh)* or if in Linux, change it to the Bourne Again shell *(bash).* Observe the prompt sign. Return to your original shell.
18. Log off to end your session.

CHAPTER 4

The vi Editor: First Look

Chapter 4 is the first of two chapters that discuss the UNIX operating system vi editor (*vi* is pronounced *vee-eye*); Chapter 6 is the second. Chapter 4 starts with an explanation of editors in general and then discusses the types of editors and their applications. After a brief description of the editors supported by UNIX, the chapter introduces the vi editor. The rest of the chapter presents the basic commands necessary to do a simple editing job in the vi editor. Basic concepts and operations with the vi editor are explained:

- The different modes of the vi editor
- Memory buffers
- The process of opening a file for editing
- Saving a file
- Quitting vi

In This Chapter

4.1 WHAT IS AN EDITOR?

Editing a text file is one of the most frequently performed computer operations. In fact, most of the things you will want to do will require some sort of file editing sooner or later.

An editor (text editor) is a tool that facilitates the creation of files or modification of existing files. Files may contain notes, memos, program source code, and so on. An editor is a watered down, vanilla sort of word processor. It does not have the typographical features (bold, center, underline, etc.) that word processors do.

At least one editor program is supplied with the software of every operating system. There are two general types of editors:

- Line editors
- Full screen editors

The Line Editor In a *line editor,* most of the changes are applied to only one line or group of lines at a time. To make a change, you must first specify the line number in the text, and then you must specify the change itself. Line editors are usually difficult to use because you cannot see the scope and context of your editing task. Line editors are good for global operations like searching, replacing, and copying large blocks of text in your file.

The Screen Editor A *screen editor* displays the text that you are editing one screen at a time and allows you to move the cursor around the screen and make changes. Any changes you make are applied to the file, and you get an immediate feedback on the screen. You can easily view the rest of the text one screen at a time. Screen editors are more user friendly than line editors and are preferred for everyday editing jobs.

4.1.1 UNIX-Supported Editors

The UNIX operating system supports a number of both line and screen editors, so you can create or alter a file easily and efficiently. To name a few, *emacs* and *ex* are line editors, and *vi* is a screen editor under UNIX.

A line editor called *ed* is an old editor supplied with the early versions of the UNIX operating systems. Currently, the ex family of editors is supplied with most UNIX operating systems. The ex editor originally provided a display facility that showed and let you work with a screen of text instead of one line at a time. In order to use this capability, you had to give ex the **vi** (for *visual*) command. However, the use of the visual mode of the ex editor became so popular that the developers of the ex editor now provide a separate vi editor. This means you can use vi without having to start the ex editor.

The Text Formatter The UNIX editors are not text-formatting programs. They do not provide functions like centering a line or setting up margins, things that are readily available with any word processing program. To format text, UNIX supports utilities like *nroff* and *troff.*

See page x for an explanation of the icons used to highlight information in this chapter.

Text formatters are usually used to prepare documents. Input to these programs are text files that you create using editors like the vi editor. Output from text formatters is paginated, and you can display the formatted text on the screen or send it to the printer.

Table 4.1 shows some of the UNIX operating system editors and their categories.

Table 4.1
Some of the Editors Supported by UNIX

Editor	Category
ed	The original line editor
ex	A more sophisticated superset of the ed editor
vi	A visual screen editor
emacs	A public-domain screen editor

4.2 THE vi EDITOR

The vi editor is a screen-oriented text editor available on most UNIX operating systems, and it provides some of the flexibility and ease of a word processing program. Because vi is based on the line editor ex, it is possible to use ex commands from within vi. When you use vi, changes to your file are reflected on the terminal screen, and the position of the cursor on the screen indicates the position within your file. More than 100 vi commands are available, providing many capabilities—and also the challenge of learning them! Do not panic. Only a few of the commands are necessary to do a simple editing task.

Two versions of vi, the *view editor* and the *vedit editor,* are tailored for specific tasks or users. These versions are the same as the vi editor except that certain flags (options) are preset. Not every system has these versions on it. By the end of the two vi chapters in this book, you will have learned to customize the vi environment according to your own needs.

The view Editor The *view editor* is the version of the vi editor with the read-only flag set in it. The view editor is useful when you only want to look at the contents of your file and not modify it. The view editor prevents you from inadvertently changing your file, and so your file remains intact.

The vedit Editor The *vedit editor* is a version of the vi editor with several flags set. The vedit editor is intended for beginners. Its flag settings make it easier to learn how to use vi.

4.2.1 The vi Modes of Operation

The vi editor has two basic modes of operation: command (edit) mode and text input mode. Some commands are applicable only to the command mode, and others work only in the text input mode. During a vi editing session, you change from one mode of vi to the other to do the editing job.

Command Mode When you start the vi editor, it comes up in the command mode. In command mode, key entries (any key or sequence of keys you press) are interpreted as commands. The keys are not echoed on the screen, but the commands associated with the pressed key or keys are carried out. While the vi editor is in the command mode, you can delete lines, search for a word, move the cursor on the screen, and perform a number of other useful operations by pressing a key or sequence of keys.

Certain commands in the command mode start with the colon (**:**), forward slash (**/**), or question mark (**?**). The vi editor displays these commands on the last line on the screen. To signal the end of the command line for these types of commands, you press [Return].

Text Input Mode In the text input mode, the keyboard becomes your typewriter. vi displays any key or sequence of the keys that you press, and input is interpreted not as commands but as text that you want to write into your file.

Status Line The bottom line on the screen, usually line 24, is used by vi to give you some feedback about your editing operation. Error messages and other informative messages are displayed on the status line. vi also displays the commands that start with **:**, **/**, or **?** on line 24.

4.3 BASIC vi EDITOR COMMANDS

The basic editing job usually involves the following operations:

- Creating a new file or modifying an existing file (open file operation)
- Entering text
- Deleting text
- Searching text
- Changing text
- Saving the file and quitting the editing session (close file operation)

The following sections walk you through a few editing sessions, explore the vi editor's way of working, and show you some of the necessary editing commands in both text input mode and command mode.

Relying solely on the textbook to learn the vi editor (or any other editor) is not a good idea. It is highly recommended that, after an initial reading of this chapter, you practice the examples and terminal session exercises on your system.

Assumptions In order not to repeat the setup of examples in every section, the general assumptions that are applicable to all examples are summarized here:

- The *current line* is the line that the cursor is on.
- The double underline (=) indicates the position of the cursor on the line.
- The `myfirst` file is used to show the operations of different keys and commands. When necessary, examples are illustrated on screens. If there are multiple screens, the top one shows the current state of the text, and the subsequent ones show the changes on the screen after the specified keys or commands are applied to the text.
- To save space, only the relevant part of the screen is shown.
- The examples are not continuous; that is, the editing changes are not carried over from one example to another.

4.3.1 Access to the vi Editor

As with any other software program, the first step with vi is to learn how to start and end the program. This section shows how to invoke the vi editor to create a small text file and then save the file.

Starting vi

To start the vi editor, type **vi**, press [Spacebar], type the name of the file (in this example, **myfirst**), and then press [Return].

If a `myfirst` file already exists, vi displays the first page (23 lines) of the file on the screen. If `myfirst` is a new file, vi clears the screen, shows the vi blank screen, and positions the cursor at the upper-left corner of the screen.The screen is filled with tildes (~) in the first column. The vi editor is in the command mode and is ready to accept your commands. Figure 4.1 shows the vi editor blank screen.

Figure 4.1
The vi Editor Blank Screen

```
~
~
~
~
"myfirst" [new file] 0 lines, 0 characters
```

1. *The status line shows the filename and the fact that it is a new file.*
2. *To save space, screens shown in this book are not shown in full size (24 lines).*

In order to input your text, you must have the vi editor in the text input mode. Make sure [Caps Lock] is off, and then press **i** (for insert). vi does not display the letter *i* but enters into the text input mode. Now type the lines shown in Figure 4.2. The vi editor displays everything you type on the screen. Use [Backspace] (or [Ctrl-h]) to erase characters, and press [Return] at the end of each line to go to the next line. Do not be overly concerned about the syntax and spelling of the text. The goal here is to create a file and save it.

Figure 4.2
The vi Editor Display Screen

```
The vi history
The vi editor is an interactive text editor that is supported
by most of the UNIX operating systems.
~
~
~
~
~
~
"myfirst" [new file]
```

Make sure [Caps Lock] is off because uppercase and lowercase letters have different meanings in the command mode.

If your file does not fill the whole screen, the vi editor fills the first column of the remaining lines with tildes (~).

Ending vi

In order to save a file that you have created or edited with vi, you must first place vi in command mode. To do so, press [Esc]. If the sound on your terminal is activated, you hear a beep that indicates vi is in the command mode. The **save file** and **quit** vi commands are both commands that start with a colon (**:**). Press **:** to put the cursor on the last line of the terminal screen. Then type **wq** (for *write* and *quit*) and press [Return]. The vi editor saves your file (which you called "myfirst") and passes control back to the shell. The shell displays the dollar sign prompt. The **$** signals that you are out of the vi editor and back to the shell and the system is ready for your next command. Figure 4.3 shows the vi editor display after you enter **:wq**.

Figure 4.3
The vi Editor Screen After the **wq** Command

```
The vi history
The vi editor is an interactive text editor that is supported
by most of the UNIX operating systems.
~
~
~
~
:wq
"myfirst" [new file] 3L, 112C written
$_
```

The vi editor feedback appears on the last line of the screen. It shows the filename followed by the number of lines and number of characters in the file.

4.3.2 Cursor Movement Keys: First Look

In order to accomplish the next terminal sessions, you need to be able to change the mode of the vi editor to command mode or text input mode and to be able to position the cursor on a specific location on the screen. The following section explains the subset of the keys in command mode that are needed to do these operations.

When you start the vi editor, it enters the command mode. Pressing the [Esc] key will place vi in command mode regardless of which mode vi was in. Table 4.2 is a partial list of the cursor movement keys. These keys are used to position the cursor. Later in this chapter these and additional cursor movement keys will be explained and practiced.

Table 4.2
Partial List of the Cursor Movement Keys

Key	Operation
h or **[Left Arrow]**	Moves the cursor position one space to the left.
j or **[Down Arrow]**	Moves the cursor position one line down.
k or **[Up Arrow]**	Moves the cursor position one line up.
l or **[Right Arrow]**	Moves the cursor position one space to the right.

1. **Remember, vi must be in command mode when you want to use the cursor movement keys. If in doubt, press the [Esc] key before using these keys.**
2. **Make sure [Caps Lock] is off, because uppercase and lowercase letters have different meanings in the command mode.**
3. **While the vi editor is in the command mode, most commands are initiated as soon as you press the key. There is no need to use [Return] to indicate the end of a command line.**

4.3.3 Text Input Mode

You must be in the text input mode in order to type text into your file. However, there are several different commands you can use to change to text input mode, and each change mode key enters the vi text input mode in a slightly different manner. The location of the text in your file depends on the position of the cursor on the screen and the key you choose to place vi in the text input mode.

Table 4.3 summarizes the keys that change the vi editor from the command mode to the text input mode. For example, let's use `myfirst`, the file created in the previous section. Suppose you have it on the screen and want to enter **999** in your file. (There is nothing special about 999. It is merely being used to illustrate the location of the text you enter relative to the position of the cursor on the screen. You can type anything you

Table 4.3
The vi Change Mode Keys

Key	Operation
i	Inserts the text you enter before the character that the cursor is on.
I	Places the text you enter at the beginning of the current line.
a	Appends the text you enter after the character that the cursor is on.
A	Places the text you enter after the last character of the current line.
o	Opens a blank line below the current line and places the cursor at the beginning of the new line.
O	Opens a blank line above the current line and places the cursor at the beginning of the new line.

wish.) Also assume the cursor is on the letter *m* of the word *most*, as indicated by the double underline. Depending on what key you choose to enter the text input mode, the 999 that you type will be put in a different location in your file.

Normally, when you enter into text input mode by using any of the change mode keys, vi does not provide feedback or a confirmation message to indicate that you are indeed in text input mode. The vi editor can be tailored to provide feedback and indicate which of the vi modes is in use. If words such as *insert*, *open*, *append*, and so on appear on the screen indicating the mode vi is in, then the vi editor on your system has been tailored to show feedback. You will learn how to tailor the vi editor to your preference in Chapter 6.

1. *The cursor position is identified with a double underline (_).*
2. *The current line is a line that has the cursor placed anywhere in it.*

For the terminal sessions in the following sections, you need to open the `myfirst` file. Open the `myfirst` file by using the following command:

`$ vi myfile [Return]` . . . . Open `myfirst` file

If you have not created a `myfirst` file in the previous section, you can create it now by typing the text from Figure 4.1. Before starting each terminal session, make sure the cursor is on the designated letter. Use the arrow keys on the keyboard (not the ones on the number pad) to place the cursor on the desired letter. Your UNIX system might not recognize the arrow keys for moving the cursor. In this case, use the **h**, **j**, **k**, or **l** keys to move and position the cursor.

Also, for each of the following terminal sessions, you have to close the file and end vi without saving the changes in the file. You want to use the original `myfirst` file for your next terminal sessions. Use the following key sequence to close the `myfirst` file:

- □ Press [Esc]. Just to make sure vi is in command mode.
- □ Press **:** to place the prompt on the status line at the bottom of the screen.
- □ Press **q! [Return]**. This command quits vi without saving the changes. The [!] key indicates you are sure you want to abandon the changes.

These steps for opening and closing the `myfirst` *file at the beginning and end of each part of the vi terminal sessions might seem excessive, but the goal is to have the original* `myfirst` *file for each exercise and to make the exercises independent from each other. If you are comfortable having the results of each of your practice sessions in the same file, then you can skip these steps opening and closing the file for each exercise.*

Inserting Text: Using i or I

Pressing either **i** or **I** places the vi editor in text input mode. The choice you make, however, puts the text you enter in a different place in the file. Pressing **i** places the text you enter before the cursor position, but pressing **I** places your text at the beginning of the current line.

Do the following things to experiment with **i**:

- □ Type **vi myfile [Return]**. This opens the `myfile` file.
- □ Press [Esc]. Just to make sure vi is in the command mode.

□ Use the cursor movement keys to place the cursor on the letter *m* of *most.*

□ Press **i**. vi enters the text input mode.

□ Press **9** three times. 999 appears before the *m.*

The cursor remains on the letter *m,* and the vi editor remains in the text input mode until you press [Esc] to return to the command mode.

```
The vi history
The vi editor is an interactive text editor that is supported
by most of the UNIX operating systems.
```

```
The vi history
The vi editor is an interactive text editor that is supported
by 999most of the UNIX operating systems.
```

Do the following things to experiment with **I**:

□ Press [Esc] to change vi to the command mode.

□ Use the cursor movement keys to place the cursor on the letter *s* of *supported.*

□ Press **I**. vi enters the text input mode and moves the cursor to the beginning of the current line.

□ Press **9** three times. 999 appears at the beginning of the current line, and the cursor moves to the *T*.

```
The vi history
The vi editor is an interactive text editor that is supported
by 999most of the UNIX operating systems.
```

```
The vi history
999The vi editor is an interactive text editor that is supported
by 999most of the UNIX operating systems.
```

The cursor remains on the *T,* and the vi editor remains in the text input mode until you press [Esc] to return to the command mode.

□ Press [Esc] to change vi to the command mode.

□ Type **:q! [Return]**. This quits vi without saving your changes.

Adding Text: Using a or A

Pressing either **a** or **A** places the vi editor in text input mode. The choice you make, however, puts the text you enter in a different place in the file. Pressing **a** places the text you enter after the cursor position at the time you pressed **a**, and pressing **A** adds the text you enter to the end of the current line.

Do the following things to experiment with **a**:

- ☐ Type **vi myfile [Return]**. This opens the `myfile` file.
- ☐ Press [Esc] just to make sure vi is in the command mode.
- ☐ Use the cursor movement keys to place the cursor on the letter *m* of *most.*
- ☐ Press **a**. vi enters the text input mode, and the cursor moves to the *o*, the next letter on the right.
- ☐ Press **9** three times. 999 appears after the *m*.

```
The vi history
The vi editor is an interactive text editor that is supported
by most of the UNIX operating systems.
```

```
The vi history
The vi editor is an interactive text editor that is supported
by m999ost of the UNIX operating systems.
```

The cursor remains on the *o,* and the vi editor is in the text input mode until you press [Esc] to return to the command mode.

Do the following things to experiment with **A**:

- ☐ Press [Esc] to change vi to the command mode.
- ☐ Use the cursor movement keys to place the cursor on the letter *o* of *most.*
- ☐ Press **A**. The vi editor enters the text input mode, and the cursor moves to the end of the current line.
- ☐ Press **9** three times. 999 appears after the . (period), the last character on the current line.

```
The vi history
The vi editor is an interactive text editor that is supported
by m999ost of the UNIX operating systems.
```

```
The vi history
The vi editor is an interactive text editor that is supported
by m999ost of the UNIX operating systems.999_
```

The cursor moves to the end of the line, and the vi editor remains in the text input mode until you press [Esc] to return to the command mode.

- ☐ Press [Esc] to change vi to the command mode.
- ☐ Type **:q! [Return]**. This quits vi without saving your changes.

Opening a Line: Using o or O

Pressing either **o** or **O** places the vi editor in text input mode. Pressing **o** opens a blank line below the current line, and pressing **O** opens a blank line above the current line.

Do the following things to experiment with **o**:

- □ Type **vi myfile [Return]**. This opens the `myfile` file.
- □ Press [Esc] just to make sure vi is in the command mode.
- □ Use the cursor movement keys to place the cursor on the letter *s* of *supported.*
- □ Press **o**. The vi editor enters the text input mode, opens a line below the current line, and moves the cursor to the beginning of the new line.
- □ Press **9** three times. 999 appears on the new line.

```
The vi history
The vi editor is an interactive text editor that is supported
by most of the UNIX operating systems.
```

```
The vi history
The vi editor is an interactive text editor that is supported
999_
by most of the UNIX operating system.
```

The cursor moves to the end of the new line, and the vi editor remains in the text input mode until you press [Esc] to return to the command mode.

Do the following things to experiment with **O**:

- □ Press [Esc] to change vi to the command mode.
- □ Use the cursor movement keys to place the cursor on the letter *s* of *supported.*
- □ Press **O**. The vi editor enters the text input mode, opens a line above the current line, and moves the cursor to the beginning of the new line.
- □ Press **9** three times. 999 appears on the new line.

```
The vi history
The vi editor is an interactive text editor that is supported
999
by most of the UNIX operating systems.
```

```
The vi history
999_
The vi editor is an interactive text editor that is supported
999
by most of the UNIX operating systems.
```

The cursor moves to the end of the new line, and the vi editor remains in the text input mode until you press [Esc] to return to the command mode.

- □ Press [Esc] to change vi to the command mode.
- □ Type **:q! [Return]**. This quits vi without saving your changes.

Avoid the use of the cursor keys (arrow keys) in the text input mode. On some systems, the cursor keys are interpreted as regular ASCII characters, and their ASCII codes will be inserted into your file if they are used in text input mode.

Using [Spacebar], [Tab], [Backspace], and [Return]

While in the text input mode, the vi editor displays the letters you type on the screen, but not all keys on the keyboard produce a displayable character. For example, when you press [Return] you intend to move the cursor to the next line, and you do not expect to see a [Return] symbol on the screen. As you know by now, depending on the vi editor's mode, keys have different meanings. Consider the use of these keys in the text input mode.

[Spacebar] and [Tab] Pressing [Spacebar] always produces a space character before the cursor position. Pressing [Tab] usually produces eight spaces. The tab size is changeable and can be set for any number of spaces. (See Chapter 6.)

[Backspace] Pressing [Backspace] moves the cursor one character to the left over the currently typed text.

[Return] Pressing [Return] when you are in text input mode always opens a new line. Depending on the position of the cursor on the current line, it opens a line above or below the current line:

- If the cursor is at the end of the line or there is no text to the right of it, then pressing [Return] opens an empty line below the current line.
- If the cursor is on the very first character of the current line, then pressing [Return] opens an empty line above the current line.
- If the cursor is somewhere on the line with text to the right of it, then pressing [Return] splits the line, and the text to the right of the cursor moves to the new line.

This explanation applies only when the vi editor is in the text input mode.

Do the following things to experiment with the [Spacebar], [Tab], and [Return] keys:

- □ Type **vi myfile [Return]**. This opens the `myfile` file.
- □ Press **i**. vi enters the text mode.
- □ Type something and practice using the [Spacebar] key. Observe the results.
- □ Practice using the [Tab] key and observe the results.
- □ Practice using the [Return] key and observe the results.
- □ Press [Esc] to change vi to the command mode.
- □ Type **:q! [Return]**. This quits vi without saving your changes.

4.3.4 Command Mode

When you start the vi editor, it enters the command mode. If vi is in the text input mode and you want to change to command mode, press [Esc]. If you press [Esc] while the vi editor is in the command mode, nothing happens, and vi remains in the command mode.

Cursor Movement Keys

In order to delete text, correct text, or insert text, you need to move the cursor to a specific location on the screen. While in command mode, you use the arrow keys (cursor control keys) to move the cursor around on the screen. On some terminals, the arrow keys do not work as explained or are not available. In these cases, you can use the **h**, **j**, **k**, and **l** letter keys to move the cursor left, down, up, and right, respectively. Pressing any of the cursor movement keys moves the cursor position one space, one word, or one line at a time.

Table 4.4 summarizes the cursor movement keys and their applications. Each key application is explained in the examples and the terminal sessions. Figure 4.4 shows the effect of these cursor movement keys on the screen.

Table 4.4
vi's Cursor Movement Keys

Key	Operation
h or **[Left Arrow]**	Moves the cursor position one space to the left.
j or **[Down Arrow]**	Moves the cursor position one line down.
k or **[Up Arrow]**	Moves the cursor position one line up.
l or **[Right Arrow]**	Moves the cursor position one space to the right.
$	Moves the cursor position to the end of the current line.
w	Moves the cursor position forward one word.
b	Moves the cursor position back one word.
e	Moves the cursor position to the end of the word.
0 (zero)	Moves the cursor position to the beginning of the current line.
[Return]	Moves the cursor position to the beginning of the next line.
[Spacebar]	Moves the cursor position one space to the right.
[Backspace]	Moves the cursor position one space to the left.

j and k The **j** or [Down Arrow] key moves the cursor one line down to the same position as it was on the previous line. The **k** or [Up Arrow] key moves the cursor one line up. When there is no line below the current line (end of the file), or the current line is the first line (top of the file), then you hear a beep, and the cursor remains on the current line. These keys do not cause the text to wrap around.

Figure 4.4

The Movement of the Cursor on the Screen

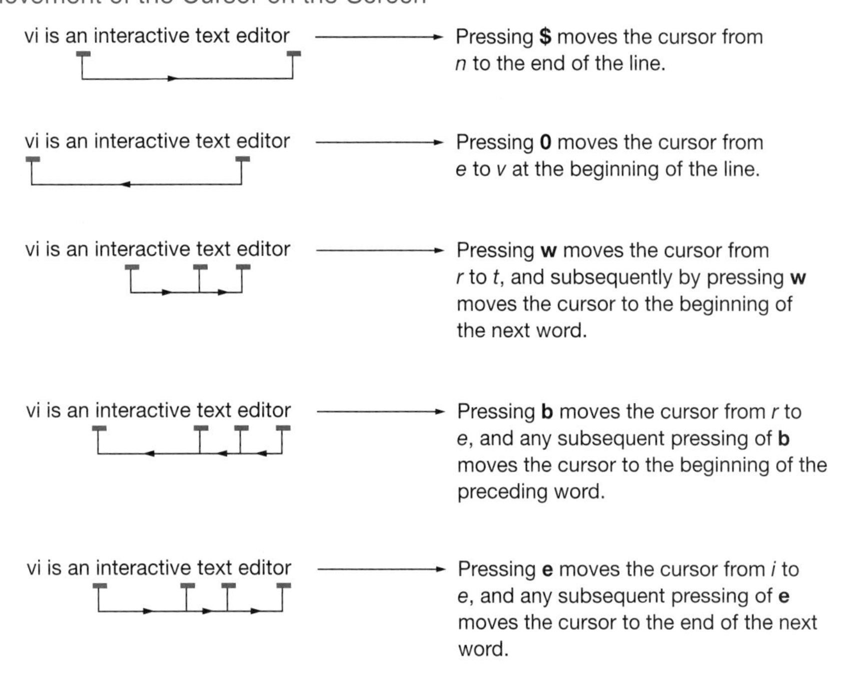

h and l Each time you press the **h** or [Left Arrow] key, the cursor moves one character to the left until it is on the first character of the current line. Then it beeps, indicating it cannot move farther to the left. The **l** (lowercase *L*) or [Right Arrow] key moves the cursor to the right in a similar manner. These keys do not cause text to wrap around.

$ and 0 Pressing **$** moves the cursor to the end of the current line. This key cannot be pressed more than once: When the cursor is at the end of the line, pressing **$** does not change anything. Pressing **0** (zero) moves the cursor to the beginning of the current line in a similar manner.

w, b, and e Each time you press **w**, the cursor moves to the beginning of the next word. Pressing **b** moves the cursor left (back) to the beginning of the preceding word, and pressing **e** moves the cursor to the end of the word. These keys cause the text to wrap around and, if necessary, move the cursor to the next line.

[Return] Each time you press [Return] in command mode, the cursor moves to the beginning of the next line below the current line until you reach the end of the file.

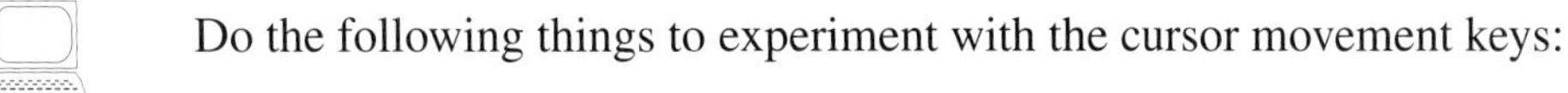

Do the following things to experiment with the cursor movement keys:

- Type **vi myfile [Return]**. Opens `myfile` file.
- Move the cursor to the left using **h** or [Left Arrow]. Observe that the cursor moves over each character and stops at the beginning of the line.
- Move the cursor to the right using **l** (ell) or [right Arrow]. Observe that the cursor movcs over each character and stops at the end of the line.
- Move the cursor to the top of the file using the **k** or [Up Arrow]. Observe that the cursor moves over each line and stops on the first line of the file.
- Move the cursor to the bottom of the file using the **j** or [Down Arrow]. Observe that the cursor moves over each line and stops on the last line of the file.

- Press **$**. Observe that the cursor moves to the end of the current line.
- Press **0**. Observe that the cursor moves to the beginning of the current line.
- Press **w**. Observe that the cursor moves forward one word.
- Press **b**. Observe that the cursor moves backward one word.
- Press **e**. Observe that the cursor moves to the end of the current word.
- Press [Return]. Observe that the cursor moves to beginning of the next line.
- Press [Spacebar]. Observe that the cursor moves one position to the right.
- Press [Backspace]. Observe that the cursor moves one position to the left.
- Practice using the cursor movement keys until you feel comfortable using them or you get tired of using them, whichever comes first.
- Press [Esc] to change vi to the command mode.
- Type **:q! [Return]**. This quits vi without saving your changes.

Text Correction

While in the command mode, you can replace (overwrite) characters and delete characters, a line, or a number of lines. You can also correct some of your mistakes by using the **undo** command. As the name implies this disregards your most recent command. These text correcting commands apply only when vi is in the command mode, and most of them do not change the vi mode. Table 4.5 summarizes the text correction keys and their applications.

Table 4.5
The vi Editor Keys for Correcting Text in Command Mode.

Key	Option
x	Deletes the character specified by the cursor position.
dd	Deletes the line specified by the cursor position.
u	Undoes the most recent change.
U	Undoes all the changes on the current line.
r	Replaces a character that the cursor is on.
R	Replaces characters starting from the cursor position. Also changes vi to the text input mode.
. (dot)	Repeats the last text changes.

Character Deletion: Using x

Suppose you have the `myfirst` file on the screen and want to correct some text in the file. The cursor is positioned on the letter *m* of the word *most.* Using **x**, you can delete characters starting from the cursor position:

- Type **vi myfile [Return]**. This opens the `myfile` file.
- Press [Esc] to make sure vi is in command mode.

□ Use the cursor movement keys to place the cursor on the letter *m* of *most*.

□ Press **x**. The vi editor deletes the *m,* and the cursor moves to *o,* the next letter to the right. The vi editor remains in command mode.

```
The vi history
The vi editor is an interactive text editor that is supported
by most of the UNIX operating systems.
```

```
The vi history
The vi editor is an interactive text editor that is supported
by ost of the UNIX operating systems.
```

□ Press **x** three more times. The vi editor deletes *o*, *s*, and *t*, respectively, as you press **x** repeatedly.

```
The vi history
The vi editor is an interactive text editor that is supported
by _of the UNIX operating systems.
```

The vi editor remains in the command mode and the cursor moves to the space before the *o*. If you want to delete more than one character in a command, you can use the **nx** command, where *n* is an integer immediately followed by the letter *x*. For example, the command **5x** deletes five characters starting from the cursor position.

□ Press **5x**. The vi editor deletes five characters and the cursor moves to the letter *h*.

```
The vi history
The vi editor is an interactive test editor that is supported
by he UNIX operating systems.
```

□ Press [Esc] to make sure vi is in the command mode.

□ Type **:q! [Return]**. This quits vi without saving your changes.

*Repetition can be used with other vi commands. For example, **dd** deletes one line, and **3dd** deletes three lines.*

Deletion and Recovery: Using dd and u

By pressing **d** twice, you can delete a line, starting from the current line. To learn about using **dd**, do the following:

□ Type **vi myfile [Return]**. This opens the `myfile` file.

☐ Press [Esc] to make sure vi is in command mode.

☐ Use the cursor movement keys to place the cursor on the letter *m* of *most.*

```
The vi history
The vi editor is an interactive text editor that is supported
by most of the UNIX operating systems.
```

☐ Press **d** twice. The vi editor deletes the current line, regardless of the cursor position on the line.

```
The vi history
The vi editor is an interactive text editor that is supported
```

☐ The vi editor is in the command mode, and the cursor moves to the beginning of the next line. Press **u**, and the vi editor undoes your last delete.

```
The vi history
The vi editor is an interactive text editor that is supported
by most of the UNIX operating systems.
```

The vi editor remains in the command mode, and the cursor moves to the beginning of the line. If you want to delete more than one line in a command, you can use the **ndd** command, where *n* is an integer followed immediately by two presses of **d**. For example, the command **3dd** deletes three lines starting from the current line.

☐ Use the cursor movement keys to place the cursor on the letter *T* of *The* on the first line.

☐ Press **3dd**. This deletes three lines starting from the current line. The vi editor remains in the command mode.

```
The vi history
The vi editor is an interactive text editor that is supported
by most of the UNIX operating systems.
```

```
=
~
~
```

☐ Press **u**. vi undoes your last delete command, and the deleted lines appear again in your file.

```
The vi history
The vi editor is an interactive text editor that is supported
by most of the UNIX operating systems.
```

- Practice using the **dd** and **u** commands.
- Press [Esc] to make sure vi is in the command mode.
- Type **:q! [Return]**. This quits vi without saving your changes.

Text Replacement: Using r, R, and U

Using **r** or **R**, you can replace a character or a group of characters, starting from the cursor position. However, pressing **R** puts the vi editor in the text input mode, and you must press [Esc] to return to the command mode.

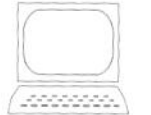

To learn about **r**, do the following:

- Type **vi myfile [Return]**. This opens the `myfile` file.
- Press [Esc] to make sure vi is in command mode.
- Use the cursor movement keys to place the cursor on the letter *m* of *most*.
- Press **r** to replace (overwrite) the character that the cursor is on.
- Press **9**. The vi editor responds by changing the *m* to *9*. The vi editor remains in the command mode, and the cursor stays on the same position.

```
The vi history
The vi editor is an interactive text editor that is supported
by most of the UNIX operating systems.
```

```
The vi history
The vi editor is an interactive text editor that is supported
by 9ost of the UNIX operating systems.
```

To learn about **R** and **U**, do the following:

- Press **R** to replace the characters starting from the cursor position. The vi editor enters the text input mode.
- Press **9** three times. The vi editor responds by adding 999 after the cursor position, overwriting *ost*. The vi editor remains in the text input mode.

```
The vi history
The vi editor is an interactive text editor that is supported
by 9999 of the UNIX operating systems.
```

- Press [Esc] to change to the command mode.
- Press **U** to undo all your changes on the current line.

The vi editor responds by restoring the current line to its previous state.

```
The vi history
The vi editor is an interactive text editor that is supported
by most of the UNIX operating systems.
```

- Practice, using the **r**, **R**, and **u**, **U** commands.
- Press [Esc] to make sure vi is in the command mode.
- Type **:q! [Return]**. This quits vi without saving your changes.

Pattern Search: Using / and ?

The vi editor provides operators to search a file for a specified pattern. The forward slash **/** and question mark **?** are the keys used to search forward and backward, respectively, through your file. If you are editing a large file, you also can use these operators to position the cursor at a specific place in the file. For example, if you want to look for the word *UNIX* in your file, press [Esc] (to make sure vi is in command mode) and then type **/UNIX** and press [Return].

When you press **/**, vi shows the **/** at the bottom of the screen and waits for the rest of the command. After you press [Return], the vi editor starts searching forward for the character string *UNIX* from the current position of the cursor. If the character string *UNIX* is in your file, vi positions the cursor on the first occurrence of it.

You can move the cursor to the next occurrence of *UNIX* (or whatever word you are looking for) by pressing **n** (for next). Every time you press **n**, vi shows the next occurrence of the pattern you are searching for, until the vi editor reaches the end of the file. Then it goes back to the beginning of the file and continues the process.

If you prefer to search backward through the file, type **?UNIX** and press [Return]. The vi editor continues to search backward as long as you press **n** after each finding of the word *UNIX*.

Repeating the Previous Change: Using . (dot)

The **.** (dot) key is used in the command mode to repeat the most recent previous text change. This feature is quite useful when you want to do a lot of repetitive changes in a file.

To experiment with **. (dot)**, try the following:

- Type **vi myfile [Return]**. This opens the `myfile` file.
- Press [Esc] to make sure vi is in command mode.
- Use the cursor movement keys to place the cursor on the letter *m* of *most.*
- Press **dd** to delete the current line. The cursor moves to the beginning of the line above.

```
The vi history
The vi editor is an interactive text editor that is supported
by most of the UNIX operating systems.
```

```
The vi history
The vi editor is an interactive text editor that is supported
```

- Press **. (dot)**. The vi editor repeats the previous text change and deletes the current line. The cursor moves to the beginning of the line above, and vi remains in the command mode.

```
The vi history
~
~
~
```

- Practice, using the **.** and **u**, **U** commands.
- Press [Esc] to make sure vi is in the command mode.
- Type **:q! [Return]**. This quits vi without saving your changes.

Exiting the vi Editor

There is only one way to enter vi, but there are several ways to exit it. The vi editor gives you a choice, depending on what you intend to do with your file after editing. Table 4.6 summarizes the vi editor quit commands.

Table 4.6
The vi Editor Quit Commands

Key	Operation
wq	Writes (saves) the contents of the file and quits the vi editor.
w	Writes (saves) the contents of the file but stays in the editor.
q	Quits the editor.
q!	Quits the editor and abandons the contents of the file.
ZZ	Writes (saves) the contents of the file and quits the vi editor.

The :wq Command Most of the time, you use the **:wq** command at the end of an editing session. This saves the file and exits the vi editor. UNIX displays the **$** prompt, indicating that the shell has returned. The **ZZ** command (uppercase *z*) works just the same: It saves your file and quits the vi editor.

The :q Command You use the **:q** command to look at the contents of a file without doing any editing and exit the vi editor. However, if you have changed something in your file and use the command to exit, vi responds by showing the following message at the bottom line on the screen—a typically terse UNIX message—and the vi editor remains on the screen:

```
No write since last change (:q! overrides).
```

The :q! Command If you change something in a file and then decide not to save the changes, use the **:q!** command to leave the vi editor. In this case, the original file remains intact and changes are abandoned.

The :w Command Use the **:w** command to save your file periodically during the course of a long editing session and to prevent losing your work accidentally. The **:w** command will accept a new filename to save your changes in a new file, if you do not want to write over the original file.

The ZZ Command Use the **ZZ** (uppercase *z*) command to quickly save the file and exit the vi editor.

*The **ZZ** command is not preceded by : and you do not press [Return] to complete the command. Just type **ZZ**, and the job is done.*

Figure 4.5 shows the vi editor modes of operation and the key sequences or commands that change the vi editor from one mode to the other.

Figure 4.5
The vi Editor Modes of Operation

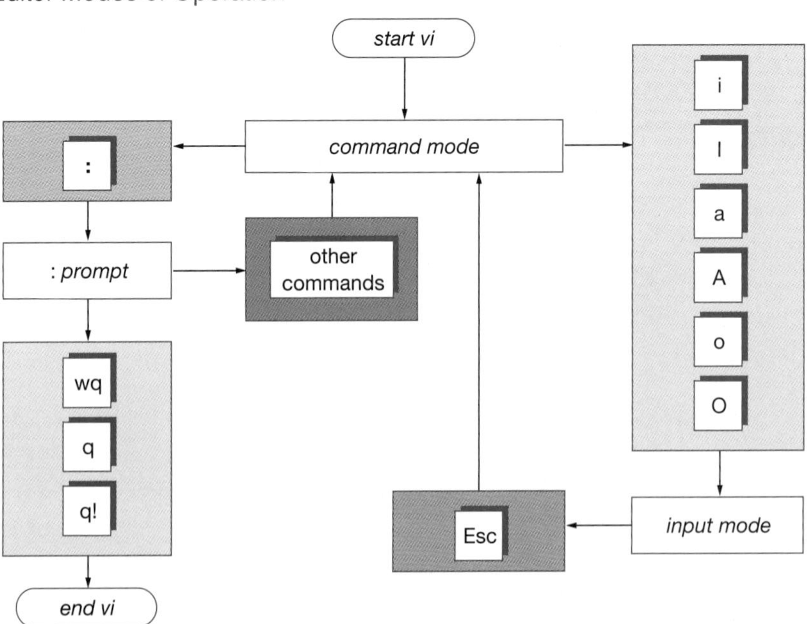

1. **Most of the commands just discussed start with :.**
2. **Pressing : positions the cursor on the last line on the screen. Then vi displays any key you press to complete the command on the same line.**
3. **Remember to press [Return] to signal the completion of a command.**

4.3.5 Linux: vi Online Help

Linux supports an enhanced version of vi called vim (vi improved). This version is upwardly compatible with vi and provides some extra command and features. One of these features is an online help feature that is described here.

In command mode, you type **:help [Return]**. This will show a general help command description similar to that shown in Figure 4.6.

Figure 4.6
vim Help Screen

```
*help.txt*
              VIM - main help file
                                                                  k
       Move around: Use the cursor keys, or "h" to go left,      h  l
                    "j" to go down, "k" to go up, "l" to go right.  j
Close this window: Use ":q<Enter>".
   Get out of Vim: Use ":qa!<Enter>" (careful, all changes are lost!).
Jump to a subject: Position the cursor on a tag between |bars| and hit CTRL-].
   With the mouse: ":set mouse=a" to enable the mouse (in xterm or GUI).
                    Double-click the left mouse button on a tag between |bars|.
        jump back: Type CTRL-T or CTRL-O.
Get specific help: It is possible to go directly to whatever you want help
                    on, by giving an argument to the ":help" command |:help|.
                    It is possible to further specify the context:
                          WHAT              PREPEND     EXAMPLE  ~
                 Normal mode commands     (nothing)    :help x
                 Visual mode commands        v_        :help v_u
                 Insert mode commands        i_        :help i_<Esc>
                 command-line commands       :         :help :quit
~
~
:help
```

You type **:q [Return]** to exit the help screen and return to your file.

In order to get help for a specific command, you type **:help** followed by the name of the command. For example, while in command mode, type **: help wq [Return]** to get help for the **wq** command. This will show a description of the **wq** command similar to that shown in Figure 4.7.

Figure 4.7
vim Help Screen for **wq** Command

```
                                                        *:wq *
:wq          Write the current file and exit (unless editing the
             last file in the argument list or the file is
             read-only)
~
~
~
:help wq
```

You type **:q [Return]** to exit the help screen and return to your file.

4.4 THE MEMORY BUFFER

The vi editor creates a temporary workspace for a file that you want to create or modify. If you are creating a new file, vi opens a temporary workspace for your file. If the specified file is an existing file, vi copies the original file into temporary workspace and the changes you make are applied to that copy and not the original file. This temporary workspace is called a *buffer* or *work buffer*. The vi editor uses several other buffers to manage your file during an editing session. If you want to keep the changes you have made, you must save the altered file (the copy in the buffers) to replace the original file. Changes are not saved automatically: You save your file by issuing a **write** command.

When you open a file for editing, vi copies the file into a temporary buffer and displays the first 23 lines of the file on the screen. A window of 23 lines of the text in the buffer is what you see on the screen (see Figure 4.8A). By moving this window up and down over the buffer, vi displays other parts of the text. When you use the arrow keys or other commands to move the window down, say, to line 10, the first nine lines on the screen are scrolled out (disappear from the screen) and you see lines 10 to 32 of the text (Figure 4.8B). Using arrow keys and other commands, you can move the window up or down over any portion of the file.

Figure 4.8
The vi Editor Temporary Buffer

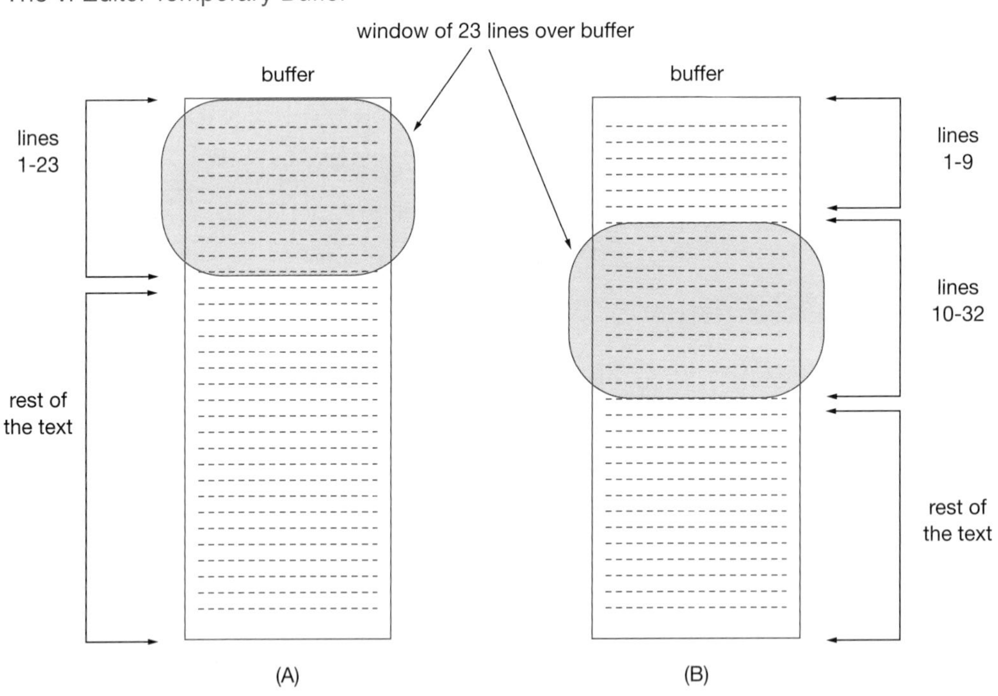

You must remember to write (save) your changes before quitting the vi editor; otherwise, your changes are discarded.

COMMAND SUMMARY

The following vi editor commands and operators have been discussed in this chapter. To refresh your memory, Figure 4.9 shows the vi editor modes of operation.

Figure 4.9
The vi Editor Modes of Operation

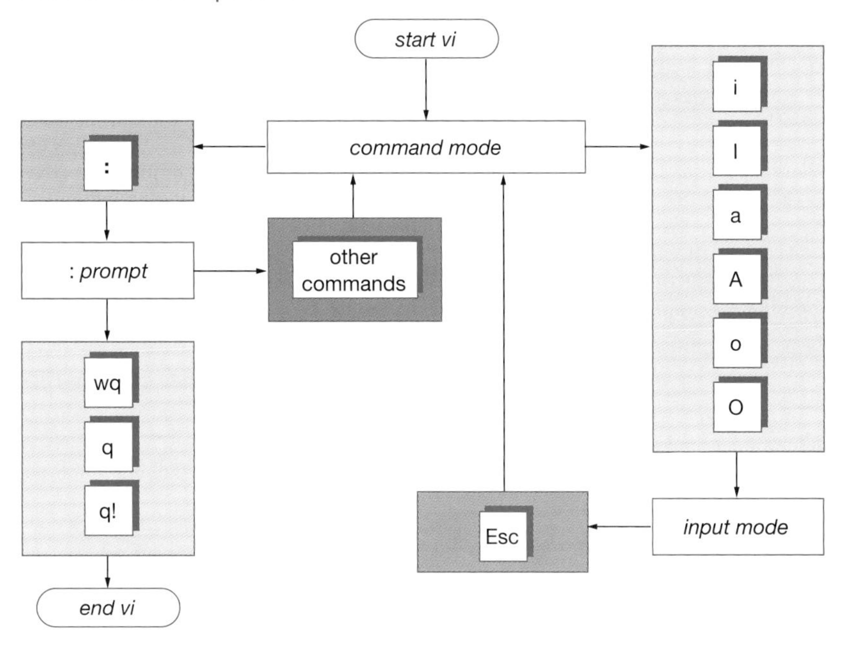

The vi editor

vi is a screen editor you can use to create files. vi has two modes: the command mode and the text input mode. To start vi, type **vi**, press [Spacebar], and type the name of the file. Several keys place vi in the text input mode, and [Esc] always returns vi to the command mode.

The change mode keys

These keys change vi from the command mode to the text input mode. Each key places vi in the text input mode in a different manner. [Esc] places vi back in the command mode.

Key	Operation
i	Places the text you enter before the character that the cursor is on.
I	Places the text you enter at the beginning of the current line.
a	Places the text you enter after the character that the cursor is on.
A	Places the text you enter after the last character of the current line.
o	Opens a blank line below the current line and places the cursor at the beginning of the new line.
O	Opens a blank line above the current line and places the cursor at the beginning of the new line.

Correcting text keys

These keys are applicable in the command mode only.

Key	Operation
x	Deletes the character specified by the cursor position.
dd	Deletes the line specified by the cursor position.
u	Undoes the most recent change.
U	Undoes all the changes on the current line.
r	Replaces a character that the cursor is on.
R	Replaces characters starting from the cursor position, and changes vi to the text mode.
.	Repeats the last text changes.

Cursor movement keys

These keys allow you to move around in your document in command mode.

Key	Operation
h or **[Left Arrow]**	Moves the cursor position one space to the left.
j or **[Down Arrow]**	Moves the cursor position one line down.
k or **[Up Arrow]**	Moves the cursor position one line up.
l or **[Right Arrow]**	Moves the cursor position one space to the right.
$	Moves the cursor position to the end of the current line.
w	Moves the cursor position forward one word.
b	Moves the cursor position back one word.
e	Moves the cursor position to the end of the word.
0 (zero)	Moves the cursor position to the beginning of the current line.
[Return]	Moves the cursor position to the beginning of the next line.
[Spacebar]	Moves the cursor position one space to the right.
[Backspace]	Moves the cursor position one space to the left.

The quit commands

With the exception of the **ZZ** command, these commands start with **:**, and you must end a command line with [Return].

Key	Operation
wq	Writes (saves) the contents of the buffer and quits the vi editor.
w	Writes (saves) the contents of the buffer but stays in the editor.
q	Quits the editor.
q!	Quits the editor and abandons the contents of the buffer.
ZZ	Writes (saves) the contents of the buffer and quits the vi editor.

The search commands

These keys allow you to search forward or backward in your file for a pattern.

Key	Operation
/	Searches forward for a specified pattern.
?	Searches backward for a specified pattern.

REVIEW EXERCISES

1. What is an editor?
2. What is a text formatter?
3. Name the editors that the UNIX operating system supports.
4. Name the vi modes.
5. Name the keys that place the vi editor in the text input mode.
6. Explain how the vi editor uses buffers.
7. Name the command that saves your files and quits the vi editor.
8. Name the command that just saves your file and remains in the vi editor.
9. Name the key that places the vi editor in the command mode.
10. Name the operator that deletes one line of text and the operator that deletes five lines of text.
11. Name the operator that deletes a character and the operator that deletes 10 characters.
12. Name the key that repeats your most recent text change.
13. Name the key that moves the cursor position to the end of the current line.
14. Name the key that moves the cursor position forward one word.
15. Name the cursor movement keys that move the cursor up, down, left, and right.
16. Name the key that appends the text you enter to the end of the current line.

17. Name the key that opens a line above the current line.

18. Name the key that opens a line below the current line.

19. Name the key that undoes the most recent changes.

20. Name the key that undoes all the changes on the current line.

Match the commands shown in the left-hand column to the explanations shown in the right-hand column.

1. x	a. Moves the cursor position one line up.
2. r	b. Deletes the character under the cursor.
3. /	c. Moves the cursor position to the beginning of the current line.
4. ?	d. Places the text you enter after the last character on the current line.
5. h	e. Moves the cursor position to the end of the current line.
6. A	f. Searches backward for a specified pattern.
7. q!	g. Quits the vi editor without saving your file.
8. wq	h. Writes (saves) your file and quits the vi editor.
9. a	i. Places the text you enter after the character that the cursor is on.
10. $	j. Moves the cursor position one space to the left.
11. 0 (zero)	k. Searches forward for a specified pattern.
12. k	l. Replaces the character that the cursor is on.

Terminal Session

In this terminal session, you will create a small text file and practice using the editing keys that vi provides. Use your imagination. Do not limit yourself to the small file in this exercise.

Try to use all the keys that are explained in this chapter.

1. Use the vi editor to create a file called `Chapter4` and type in the text shown on screen 1.

2. Save this file.

Screen No. 1

```
The vi history
The vi editor was developed at the University of california, berkeley
as part of the berkeley unix system.
~
~
"Chapter4" [new file ]
```

3. Open the test file again and add text to make it look like the text shown on screen 2.

4. Save this file again.

Screen No. 2

```
The vi history
The vi editor was developed at the University of california
berkeley as part of the berkeley unix system.
At the beginning the vi editor was part of another editor
The vi part of the ex editor was often used and became very.
This popularity forced the developers to come up with a separate
vi editor.
now the vi editor is independent of the ex editor and is available on
most of the UNIX operating system.
The vi editor is a good editor for everyday editing jobs.
```

5. Open the test file once more and edit the text to make it look like the text on screen 3.
6. Search for the word **vi** using the **/** (forward search). Use **n** to find the next occurrence of the word **vi**.
7. Search for the word **vi** using the **?** (backward search). Use **n** to find the next occurrence of the word **vi**.
8. Place the cursor at the beginning of the file and delete five lines. Undo your delete.
9. Place the cursor at the beginning of the second line and delete 10 characters. Undo your delete.
10. Use **r** to replace the character at the cursor position. Undo your action.
11. Use **R** to change the word **developers** to **creators**. Undo your action.
12. If you have a Linux system, practice the following:
    a. Get help for the **help** command.
    b. Get help for the **ZZ** command.
    c. Get help for the **search** command.
13. Save your file for later exercise sessions using the **ZZ** command.

Screen No. 3

```
The vi history
The vi editor was developed at the University of California
Berkeley as part of the Berkeley UNIX system.
At the beginning the vi (visual) editor was part of the ex editor
and you had to be in the ex editor to use the vi editor.
The vi part of the ex editor was often used and became
very popular. This popularity forced the developers to come up
with a separate vi editor.
Now the vi editor is independent of the ex editor and is available
on most of the UNIX operating systems.
The vi editor is a good, efficient editor for everyday editing jobs
although it could be more user friendly.
~
~
~
~
```

CHAPTER 5

Introduction to the UNIX File System

This is the first of two chapters that discuss the file structure of the UNIX system; Chapter 7 is the second. Chapter 5 describes the basic concepts of files and directories and their arrangement in a hierarchical tree structure. It defines the terminology used in the UNIX file system. It discusses commands that facilitate the manipulation of the file system, explains the naming conventions for files and directories, and shows a practical view of the file system and its associated commands in the terminal session exercises.

In This Chapter

5.1 DISK ORGANIZATION

The information a computer uses is stored in files. The memos you write, the programs you create, and the text you edit are all stored in files. But where are the files stored? How do you keep track of them? You give each file a name, and the files are usually saved on a particular section of the hard disk. But considering the capacity of disks, how do you keep track of the whereabouts of the files on the disk? You divide your disk into smaller units and subunits, name each of them, and store related information in the same unit or subunit.

UNIX follows the same procedure in handling your disk. While you are working on a computer, the files you work with are stored in the computer's random access memory (RAM) or what in general is called the *main memory.* UNIX uses RAM for short-term storage. Your files are usually stored on hard disk, for permanent, long-term storage. Hard disk is the most common file storage medium. However, you can store your files on other storage media such as floppy disks, zip disks, or tape.

UNIX allows you to divide your hard disk into many units (called *directories*), and subunits (called *subdirectories*), thereby nesting directories within directories. UNIX provides commands to create, organize, and keep track of directories and files on the disk.

5.2 FILE TYPES UNDER UNIX

For the UNIX operating system, a file is a sequence of bytes. UNIX does not support other structures (such as records or fields), as some other operating systems do. UNIX has three categories of files:

Regular Files Regular files contain sequences of bytes that could be programming code, data, text, and so on. The files you create using the vi editor are regular files, and most of the files you manipulate are this type of file.

Directory Files In most respects, a directory file is a file like any other file, and you name it as you name any other file. It is not, however, a standard ASCII text file. The directory file is a file that contains information (like the file name) about other files. It consists of a number of such records in a special format defined by your operating system.

Special Files Special files (device files) contain specific information corresponding to peripheral devices such as printers, disks, and so on. UNIX treats I/O (input and output) devices as files, and each device in your system—the printer, floppy disk, terminal, and so on—has a separate file.

See page x for an explanation of the icons used to highlight information in this chapter.

5.3 ALL ABOUT DIRECTORIES

Directories are an essential feature of the UNIX file system. The directory system provides the structure for organizing files on a disk. To visualize a disk and its directory structure, think of your disk as a file cabinet. A file cabinet may have several drawers, which can be compared to disk directories. A drawer may be divided into several sections, which are comparable to subdirectories.

In UNIX, the directory structure is organized in levels and is known as a hierarchical structure. This structure allows you to organize files so you can easily find any particular one. The highest level directory is called the *root* and all other directories branch directly or indirectly from it. Directories do not contain the information contained in your files, but instead provide a reference path to allow you to organize and find your files. Figure 5.1 shows the root and some other directories.

Figure 5.1
Directory Structure

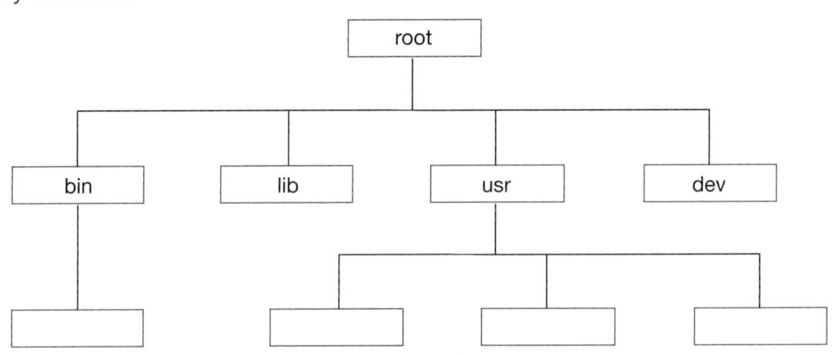

One example of a hierarchical structure is your family tree. A couple may have a child, that child may have several children, and each of those children may have more children. The terms *parent* and *child* describe the relationship between levels of the hierarchy. Figure 5.2 shows this relationship. Only the root directory has no parents. It is the ancestor of all the other directories.

Figure 5.2
Parent and Child Relationship

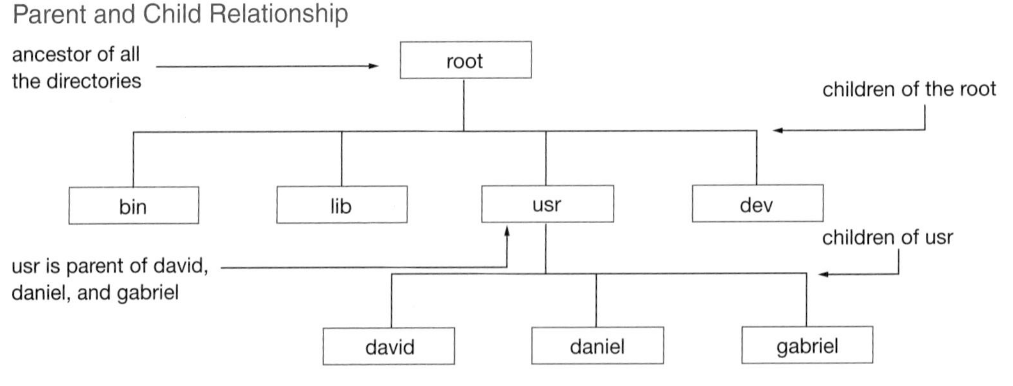

A hierarchical directory structure is often illustrated as a tree. The tree representing a file structure is usually pictured upside down, with its root at the top. Using this tree analogy, the tree's root represents the root directory, the branches the other directories, and the leaves are the files.

5.3.1 Important Directories

Most of the directories that hold UNIX system files are standard. Other UNIX systems such as Linux will have identical directories with similar contents. These directories are usually accessible to ordinary users to list files or read files; however, you probably cannot edit, copy, or delete them. As an ordinary user, you have full access to your own directories and files that you create in your home directory, but you have limited access to other files. This topic will be addressed later in this chapter and Chapter 13.

Following are summaries of some of the more important directories on your UNIX system.

/
This is the root directory. It is the highest-level directory and all other directories branch from it.

/usr
This directory holds users' home directories. In other UNIX systems including Linux, this can be the **/home** directory. This directory holds many other user-oriented directories including the following:

/usr/docs
This directory holds various documents.

/usr/man
This directory holds man (online manual) pages.

/usr/games
This directory holds game programs.

/usr/bin
This directory holds user-oriented UNIX programs.

/usr/spool
This directory has several subdirectories such as **mail**, which holds mail files, and **spool**, which holds files to be printed.

/usr/sbin
This directory holds system administration files. You must be a privileged user (root user) to have access to most of these files.

/bin
This directory holds many of the basic UNIX program files. bin stands for *binaries*, and these files are executable files.

/dev
This directory holds device files. These are special files that represent the physical computer components such as printer or disk. UNIX treats everything as a file. For example, your terminal is one of the ***/dev/tty*** files. A special device is ***/dev/null***, the null device (sometimes called the bit bucket). All information sent to null device is deleted.

/sbin
This directory holds system files that usually are run automatically by the UNIX system.

/etc
This directory and its subdirectories hold many of the UNIX configuration files. These files are usually text, and they can be edited to change the system's configuration. Of course, you must be a privileged user to have access to edit these files.

5.3.2 The Home Directory

The system administrator creates all user accounts on the system and associates each user account with a particular directory. This directory is the *home directory.* When you log on to the system, you are placed automatically into your home directory. From this single directory (your home directory), you can expand your directory structure according to your needs. You can add as many subdirectories as you like, and by dividing subdirectories into additional subdirectories, you can continue expanding your directory structure. Figure 5.3 shows that the directory called `usr` has three subdirectories called `david`, `daniel`, and `gabriel`. The directory `david` contains three files, but the other directories are empty.

Figure 5.3
Directories, Subdirectories, and Files

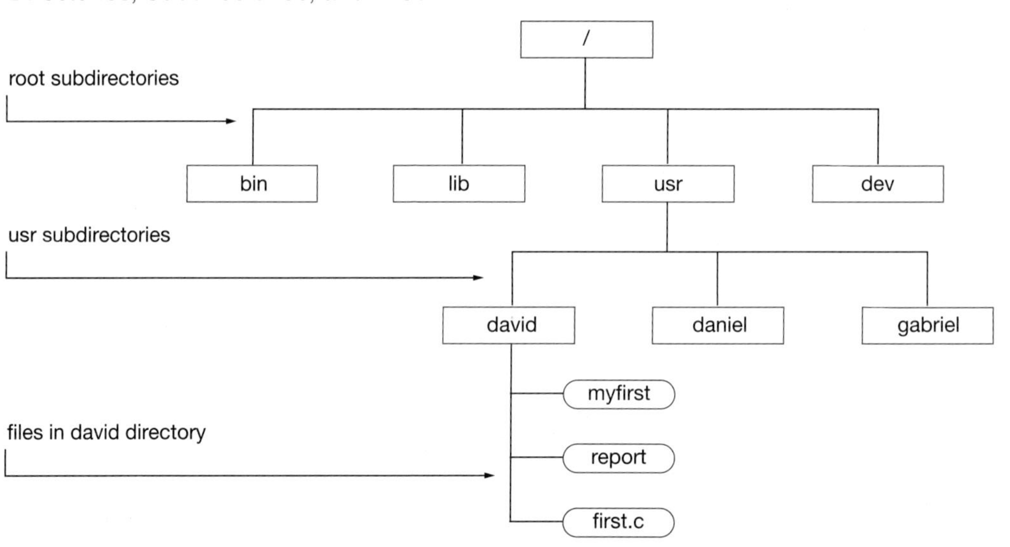

1. *Figure 5.3 is not the standard UNIX file structure, and file structure setup varies from one installation to another.*
2. *Your login name and your home directory name are usually the same and are assigned by the system administrator.*
3. *The root directory is present in all UNIX file structures.*
4. *The name of the root directory is always the forward slash (/).*

5.3.3 The Working Directory

While you are working on the UNIX system, you are always associated with a directory. The directory you are associated with or working in is called the *working directory* or the *current directory*. Several commands allow you to view or change your working directory. These commands will be covered later in this chapter.

5.3.4 Understanding Paths and Pathnames

Every file has a pathname. The pathname locates the file in the file system. You determine a file's pathname by tracing a path from the root directory to the file, going through all intermediate directories. Figure 5.4 shows a hierarchy and the pathnames of its directories and files. For example, using Figure 5.4, if your current directory is *root,* then the path to a file (say, `myfirst`) under the `david` directory is `/usr/david/myfirst`.

Figure 5.4
Pathnames in a Directory Structure

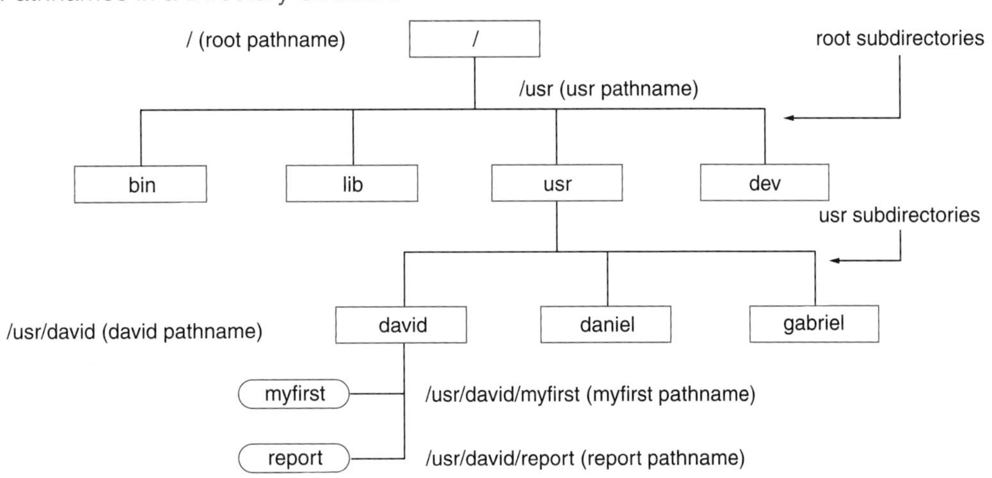

At the end of each path is an ordinary file (called *file*) or a directory file (called *directory*). Ordinary files are at the ends of paths and cannot have further directories—or, using the tree analogy, leaves cannot have branches. Directory files are the points in the file structure that can support other paths, just as tree branches can have other branches.

1. *The forward slash (/) at the very beginning of a pathname stands for the root directory.*
2. *The other slashes serve to separate the names of the other directories and files.*
3. *The files in your working directory are immediately accessible. To access files in another directory you need to specify the particular file by its pathname.*

Absolute Pathname An *absolute pathname* (full pathname) traces a path from the root to the file. An absolute pathname always begins with the name of the root directory, forward slash (/). For example, if your working directory is `usr`, the absolute pathname of the file called `myfirst` under the directory `david` is `/usr/david/myfirst`.

1. *The absolute pathname specifies exactly where to find a file. Thus, it can be used to specify file location in the working directory or any other directory.*
2. *Absolute pathnames always start from the root directory and therefore have a forward slash (/) at the beginning of the pathname.*

Relative Pathname A *relative pathname* is a shorter form of the pathname. It traces a path from the working directory to a file. Like the absolute pathname, the relative pathname can describe a path through many directories. For example, if your working directory is `usr`, the relative pathname to the file called `REPORT` under the `david` directory is `david/REPORT`.

There is no initial forward slash (/) for a relative pathname. It always starts from your current directory.

5.3.5 Using File and Directory Names

Every ordinary and directory file has a filename. UNIX gives you much freedom in naming your files and directories. The maximum length of the filename depends on the UNIX version and system manufacturer. All UNIX systems allow a filename to be at least 14 characters, and most support much longer filenames, up to 255 characters.

You name a file using a combination of characters or numbers. The only exception is the root directory, which is always named / (forward slash) and is referred to by this single character. No other file can use this name.

Certain characters have a special meaning to the shell you are using. If those characters are used in a filename, the shell interprets them as part of a command and acts on them. Although there are ways to override this interpretation of special characters, a better choice is to avoid using them. In particular, avoid using the following characters in filenames:

< >   less than and greater than signs
( )   open and close parentheses
[ ]   open and close brackets
{ }   open and close braces
*   asterisk or star
?   question mark
"   double quotation mark
'   single quotation mark
–   minus sign
$ dollar sign
^   caret

You can avoid confusion if you choose characters for filenames from the following list:

(A–Z)   uppercase letters
(a–z)   lowercase letters
(0–9)   numbers
(_)   underscore
(.)   dot (period)

UNIX uses spaces to tell where one command or filename ends and another one begins. Use a period or underline where you normally would use a space. For example, if you want to give the name `MYNEWLIST` to a file, you may call it `MY_NEW_LIST` or `MY.NEW.LIST` instead.

The filename you choose should mean something. Names like *junk, whoopee,* or *ddxx* are correct filenames, but they are poor choices because they do not help you recall what you stored in the file. Choose a filename that is as descriptive as possible and associate it with the contents of the file. The following filenames have correct syntax and also convey information about the contents of the file:

```
REPORTS        Jan_list  my_memos
shopping.lis   Phones    edit.c
```

The UNIX operating system is case sensitive: Uppercase letters are distinguished from lowercase letters. You can use a mixture of uppercase and lowercase letters in any sequence within your filename. Keep in mind, however, that files named `MY_FILE`, `My_File`, and `my_file` are considered to be three different files.

UNIX makes no distinction between names that can be assigned to ordinary files and those that can be assigned to directory files. Hence, it is possible to have a directory and a file with the same name, for example, the `lost+found` directory could have a file in it called `lost+found`.

Referring back to the parent and child analogy, no two files in the same directory can have the same name, like children of one parent. It makes good sense for parents to give their children different names, but in UNIX it is mandatory. However, like children of different parents, files in different directories can have the same names.

You must avoid using spaces in any part of a filename.

Filename Extensions The filename extension helps to further categorize and describe the contents of a file. Filename extensions are part of the filename following a period and in most cases are optional. Some programming language compilers, like those for C, depend on a specific filename extension. (Compilers are explained in Chapter 10.) In Figure 5.5, `first.c` and `first.cpp` in the `source` directory have typical file extensions, (**`.c`** and **`.cpp`** for the C and C++ programming languages respectively).

The following examples show some filenames with extensions:

```
report.c       report.o
memo.04.10
```

Note that the use of more than one period in a file extension is allowed in UNIX.

Figure 5.5
An Example of a Directory Structure

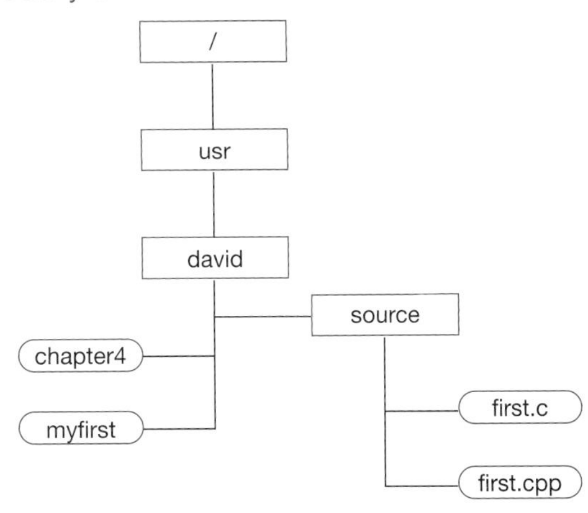

5.4 DIRECTORY COMMANDS

Now that you are familiar with some of the basic file concepts and definitions, it is time to learn to work with files and directories. The following examples and command sequences show the use of commands that let you manipulate your files and directories.

In the following examples, assume that your login name is `david`, Figure 5.5 is your directory structure, and your home directory is `david`.

5.4.1 Displaying a Directory Pathname: The pwd Command

The **pwd** (print working directory) command displays the absolute pathname of your working (current) directory. For example, when you first log in on a UNIX system, you are in your home directory. To display the pathname of your home directory, which at this point is also your working directory, you use the **pwd** command just after you log in.

Log in and show the pathname of your home directory.

```
login: david [Return] . . . . . Enter your login name (david)
password: . . . . . . . . . . . Enter your password.
Welcome to UNIX!
$ pwd [Return]  . . . . . . . . Display your HOME directory path.
/usr/david
$_  . . . . . . . . . . . . . . Prompt for next command.
```

1. `/usr/david` *is your home directory pathname.*
2. `/usr/david` *is also your current or working directory pathname.*
3. `/usr/david` *is an absolute pathname because it begins with /, tracing the path of your home directory from the root.*
4. `david` *is your login name and your home directory name.*

Locating a File in Your Working Directory Your working directory is `david`, and Figure 5.5 shows that you have two files plus a directory called `source`, which has two files in it. You want to locate the file called `myfirst`. The pathname for `myfirst`, which is in the `david` directory, is `/usr/david/myfirst`. This is the absolute pathname to the file. However, when a file is in your working directory, you do not need a pathname to refer to it. The name of the file (in this case, `myfirst`) by itself is sufficient.

Locating a File in Another Directory When a file is in a directory different from your working directory, you need to specify which directory the file is in. Suppose your working directory is `usr`. The pathname for the file called `first.c` in your source directory is `david/source/first.c`.

`david/source/first.c` *is what is called a relative pathname. It does not start from the root directory.*

5.4.2 Changing Your Working Directory: The cd Command

You are not going to work in your home directory all the time. You are likely to change your working directory from one directory to another. The **cd** (change directory) command makes the specified directory your working directory.

For example, to change your working directory to the `source` directory, use the following command sequence:

```
$ pwd [Return]  . . . . . . Check your current directory.
/usr/david
$ cd source [Return]  . . . Change to source directory.
$ pwd [Return]  . . . . . . Display your working directory.
/usr/david/source
$_  . . . . . . . . . . . . Prompt for the next command.
```

Assuming you have permission, you can change your working directory to `/dev` by using the following command sequence.

```
$ cd /dev [Return]. . . . . Change to /dev directory.
$ pwd [Return]. . . . . . . Check your working directory.
/dev. . . . . . . . . . . . Your current directory is /dev.
$_  . . . . . . . . . . . . Prompt for next command.
```

Returning to Your HOME Directory When you have levels of directories and your working directory happens to be a few levels deep in the nested directory structure, then it is convenient to be able to return to your home directory without too much typing. You use the **cd** command with **$HOME** (a variable that holds your home directory pathname) as the directory name. You can also type just the **cd** command followed by [Return] since the default is your home directory.

Use the following command sequence to practice the use of the **cd** command:

```
$ cd $HOME [Return] . . . . Return to HOME directory.
$ cd /bin [Return]  . . . . Change to /bin directory.
$ pwd [Return]  . . . . . . Check your working directory.
/bin  . . . . . . . . . . . Your current directory is /bin.
$ cd [Return] . . . . . . . No directory name is specified; the default
                            is your HOME directory.
$ pwd [Return]  . . . . . . Check your working directory.
/usr/david
$ cd xyz [Return] . . . . . Change to directory named xyz. It does
                            not exist, so you get an error message.
xyz: not a directory
$_. . . . . . . . . . . . . Prompt for next command.
```

5.4.3 Creating Directories

The very first time you log on to the UNIX system, you begin work from your home directory, which is also your working directory. Probably no file or subdirectory exists in your home directory at this point, and you will want to build your own subdirectory system.

Advantages of Creating Directories

There are no restrictions on directory structure in UNIX. If you want to keep all your files under your home directory, you can. Although creating an effective directory structure takes some time, it also provides many advantages. The advantages become apparent, particularly when you have a large number of files. The following lists some of the advantages of using directories:

- Grouping related files in one directory makes it easier to remember and access them.
- Displaying a shorter list of your files on the screen enables you to find a file more quickly.
- You can use identical filenames for files that are stored in different directories.
- Directories make it feasible to share a large-capacity disk with other users (perhaps other students) with a well-defined space for each user.
- You can take advantage of the UNIX commands that manipulate directories.

Using directories requires careful planning. If you group your files logically into manageable directories, your plan will pay off handsomely. If you haphazardly create and fill directories, then you might have difficulty finding your files or you might require more time to recognize them.

Directory Structure

Let's start with the directory structure presented in Figure 5.6. At this point, this structure should be similar to your directory structure. You created `myfirst` and `chapter4` files in your HOME directory when you were practicing with the vi editor, and there are no subdirectories. Depending on your system configuration and administration requirements, you might have other files or subdirectories in your HOME directory. These other files are usually default files and subdirectories that are provided by your system administrator, not ones that you have created.

Figure 5.6
Your Directory Structure at the Beginning

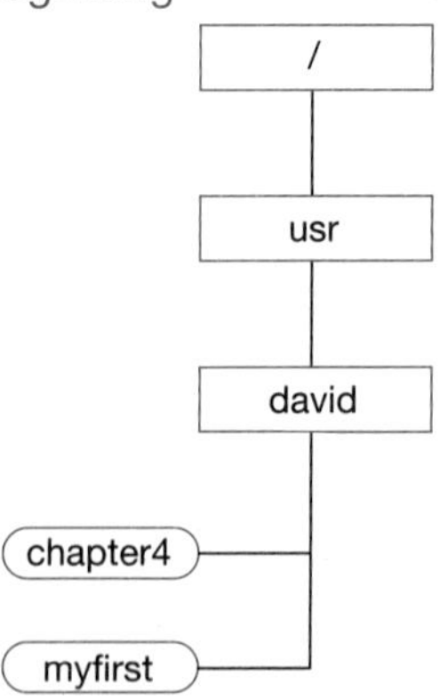

Directory Creation: The mkdir Command

The **mkdir** (make directory) command creates a new subdirectory under your working directory or any other directory you specify as part of the command. For example, the following command sequences show how to create subdirectories under your home directory or other directories.

To create a directory called memos under your HOME directory, use the following command sequence:

```
$ cd [Return]  . . . . . . . . . .  Make sure you are in your HOME
                                    directory.
$ mkdir memos [Return]  . . . . .   Create a directory called memos.
$ pwd [Return]  . . . . . . . . .   Check your working directory.
/usr/david
$ cd memos [Return]  . . . . . .    Change to memos directory.
$ pwd [Return]  . . . . . . . . .   Check your working directory.
/usr/david/memos  . . . . . . . .   Your current directory is memos.
$_  . . . . . . . . . . . . . . .   Prompt for next command.
```

Figure 5.7 shows your directory structure after adding the memos subdirectory.

Figure 5.7

Your Directory Structure After Adding the memos Subdirectories

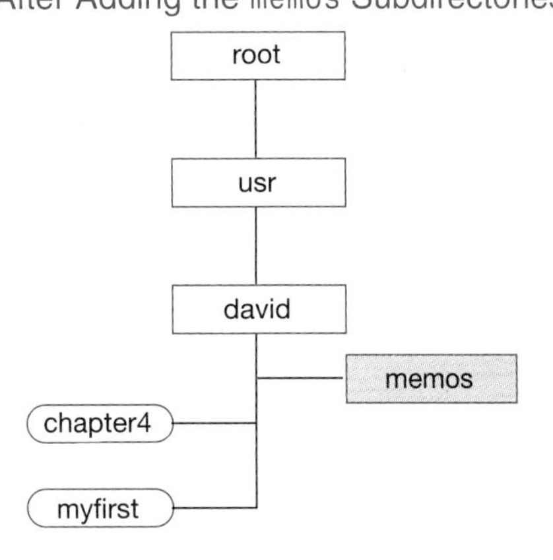

While you are in your HOME directory, create a new subdirectory called important in the memos directory.

```
$ cd [Return]  . . . . . . . . . .  Make sure you are in your HOME
                                    directory.
$ mkdir memos/important [Return]    Specify the important directory
                                    pathname.
$ cd memos/important [Return]  .    Change to important directory.
$ pwd [Return]  . . . . . . . . .   Check your working directory.
/usr/david/memos/important
$_  . . . . . . . . . . . . . . .   Now your working directory is
                                    important.
```

Figure 5.8 shows your directory structure after adding memos and important subdirectories.

Using the same command sequence, Figure 5.9 shows how to create a directory called source under your HOME directory.

Figure 5.10 shows your directory structure after adding the source subdirectory.

A directory structure can be created according to your specific needs.

Figure 5.8
Your Directory Structure After Adding the `memos` and `important` Subdirectories

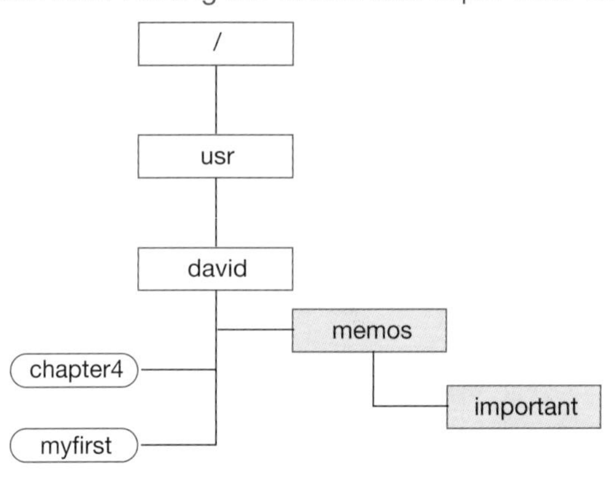

Figure 5.9
Creating the `source` Directory

```
$ cd
$ mkdir source
$ pwd
/usr/david
$ cd source
$ pwd
/usr/david/source
$_
```

Figure 5.10
Your Directory Structure After Adding the `source` Directory

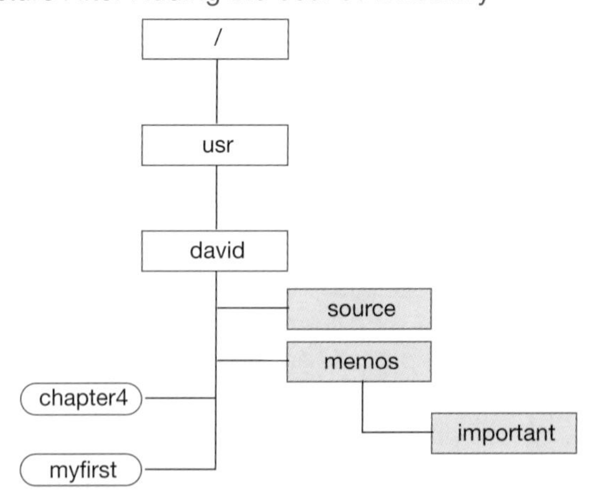

-p Option You can create a whole directory structure using a single command line. You use the **-p** option to create levels of directories under your current directory. For example, suppose you want to create a directory structure three levels deep, starting in your HOME directory. The following command sequence shows how to do

Figure 5.11

Your Directory Structure After Adding the Three-Levels Deep Subdirectories

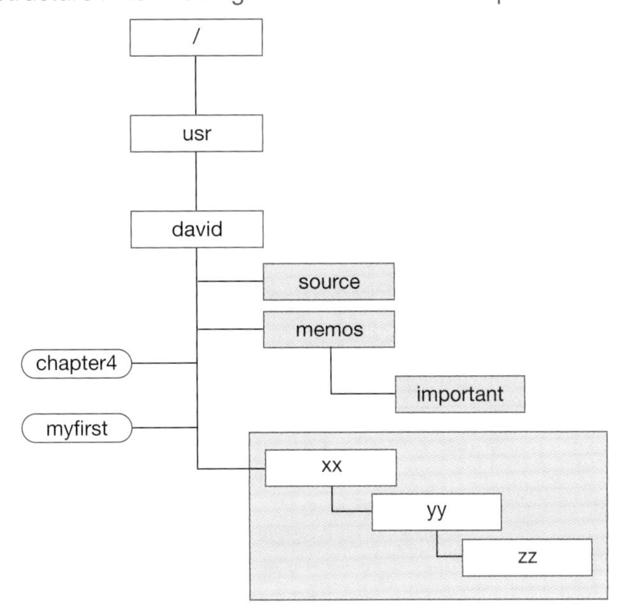

it, and Figure 5.11 depicts the directory structure after this command sequence has been applied.

While in your HOME directory, create a directory structure with three levels by doing the following:

$ **cd [Return]**	Make sure you are in your HOME directory.
$ **mkdir -p xx/yy/zz [Return]**	Create a directory called xx; in xx create a directory called yy, and in yy create a directory called zz.
$_	Ready for next command.

--parents Option The alternative option in Linux is **--parents**. Like the **-p** option, **--parents** creates levels of directories under your current or the specified directory. For example, the command line for using the **--parents** is

$ **mkdir --parents xx/yy/zz [Return]** . .	Create a directory called xx; in xx, create a directory called yy; and in yy, create a directory called zz.

Figure 5.11 shows your directory structure after adding the three-level deep directories named xx, yy, and zz.

1. **The parent directory must be nonexistent. In this example, you should not have a directory called xx in your current directory.**
2. **You do not have to be in the parent directory to create a subdirectory. As long as you give the pathname for the new directory, you can issue the command from any level of the directories.**

5.4.4 Removing Directories: The rmdir Command

Sometimes you find that you have no more use for a directory, or that you created a directory by mistake. In both cases, you want to remove the unwanted directory, and UNIX has the command for you!

The **rmdir** (remove directory) command removes (deletes) the specified directory. However, it removes only empty directories—directories that contain no subdirectories or files other than the dot (.) and dot dot (..) directories (which are explained later in this chapter).

To remove the `important` directory from your `memos` directory, use the following command sequence:

```
$ cd [Return] . . . . . . . . . . . Make sure you are in your
                                    HOME directory.
$ cd memos [Return] . . . . . . . . Change your working
                                    directory to memos.
$ pwd [Return] . . . . . . . . . .  Make sure you are in memos.
/usr/david/memos
$_ . . . . . . . . . . . . . . . .  Yes, you are in memos.
$ rmdir important [Return] . . . .  Remove the important
                                    directory.
$_ . . . . . . . . . . . . . . . .  Ready for next command.
```

1. *You were able to remove the* `important` *subdirectory because it was an empty directory.*
2. *You must be in a parent directory to remove a subdirectory.*

From the `david` directory, try to remove the `source` subdirectory by doing the following:

```
$ cd [Return] . . . . . . . . . . . Change to david directory.
$ rmdir important [Return] . . . .  Remove the source directory.
rmdir: source: Directory not empty
$ rmdir xyz [Return] . . . . . . .  Remove a directory called
                                    xyz.
rmdir: xyz: Directory does not exist
$_ . . . . . . . . . . . . . . . .  Ready for next command.
```

1. *You could not remove the* `source` *subdirectory because it was not an empty directory. There are files in it.*
2. **rmdir** *returns an error message if you give a wrong directory name or if it cannot locate the directory name in the specified pathname.*
3. *You must be in the parent directory or a higher level of directory to remove subdirectories (children).*

5.4.5 Listing Directories: The ls Command

The **ls** (list) command is used to display the contents of a specified directory. It lists the information in alphabetical order by filename, and the list includes both filenames and directory names. When no directory is specified, the current directory is listed. If a filename is specified instead of a directory name, then **ls** shows the filename with any other information requested.

Figure 5.12 is used in the examples and command sequences as the directory structure, and subsequent figures show the effect of the example commands on the files and directories. Please follow these figures for better understanding of the examples.

Figure 5.12
The Directory Structure Used for Command Examples

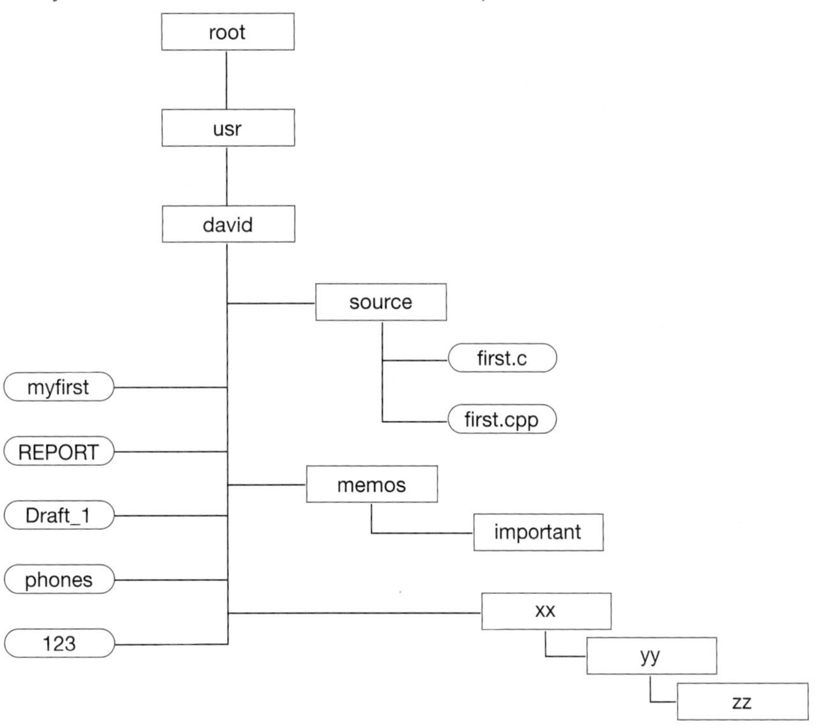

1. *Remember, a directory listing contains only the names of the files and subdirectories. Other commands let you read the contents of a file.*
2. *If no directory name is specified, the default is your current directory.*
3. *A filename does not indicate whether it refers to a file or a directory.*
4. *By default, the output is sorted alphabetically. Numbers come before letters, and uppercase letters before lowercase letters.*

Your directory structure is different and you do not have this many files and directories. To make your directory structure similar to the one presented here, you have to create new directories and files. First, create directories using the **mkdir** command as you did in previous section. Second, use the vi editor to create the files you need. There is no need to type anything in the files. You need to open and save files with the desired filenames. For example, to create a file named `REPORT`, you type

$ **vi REPORT** . . . Open/create a file called REPORT

The vi editor will open a blank screen.

```
~
~
~
"REPORT"  [new file] 0 lines, 0 characters
```

Now exit vi using the **wq** (write and quit command).

□ Press [Esc] to make sure vi is in the command mode.

□ Type **:wq** [Return]. This closes vi and saves the file.

```
~
~
~
:wq
"REPORT"  [new file] 0L, 0C written
$_
```

Repeat this command sequence to create other files. To create a file under a specific directory, first change to that directory using the **cd** command and then invoke vi to create your file.

There are other ways to create small files quickly. We will explore the other methods in later chapters.

Assuming your current directory is `david`, show the contents of your HOME directory by typing **ls** [Return].

```
$ ls
123
Draft_1
REPORT
memos
myfirst
phones
source
xx
$_
```

In some systems, the output of the **ls** command is not vertical in one column and the default format is set to display filenames across the screen.

```
$ ls
123  Draft_1  REPORT  memos  myfirst  phones  source  xx
$_
```

You may want to list the contents of directories other than your current directory. You can also list a single file to check whether it exists in the specified directory. While in your HOME directory `david`, list files in the `source` directory by doing the following:

```
$ cd [Return] . . . . . . . Make sure you are in the david directory.
$ ls source [Return]  . . . While in david, display list of files in the
                            source directory.
first.c
first.cpp
$_ . . . . . . . . . . . .  Ready for next command.
```

While in your HOME directory, check whether `first.c` exists in the `source` directory.

```
$ ls source/first.c [Return] . . Display the first.c filename in
                                 the source directory to see whether
                                 it exists. It does exist, so the file-
                                 name is displayed.
first.c
$ ls xyz [Return] . . . . . . .  Display a file called xyz if it exists.
                                 If it does not exist, you get the error
                                 message.
source/xyz: No such a file or directory
$_ . . . . . . . . . . . . . .  You get the prompt sign again.
```

ls Options

When you need more information about your files or you want the listing in a different format, use the **ls** command with options. These versatile options provide capabilities for listing your files in column format, showing a file's size, or distinguishing between filenames and directory names.

Table 5.1 shows most of the **ls** command options. Let's use some of these options and observe their outputs on the screen.

1. *Every option letter is preceded by a minus sign.*
2. *There must be a space between the command name and the option.*
3. *You can use pathnames to list files in a directory other than your working directory.*
4. *You can use more than one option in a single command line.*

The most informative option is the **-l** (long format) option. The listing produced by the **ls** command and **-l** option shows one line for each file or subdirectory and displays several columns of information for each file.

List the files in your current directory in long format by doing the following:

- □ Type **cd** and press [Return].
- □ Type **ls -l** and press [Return].

The first output line in Figure 5.13, "total 11," shows the total size of the displayed files. The size indicates the number of blocks, which is usually 512 bytes. Your output might be showing different "total," depending on the number and size of files in your directory.

Table 5.1
The **ls** Command Options

Option		Operation
UNIX	**Linux Alternative**	
-a	**--all**	Lists all files, including hidden files.
-C	**--format=vertical** **-- format=horizontal**	Lists files in multicolumn format. Entries are sorted down the columns.
-F	**--classify**	Places a forward slash (/) after each filename if that file is a directory, and an asterisk (*) if it is an executable file.
-l	**--format=single-column**	Lists files in a long format, showing detailed information about the files.
-m	**--format=commas**	Lists files across the page, separated by commas.
-p		Places a forward slash (/) after each file name if that file is a directory.
-r	**--reverse**	Lists files in reverse alphabetic order.
-R	**--recursive**	Recursively lists the contents of the directory.
-s	**--size**	Show size of each file in blocks.
-x	**--format=horizontal** **--format=across**	Lists files in multicolumn format. Entries are sorted across the line.
	--help	Displays a usage message.

Figure 5.13
The **ls** Command and the **-l** Option

```
$ cd
$ ls -l
total 11
-rw-r--r-- 1 david  student 1026 Jun 25  12:28  123
-rw-r--r-- 1 david  student  684 Jun 25  12:2   Draft_1
-rw r--r-- 1 david  student  342 Jun 25  12:28  REPORT
drw-r--r-- 1 david  student   48 Jun 25  12:28  memos
-rw-r--r-- 1 david  student  342 Jun 25  12:28  myfirst
-rw r--r-- 1 david  student  342 Jun 25  12:28  phones
drw-r--r-- 1 david  student   48 Jun 25  12:28  source
drw-r--r-- 1 david  student   48 Jun 25  12:28  xx
$_
```

Figure 5.14
The **ls** Command–Long Format

```
     1       2  3       4         5       6          7

-rwx rw- --- 1 david    student    342  Jun 25 12:28  myfirst
```

Column 1 . . . . Consists of 10 characters. The first character indicates the file type and the rest indicate the file access mode.

Column 2 . . . . Consists of a number indicating the number of links.

Column 3 . . . . Indicates the owner's name.

Column 4 . . . . Indicates the group name.

Column 5 . . . . Indicates the size of the file in bytes.

Column 6 . . . . Shows the date and time of last modification.

Column 7 . . . . Shows the name of the file.

Figure 5.14 gives you a general idea about what is in each column. Look at each column and see what type of information it conveys.

File Type The first column consists of 10 characters and the first character in each line indicates the type of file. The following list summarizes the file types:

- \- indicates an ordinary file
- d indicates a directory file
- b indicates a block-oriented special (device) file, such as a disk
- c indicates a character-oriented special (device) file, such as a printer

File Access Mode The next nine characters of the first column, which consist of three sets of the letters *r, w, x,* and/or hyphens (-), describe the access mode of each file. They tell how each user in the system can access a particular file, and they are called the *file access,* or *permission mode*. Within each set, the letters *r, w, x,* and hyphens (-) are interpreted as described in Table 5.2.

Table 5.2
File Permission Characters

Key	Permission Settings
r	Read permission granted.
w	Write permission granted.
x	Execute permission granted (permission to run the file as a program).
- (hyphen)	Permission is not granted.

1. *The execute permission* (x) *makes sense if the file is an executable file (a program).*
2. *If a hyphen appears instead of a letter, then permission is denied.*
3. *If the file is a directory file, then* x *is interpreted as permission to search the directory for a specific file.*

Each set of letters grants or denies permission to a different group of users. The first set of the *rwx* letters grants read, write, and execute permission to *owner* (user); the second *rwx* is for *group*; and the third one is for *others*. By setting the access letters for different groups of users, you can control who can have permission for your files and what type of access they have. For example, suppose you are in `david` directory, and you issue the following command:

```
$ ls -l myfirst [Return]   . . . . List myfirst in long format.
-rwx rw----  1 david  student  342  Jun 25  12:28  myfirst
```

The output shows that `myfirst` is an ordinary file, since the first character is a hyphen. The first group of letters (`rwx`) means the *owner* has read, write, and execute permission. The second group of letters (`rw-`) means the *group* has read and write permission, but execute permission is denied. The last three characters (- - -), three hyphens, show that *others* are denied all access rights.

Number of Links The second column shows a number that indicates the number of links. Links (ln command) are discussed in Chapter 7. In our example, the number of links for `myfirst` is 1.

File Owner The third column shows the owner of the file. Usually this name is the same as the user ID of the person who created it, in this case, **david**.

File Group The fourth column shows the user group. Every UNIX user has a user ID and a group ID. Both are assigned by the system administrator. For example, people who are working on a project get the same group ID. In our example, the file's group is **student**.

File Size The fifth column shows the file size. This is the number of bytes (characters) in the file. In our example, the file size is 342 bytes.

Date and Time The sixth column shows the date and time of the last modification. In our example, `myfirst` was last modified on June 25, at 12:28.

Filename The seventh column (finally) shows the name of the file, in our example, `myfirst`.

While you are in `david`, list the files in the `source` directory in long format by doing the following:

```
$ cd [Return]   . . . . . . . Make sure you are in your HOME directory.
$ ls -l source [Return]   . List files in source directory.
-rwx rw---  1  david  student  342  Jun  25  12:28  first.c
$_   . . . . . . . . . . . . Back to the prompt.
```

1. *You are in your home directory, listing files in the* `source` *directory.*
2. *You have only one file in the* `source` *directory.*
3. *Your file,* `first.c`*, is an ordinary file, indicated by a hyphen (-).*
4. *You have read, write, and execute permission (rwx). User group has read and write permission, indicated by (rw-). Permission is denied to others, indicated by (---).*
5. `first.c` *has one link.*
6. *The owner name is* **david**, *and the group name is* **student**.
7. *The* `first.c` *file is 342 bytes large.*
8. `first.c` *was last modified on June 25 at 12:28.*

To display the names of the files in your HOME directory in reverse order, type **ls -r** and press [Return]. (See Figure 5.15).

Figure 5.15
The **ls** Command and the **-r** Option

```
$ ls -r
xx
source
memos
myfirst
phones
REPORT
Draft_1
123
$_
```

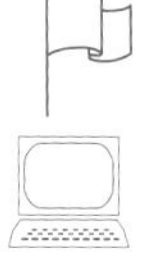

Notice the option is the lowercase r.

To display the contents of your current directory in column format, type **ls -C** and press [Return]. (See Figure 5.16.)

Figure 5.16
The **ls** Command and the **-C** Option

```
$ ls -C
123       REPORT         myfirst        source
Draft_1   memos          phones         xx
$_
```

The columns are alphabetically sorted down two columns. This is the default output format.

To display the contents of your current directory separated by commas, type **ls -m** and press [Return]. (See Figure 5.17.)

Figure 5.17
The **ls** Command and the **-m** Option

```
$ ls -m
123, Draft_1, REPORT, memos, myfirst, phones, source, xx
$_
```

Invisible Files

A filename beginning with a period refers to an *invisible file* or a *hidden file,* and directory listing commands normally do not display them. Startup files are usually invisible (named with the dot at the beginning) so that they do not clutter your directory. (Startup files are discussed in Chapter 8.)

You can create your own invisible file at HOME, or any other subdirectories you wish. Just start the filename with [.]. Two special invisible entries, a single and double dot (. and ..), appear in every directory except the root directory.

The . and .. Directory Entries The **mkdir** (make directory) command automatically puts two entries in every directory you create. They are a single and double period, representing the current and one level higher directories, respectively. Using the parent and child analogy, dot dot (..) represents the parent directory and dot (.) the child directory.

These directory abbreviations can be used in UNIX commands to refer to the parent and current directories when the pathname is required.

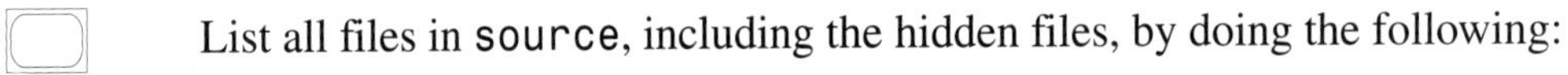

List all files in `source`, including the hidden files, by doing the following:

- ☐ Type **cd source** and press [Return] to change to the `source` directory.
- ☐ Type **ls -a** and press [Return] to list all files, including the invisible files:

```
.
..
first.c
```

1. *At this point, dot (.) means your current directory* (`/usr/david/source`).
2. *At this point, dot dot (..) means your parent directory* (`/usr/david`).

Change your working directory to the parent directory by doing the following:

```
$ cd .. [Return] . . . . . .   Change to the parent directory
                               (/usr/david).
$ pwd [Return] . . . . . . .   Check where you are.
/usr/david
$_ . . . . . . . . . . . . .   Prompt is back, and you are in
                               /usr/david directory.
```

1. *At this point, the dot (.) represents the current directory, which is* (`/usr/david`).
2. *At this point, the dot dot (..) represents the parent directory, which is* (`/usr`).

List the files in the parent directory of `david`, separated by commas, by doing the following:

```
$ cd [Return] . . . . . . .    Back to david.
$ ls -m .. [Return] . . . .    List the files in the parent directory
                               of david, which is usr; show file-
                               names across the screen, separated by
                               commas.
david, daniel, gabriel
$_ . . . . . . . . . . . . .   Ready for the next command.
```

You probably will get a long list of users IDs in your system and not the three names presented here in our example.

Using Multiple Options

You can use more than one option in a single command line. For example, if you want to list all files, including invisible (**-a** option) files, in long format (**-l** option), and with the filenames in reverse alphabetic order (**-r** option), you type **ls -alr** or **ls -a -l -r** and press [Return].

1. *You can use one hyphen to start options, but there should be no space between the option letters.*
2. *The sequence of the option letters in the command line is not important.*
3. *You can use one hyphen for each option, but there must be a space between option letters.*

List your HOME directory across the screen and indicate each directory name with a slash (/).

```
$ cd
$ ls -m -p
123, Draft_1, REPORT, memos/, myfirst, phones, source/, xx/
$_
```

Two options are used; **-m** *to produce filenames across the screen, and* **-p** *to place a slash (/) at the end of the directory filenames.*

To show all filenames, separated by commas, and to indicate the directory files with a slash and executable files with an asterisk, do the following:

```
$ cd
$ ls -amF
./, ../, 123, Draft_1, REPORT, memos/, myfirst, phones, source/, xx/
$_
```

1. *Three options are used,* **-a** *to show hidden files,* **-m** *to produce filenames across the screen separated by columns, and* **-F** *to indicate directories and executable files by placing a slash (/) or an asterisk at the end of the filenames respectively.*
2. *The two invisible files are directory files, indicated by the slash at the end of the filenames.*

List all the files in your HOME directory, in column format, in reverse order.

```
$ ls -arC
xx          phones      memos       REPORT      123     ..
source      myfirst     Draft_1     .
$_
```

List the files in david, separated by commas, and show the size of each file.

```
$ ls -s -m
total 11, 3 123, 2 Draft_1, 1 REPORT, 1 memos, 1 myfirst, 1 phones,
1 source, 1 xx
$_
```

1. *The first field (total 11) shows total size of files, usually in blocks of 512 bytes.*
2. *The option* **-s** *produces the file size; each file is at least 1 block (512 bytes), regardless of how small the file may be.*

List all files (including hidden files) in david in column format and also show the file sizes.

```
$ ls -a -x -s
Total 15      1.          1..         3 123
3 Draft_1     1 REPORTS   1 memo      1 myfirst
1 phones      1 source    2 xx
$_
```

1. *The total size is 13 blocks, since the size of the two hidden files is added.*
2. *The* **-x** *option formats the columns in a slightly different manner than* **-C**. *Each column is alphabetically sorted across rather than down the page.*

Show the directory structure under david in column format.

```
$ cd
$ ls -R -C
123           Draft_1    REPORT    memos    myfirst    phones
source        xx
./memos:
./source:
first.c
./xx:
yy
./xx/yy
zz
./xx/yy/zz:
$_
```

The command's options in this example are the uppercase letters *R* and *C*.

1. *The* **-R** *option lists the filenames in the current directory* `david`, *which has three subdirectories:* `memos`, `source`, *and* `xx`.
2. *Each subdirectory encountered is shown by its pathname followed by the* (`:`) (`./memos:`), *and then lists the files in that directory.*
3. *The pathnames are relative pathnames, starting from your current directory (the current directory sign is the dot at the beginning of the pathnames).*

As with many other Linux commands, you can use the **--help** option to get the list of options for the **ls** command.

```
$ ls --help [Return] . . . . . . . . . Display list of options.
              list of options will be displayed
$ _  . . . . . . . . . . . . . . . . . Back to the prompt.
```

Notice the Linux alternative options listed in Table 5.1. The option **--format** needs an argument such as **horizontal** or **commas**. There is no space on either sides of the = sign. Use the **man** command to look at more alternative options available under Linux.

The following command sequences show examples using the Linux alternative options:

```
$ ls --all [Return] . . . . . . . . . . Same as ls -a
$ ls --classify [Return] . . . . . . . Same as ls -F
$ ls --format=single-column [Return] . Same as ls -l
$ ls --format=commas [Return] . . . . . Same as ls -m
$ _  . . . . . . . . . . . . . . . . . Prompt for your next
                                       command.
```

5.5 DISPLAYING FILE CONTENTS

So far in this book you have learned file manipulation commands to scan directories to locate files and look at a list of filenames. What about looking at the content of a file? You can always print a file to obtain a hard copy of its contents, or use the vi editor to open a file and look at it on the screen. You also can use the **cat** command for this purpose.

5.5.1 Displaying Files: The cat Command

You can use the **cat** (for concatenate) command to display a file (or files), to create files, and to join files. In this chapter, only the display capability of the **cat** command is discussed. For example, you type the following command line to display a file called `myfirst`:

```
$ cat myfirst [Return] . . . . . . . . Display the myfirst file.
```

If `myfirst` exists in your current directory (if no pathname is specified, the default is your current directory), the **cat** command displays the contents of `myfirst` on the screen (the standard output device).

If you specify two filenames, then you see the contents of the two files, one after the other, in the same sequence specified on the command line. For example, you type the following command line to display the contents of two files, `myfirst` and `yourfirst`, on the screen. Notice the filenames are separated by a space.

`$ cat myfirst yourfirst [Return]` . . . Display the `myfirst` and `yourfirst` files.

1. *Filenames on the command line are separated by at least one space.*
2. *The **cat** command is usually used to display small (one screen) files.*

If the file is long, all you see on the screen at first are the last 23 lines of the file; the rest of the lines scroll up before your eyes. Unless you are a fast reader, this is not of much use. You can stop the scrolling process by pressing [Ctrl-s]. To continue scrolling, press [Ctrl-q].

It is rather inconvenient to look at the contents of a file in this manner. Please be patient; the best is yet to come. UNIX has other commands that show a long file a page at a time.

Every [Ctrl-s] must be canceled by a [Ctrl-q]. Otherwise the screen remains locked, and the input from your keyboard is ineffective.

5.6 PRINTING FILE CONTENTS

You can look at the contents of a file on the screen by using the vi editor or the **cat** command. However, there are times when you want a hard copy of your file.

UNIX provides commands to send your file to the printer, give you the status of your print job, and let you cancel your print job if you change your mind. Most UNIX systems provide the **lp** and **lpr** commands to print files. Let's investigate these commands.

5.6.1 Printing: The lp Command

The **lp** command sends a copy of a file to the printer for producing a hard (paper) copy of the file. For example, if you want to print the contents of the `myfirst` file, you type **lp myfirst** and press [Return].

UNIX confirms your request by displaying a request ID similar to the following message:

```
request id is lp1-8054 (1 file)
```

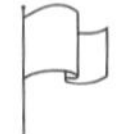

As always, there is a space between the command (lp) and the argument (filename).

If you specify a filename that does not exist, or one that UNIX cannot locate, then **lp** returns a message similar to the following:

```
lp: can't access file "xyz"
lp: request not accepted
```

You can specify several files on one command line.

```
$ lp myfirst REPORT phone [RETURN]  . . Print myfirst, REPORT,
                                        and phone files.
request id is lp1-6877 (3 files)
$_ . . . . . . . . . . . . . . . . . .  Ready for the next
                                        command.
```

1. *Filenames are separated by at least one space.*
2. *Only one banner page (first page) is produced for this request. However, each file is printed beginning at the top of a page.*
3. *The files are printed in the order in which they appear on the command line.*

If no filename is specified on the command line, the standard input is assumed. In this case, what you type (assuming the keyboard is the standard input device) will be printed. You signal the end of your input from the keyboard by pressing [Ctrl-d].

The following command sequence shows how the **lp** command is used to print your input from the keyboard:

```
$ lp [Return] . . . . . . . . .  No filename is specified.
_ . . . . . . . . . . . . . . .  Prompt waiting for your input.
                                 Type the following text:
Hello,
This is a test for checking the lp command.
[Ctrl-d]
        request id is HP_Printer-1016 (standard input)
$_  . . . . . . . . . . . . . .  Ready for the next command.
```

Each print request is associated with an id number. You use the id number to refer to the print job, such as when you want to cancel a print request.

lp Options

Table 5.3 shows the options you can use to make your print request more specific.

Table 5.3
The **lp** Command Options

Option	Operation
-d	Prints on a specific printer.
-m	Sends mail to the user mailbox after completion of the print request.
-n	Prints specified number of copies of the file.
-s	Suppresses feedback messages.
-t	Prints a specified title on the banner page (first page) of the output.
-w	Sends a message to the user's terminal after completion of the print request.

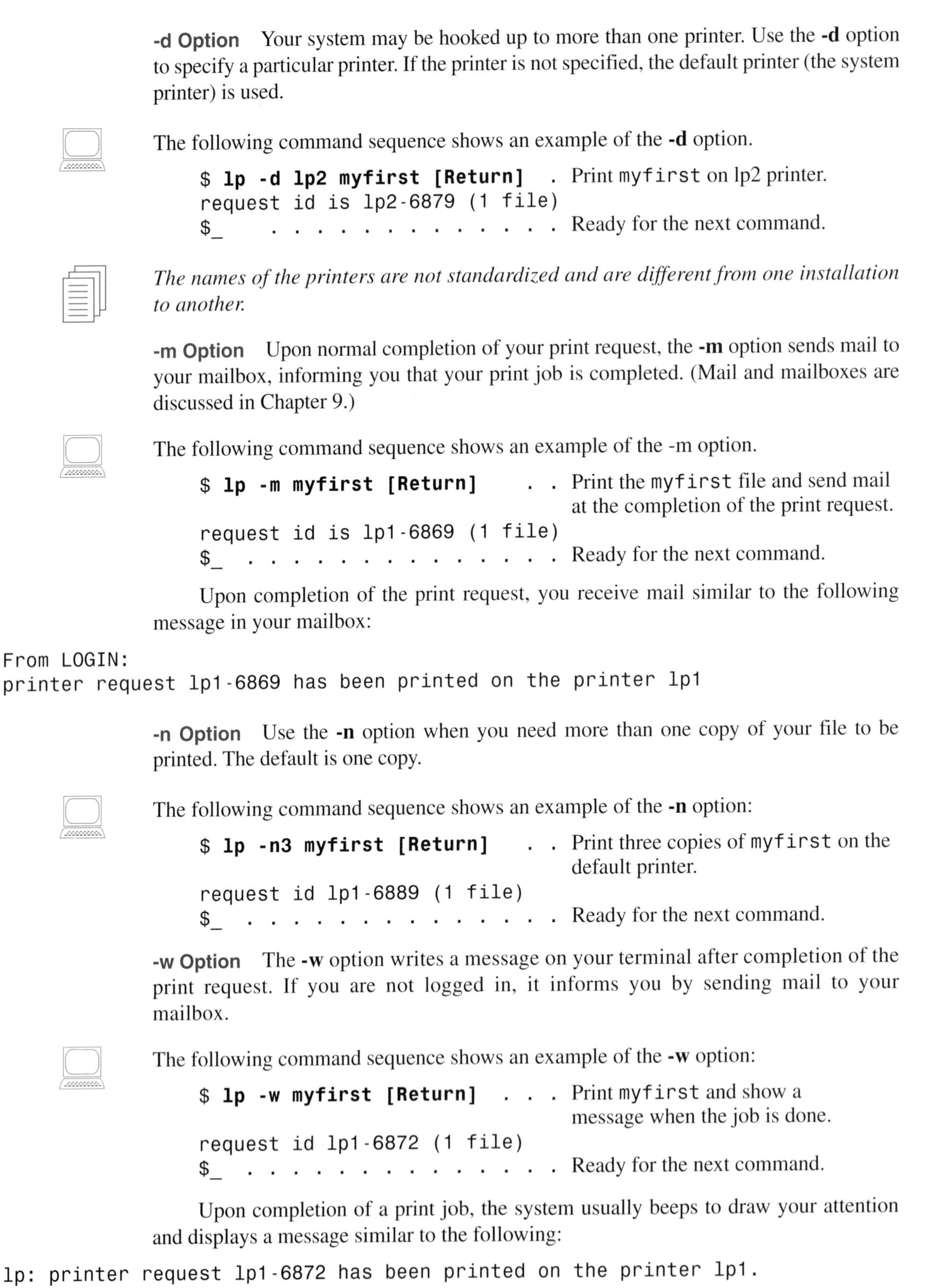

-d Option Your system may be hooked up to more than one printer. Use the **-d** option to specify a particular printer. If the printer is not specified, the default printer (the system printer) is used.

The following command sequence shows an example of the **-d** option.

```
$ lp -d lp2 myfirst [Return]   . Print myfirst on lp2 printer.
request id is lp2-6879 (1 file)
$_ . . . . . . . . . . . . .     Ready for the next command.
```

The names of the printers are not standardized and are different from one installation to another.

-m Option Upon normal completion of your print request, the **-m** option sends mail to your mailbox, informing you that your print job is completed. (Mail and mailboxes are discussed in Chapter 9.)

The following command sequence shows an example of the -m option.

```
$ lp -m myfirst [Return]     . . Print the myfirst file and send mail
                                 at the completion of the print request.
request id is lp1-6869 (1 file)
$_ . . . . . . . . . . . . . . . Ready for the next command.
```

Upon completion of the print request, you receive mail similar to the following message in your mailbox:

```
From LOGIN:
printer request lp1-6869 has been printed on the printer lp1
```

-n Option Use the **-n** option when you need more than one copy of your file to be printed. The default is one copy.

The following command sequence shows an example of the **-n** option:

```
$ lp -n3 myfirst [Return]    . . Print three copies of myfirst on the
                                 default printer.
request id lp1-6889 (1 file)
$_ . . . . . . . . . . . . . . . Ready for the next command.
```

-w Option The **-w** option writes a message on your terminal after completion of the print request. If you are not logged in, it informs you by sending mail to your mailbox.

The following command sequence shows an example of the **-w** option:

```
$ lp -w myfirst [Return]   . . . Print myfirst and show a
                                 message when the job is done.
request id lp1-6872 (1 file)
$_ . . . . . . . . . . . . . . . Ready for the next command.
```

Upon completion of a print job, the system usually beeps to draw your attention and displays a message similar to the following:

```
lp: printer request lp1-6872 has been printed on the printer lp1.
```

-t Option the **-t** option prints the specified string on the banner page (first page) of the output.

The following command sequence shows an example of the **-t** option:

```
$ lp -t hello myfirst [Return]  . . . Print myfirst and print
                                      "hello" on the banner page.
request id lp1-6889 (1 file)
$_ . . . . . . . . . . . . . . . . .  Ready for the next command.
```

5.6.2 Printing: The lpr Command in Linux

Linux is based on BSD (Berkeley Software Distribution) and some of the utilities and commands provided are different from UNIX. For example, the **lpr** command is used to print specified files. If no filename is specified, the standard input device is assumed.

lpr Options

The **lpr** command provides some of the same options available for the **lp** command and some different options. Table 5.4 shows some of the available options. Use the **man** command to obtain a full list of the options.

Table 5.4
The **lpr** Command Options

Option	Operation
-p	Prints on a specific (named) printer.
-#	Prints specified number of copies of the file.
-T	Prints a specified title on the banner page (first page) of the output.
-m	Sends mail to the user mailbox after completion of the print request.

1. *Most UNIX and Linux systems provide both* **lp** *and* **lpr** *commands.*
2. *Whenever possible, the use of the* **lp** *command is preferred.*

The following command sequences shows examples of using the **lpr** options:

```
$ lpr -p lp2 myfirst [Return] . . . . Print myfirst on lp2 printer.
$ lpr -m [Return] . . . . . . . . . . Print the myfirst file and
                                      send mail at the completion
                                      of the print request.
$ lpr -#3 myfirst [Return]  . . . . . Print three copies of
                                      myfirst file.
$ lpr -T hello myfirst [Return] . . . Print myfirst and print
                                      "hello" on the banner page.
$_ . . . . . . . . . . . . . . . . .  Ready for the next command.
```

5.6.3 Cancelling a Printing Request: The cancel Command

The **cancel** command cancels requests for print jobs made with the **lp** command. You use the **cancel** command to cancel unwanted printing requests. If you send a wrong file to the printer or decide not to wait for a long printing job, then the UNIX **cancel** command helps. To use the **cancel** command, you need to specify the ID of the printing job, which is provided by **lp,** or the printer name.

The following command sequences illustrate the use of the **cancel** command:

```
$ lp myfirst [Return] . . . . . . .  Print myfirst on the default
                                     printer.
request id lp1-6889 (1 file)
$_ . . . . . . . . . . . . . . . . . Ready for the next command.
$ cancel lp1-6889 [Return]  . . . .  Cancel the specified printing
                                     request.
Request "lp1-6889" canceled
$_ . . . . . . . . . . . . . . . . . Ready for the next command.
$ cancel lp1-6889 [Return]  . . . .  Cancel the current requests
                                     on the printer lp1.
request "lp1-6889" canceled
$ cancel lp1-85588 [Return] . . . .  Cancel the print request; wrong
                                     print request id; the system
                                     displays an error message.
cancel request "lp1-85588" nonexistent
$ cancel lp1 [Return] . . . . . . .  Cancel the request that is
                                     currently on the printer lp1; if
                                     there is no printing job on the
                                     printer, the system informs you.
cancel: printer "lp1" was not busy
$_ . . . . . . . . . . . . . . . . . Ready for the next command.
```

1. *Specifying printing request ID cancels the printing job even if it is currently printing.*
2. *Specifying the printer name only cancels the request that is currently printing on the specified printer. Your other printing jobs in the queue will be printed.*
3. *In both cases, the printer is freed to print the next job request.*

5.6.4 Getting the Printer Status: The lpstat Command

You use the **lpstat** command to obtain information about printing requests and the status of the printers. You can use the **-d** option to find out the name of the default printer on your system.

Try the following command requests to practice use of this command:

```
$ lp source/first.c [Return]  . . .  Print first.c in the source
                                     directory.
request id lp1-6877 (1 file)
$ lp REPORT [Return]  . . . . . . .  Print REPORT file on the
                                     default printer.
request id lp1-6878 (1 file)
$ lpstat [Return] . . . . . . . . .  Show status of the printing
                                     requests.
```

```
lp1-6877   student    4777      jun 11 10:50 on lp1
lp1-6877   student    4777      jun 11 10:50 on lp1
$ cancel lp1-6877 [Return]  . . . . Cancel specified printing
                                    request.
request "pl1-6877" cancelled
$ lpstat -d [Return]  . . . . . . . Show the name of the default
                                    printer.
system default destination: lp1
$ cancel lp1 [Return] . . . . . . . Cancel printing request that is
                                    currently printing on lp1.
```

If you do not have any printing request in the queue or currently printing on the printer, the **lpstat** *does not show anything, and the* **$** *prompt is displayed.*

5.7 DELETING FILES

You know how to create files and directories. You know how to remove empty directories. But what if the directory is not empty? You must remove all the files and subdirectories. How do you remove (delete) files?

Use the **rm** (remove) command to delete files that you do not want to keep anymore. You specify the filename to delete the file from your working directory, or specify the pathname to the file you intend to delete if it is in another directory. Figure 5.18 shows how your directory structure looks after the deletion of some files.

Figure 5.18
The Directory Structure After the File Deletions

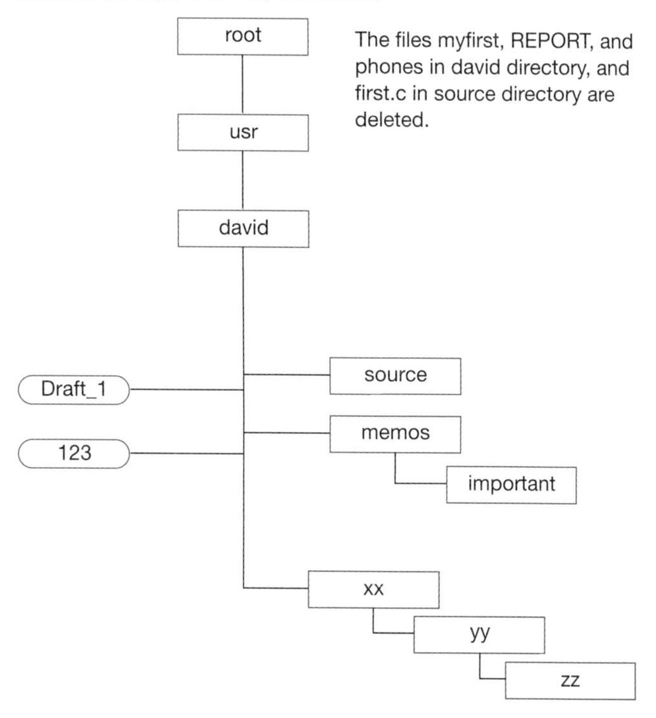

The following command sequence shows how to use the **rm** command:

```
$ cd [Return] . . . . . . . . .  Change to your HOME directory.
$ rm myfirst [Return] . . . . .  Delete myfirst from your HOME
                                 directory.
$ rm REPORTphones [Return]   .   Delete two files, REPORT and phones.
$ rm xyz [Return] . . . . . . .  Delete xyz; if the file does not exist,
                                 the system complains by showing an
                                 error message.
rm: file not found
$_ . . . . . . . . . . . . . .   Ready for next command.
```

The rm command does not give you any warning, and when a file is deleted, it is deleted for good!

rm Options

Like most UNIX commands, **rm** options modify the capabilities of the **rm** command. Table 5.5 summarizes the **rm** options.

Table 5.5
The **rm** Command Options

Option		Operation
UNIX	**Linux Alternative**	
-i	**--interactive**	Asks for confirmation before deleting any file.
-r	**--recursive**	Deletes the specified directory and every file and subdirectory in it.
	--help	Displays a usage message.

-i Option The **-i** option gives you more control over the delete operation. If you use the **-i** option, **rm** prompts you for confirmation before deleting each file. You press [y] for yes, if you are sure you want to delete the specified file, or [n], if you do not want to delete the file. This is the safest way to remove files.

The following command sequence shows examples of the use of the **-i** option:

```
$ pwd [Return] . . . . . . . . .  Check where you are.
/usr/david
$ ls source [Return] . . . . . .  List files in the source directory.
first.c
$ rm -i first.c [Return] . . . .  Delete first.c; the system displays
                                  the confirmation prompt before
                                  deletion. Press [y] for yes.
rm: remove first.c? y
$ ls source [Return] . . . . . .  Check whether the file was deleted.
$_ . . . . . . . . . . . . . . .  No files in the source directory.
```

-r Option The **-r** option deletes every file and subdirectory in a directory. You can delete an entire directory structure using **rm** with the **-r** option. Commands like this are what make UNIX an operating system for grownups!

1. **Please read the next command sequences carefully. Using the rm -r * command will delete all your directories and files starting from your working directory. These command sequences are provided here to show the dramatic and sometimes disastrous effect of this command line.**
2. **If you want to try the rm -r * command, make sure that you are not in one of the top-level directories and that you have copied your files into other directories.**

Figure 5.19
The Directory Structure After the Entire Structure Is Removed

Let's look at a command sequence using the **-r** option. Figure 5.19 shows your directory structure afterward. In this example the asterisk sign (*) is a metacharacter and means "all files." The metacharacters are explained in Chapter 7.

```
$ cd [Return]       . . . . . . . . Change to HOME directory.
$ rm -r * [Return]  . . . . . . . . Remove all there is under david
                                    (HOME) directory.
$ ls [Return]       . . . . . . . . List files in david.
$_  . . . . . . . . . . . . . . . . Sorry, nothing under david; the files
                                    are all deleted.
```

1. *Use the* **-i** *option to get the confirmation prompt.*
2. *Use the* **-r** *option sparingly, and only when it is absolutely necessary.*
3. *Use* **rmdir** *to remove directories.*

5.7.1 Before Removing Files

Under UNIX, deleting files and removing directories is quite easy. However, unlike other operating systems, UNIX does not give you any feedback or warning messages. Before you know it, the files are deleted, and the remove command is irreversible. Thus, before typing **rm,** consider the following points:

1. **Make sure it is not two o'clock in the morning when you start a major delete operation.**
2. **Make sure you know which file you want to delete, and what the content of that file is.**
3. **Think twice before pressing [Return] to complete the command.**

COMMAND SUMMARY

This chapter has introduced the following UNIX commands.

cancel (cancel print requests)

You can use this command to cancel print requests that are in queue waiting to be printed or are currently being printed.

cd (change directory)

This command changes your current directory to another directory.

lp (line printer)

This prints (provides hard copy of) the specified file.

Option	Operation
-d	Prints on a specific printer.
-m	Sends mail to the user mailbox after completion of the print request.
-n	Prints specified number of copies of the file.
-s	Suppresses feedback messages.

lpr (line printer)

This command prints the specified file. **lpr** reads from the standard input if no filename is specified.

Option	Operation
-p	Prints on a specific (named) printer.
-#	Prints specified number of copies of the file.
-T	Prints a specified title on the banner page of the output.
-m	Sends mail to the user mailbox after completion of the print request.

lpstat (line printer status)

This provides information about your printing request jobs, including printing a request ID number that you can use to cancel a printing request.

Option	Operation
-d	Prints the name of the system default printer for print requests.

ls (list)

This command lists the contents of your current directory, or any directory you specify.

Option		Operation
UNIX	Linux Alternative	
-a	--all	Lists all files, including the hidden files.
-C	-- format=vertical -- format=horizontal	Lists files in multicolumn format. Entries are sorted down the columns.
-F	--classify	Places a forward slash (/) after each filename if that file is a directory, and an asterisk (*) if it is an executable file.
-l	--format=single-column	Lists files in a long format, showing detailed information about the files.
-m	--format=commas	Lists files across the page, separated by commas.
-p		Places a forward slash (/) after each filename if that file is a directory.
-r	--reverse	Lists files in reverse alphabetical order.
-R	--recursive	Recursively lists the contents of the directory.
-s	--size	Show size of each file in blocks.
-x	--format=horizontal --format=across	Lists files in multicolumn format. Entries are sorted across the line.
	--help	Displays a usage message.

mkdir (make directory)

This command creates a new directory in your working directory, or in any other directory you specify.

Option	Operation
-p	Lets you create levels of directories in a single command line.

pwd (print working directory)

This command displays the pathname of your working directory or any other directory you specify.

rm (remove)

This command removes (deletes) files from your current directory, or any other directory you specify.

Option		Operation
UNIX	Linux Alternative	
-i	**--interactive**	Asks for confirmation before deleting any file.
-r	**--recursive**	Deletes the specified directory and every file and subdirectory in it.
	--help	Displays a usage message.

rmdir (remove directory)

This command deletes the specified directory. The directory must be empty.

REVIEW EXERCISES

1. What is the difference between a directory file and an ordinary file?

2. Can you use the / (slash) character in a filename?

3. What are the advantages of organizing your files in directories?

4. What is the difference between a relative and an absolute pathname?

5. Match the commands shown in the left column with the explanations shown in the right column.

1. ls — a. Displays the contents of the xyz file on the screen.
2. pwd — b. Deletes the `xyz` file.
3. cd — c. Asks for confirmation before deleting a file.
4. mkdir xyz — d. Prints the `xyz` file on the default printer.

5.	ls -l	e.	Deletes the `xyz` directory.
6.	cd ..	f.	Cancels the printing jobs on the lp1 printer.
7.	ls -a	g.	Displays the status of the default printer.
8.	cat xyz	h.	Lists contents of the current directory.
9.	lp xyz	i.	Creates a directory `xyz` in the current directory.
10.	rm xyz	j.	Displays the pathname of the current directory.
11.	rmdir xyz	k.	Lists the current directory in long format.
12.	cancel lp1	l.	Changes the working directory to the parent of the current directory.
13.	lpstat	m.	Lists all files, including the invisible files.
14.	rm -i	n.	Changes the current directory to the HOME directory.

6. Determine whether each one of the following is an absolute pathname, relative pathname, or a filename:

a. `REPORT`

b. `/usr/david/temp`

c. `david/temp`

d. `..(dot dot)`

e. `my_first.c`

f. `lists.01.07`

7. What is the command to

a. delete a file?

b. delete a directory?

c. get a confirmation message before delete?

d. print a file?

e. cancel a print request?

f. check the status of a printer?

g. redirect your print job to another printer?

h. print more than one copy of your document?

i. list your files?

j. list all files including the hidden files?

k. list files in long format?

l. change to your HOME directory?

m. change to another directory?

n. create a directory?

o. create a two-level directory structure?

p. change to the root directory?

q. display a file on the screen?

r. display two files on the screen?

8. Using the following directory structure:

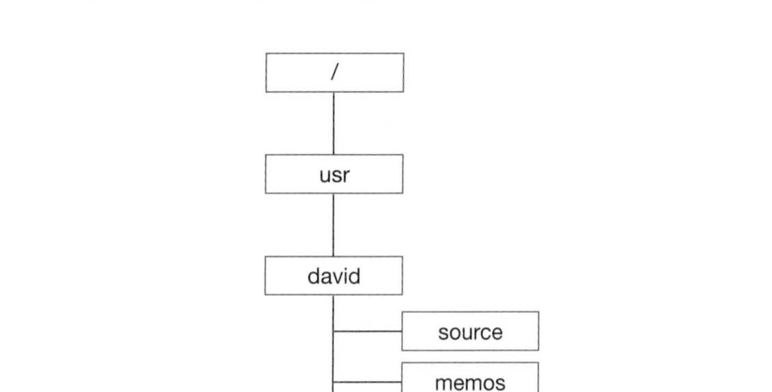

determine the *absolute* pathname for the following files and directories:

a. `123`

b. `source`

c. `xx`

d. `yy`

e. `zz`

Assuming your current directory is `david`, determine the *relative* pathnames for the following files and directories:

f. `123`

g. `source`

h. `important`

i. `yy`

j. `zz`

Assuming your current directory is `xx`, determine the *relative* pathnames for the following files and directories:

k. `yy`

l. `zz`

9. Having the following list of files determine the access mode for each file.

```
drwxrwxrwx    11
-rwxrw-rw-    counter
-rw-------    dead.letter
-rw-rw-rw-    enable
-rwxrwxrwx    xyz
-rwx------    HELLO
-rwx--x--x    Memos
```

Terminal Session

Try the following commands. Observe the output and the command feedback (error messages and so on) on the screen.

Create a directory structure in your HOME directory and try different commands until you feel comfortable with directories and file manipulation commands.

1. Show your current directory.
2. Change to your HOME directory.
3. Identify your HOME directory.
4. List the contents of your current directory.
5. Create a new directory called `xyz` under your current directory.
6. Create a file called `xyz` in the `xyz` directory.
7. Identify the directories in your working directory.
8. Show the contents of your current directory:
   a. In reverse alphabetical order
   b. In long format
   c. In horizontal format
   d. Showing the invisible files in your current directory
9. Print the `xyz` file in the `xyz` directory.
10. Check the printer status.
11. Print to `xyz` and then cancel the printing request.
12. Delete `xyz` in your `xyz` directory.
13. Delete the `xyz` directory in your current directory.

CHAPTER 6

The vi Editor: Last Look

The discussion of the vi editor began in Chapter 4 and continues in this chapter. Chapter 6 describes more of the vi editor's power and flexibility and introduces more advanced commands. It explains the commands' scopes and their usage in combination with the other commands, discusses the vi editor's manipulation of temporary buffers, and shows some of the ways the vi editor can be customized to your needs. By the end of this chapter, you will be well equipped to use vi to do your editing jobs.

In This Chapter

6.1 MORE ABOUT THE vi EDITOR

The vi editor is part of the *ex family* of editors. vi is the screen-oriented part of the ex editor, and it is possible to switch between the vi and ex editors. In fact, vi commands that start with **:** are ex editor commands. While in vi's command mode, pressing **:** displays the colon prompt at the bottom of the screen and causes vi to wait for your command. When you complete your command line by pressing [Return], ex executes the command and, upon completion, returns control to vi.

To change to the ex editor, type **:Q** and press [Return]. If you have intentionally or accidentally changed to the ex editor, type **vi** to get back to the vi editor, or type **q** to exit the ex editor and return to the shell prompt.

6.1.1 Invoking the vi Editor

In Chapter 4, you learned how to start vi, how to save a file, and how to quit vi. Expanding on those commands, let's explore other ways the vi editor can be invoked and ended.

You can start the vi editor without providing a filename. In this case, you use the write (**:w**) or write and quit (**:wq**) command to name your file.

The following command sequences show you how to invoke vi without a filename and later to name and save your file:

- Type **vi** and press [Return] to invoke the vi editor without a filename.
- Type **:w myfirst** and press [Return] to save the contents of the temporary buffer into the `myfirst` file and stay in the vi editor.
- Type **:wq myfirst** and press [Return] to save the contents of the temporary buffer into the `myfirst` file and quit the vi editor.

If the file you are currently editing does not have a name and you type **:w** or **:wq** without giving a filename, vi displays the following message:

```
No current filename
```

The vi editor normally prevents overwriting an existing file. Thus if you type **:w myfirst** and press [Return], and `myfirst` already exists, then vi warns you by showing the following message:

```
"myfirst" File exists - use ":w! to overwrite"
```

If you want to overwrite an existing file, use the **:w!** command.

The write command (**:w**) is also useful if you want to save your file or part of it under another name and keep the original file intact. The following command sequences show ways you can name a file or change the name of the current editing file:

- Type **vi myfirst** and press [Return] to invoke vi and copy the `myfirst` file into the temporary buffer.

See page x for an explanation of the icons used to highlight information in this chapter.

- Type **:w yourfirst** and press [Return] to save the contents of the temporary buffer (`myfirst`) into the `yourfirst` file. Your current editing file remains `myfirst`. The display shows this message:

```
"yourfirst" [New file] 3 lines, 103 characters
```

- Type **:wq yourfirst** and press [Return] to save the temporary buffer into the `yourfirst` file and quit the vi editor. The original `myfirst` file remains intact.

6.1.2 Using the vi Invocation Options

The vi editor provides flexibility from the very start. You can invoke vi with certain invocation options that you type as part of the command line.

The Read Only Option The **-R** (for read only) option makes a file a read only file and allows you to step through the contents of the file without making accidental changes. To use this option with the `myfirst` file, type **vi -R myfirst** and press [Return]. The vi editor shows the following message on the bottom line of the screen:

```
"myfirst" [Read Only] 3 lines, 106 characters
```

If you try to save the read only file using the **:w** or the **:wq** command, vi displays a message indicating this is a read only file:

```
"myfirst" File is read only
```

Use the **-R** option with the `myfirst` file.

`$ vi -R myfirst [Return]` . . . . Use the read only option.

The vi editor opens the `myfirst` file and shows a message that indicates this is a read only file.

```
The vi history
The vi editor is an interactive text editor that is supported
by most of the UNIX operating systems.
~
~
~
"myfirst" [readonly] 3L, 116C
```

If you try to save a read only file, using the **:w** or **:wq** command, vi displays a message indicating this is a read only file.

```
The vi history
The vi editor is an interactive text editor that is supported
by most of the UNIX operating systems.
~
~
~
:wq
'readonly' option is set (use ! to override)
```

If you want to save a read only file, you must force the write operation by using the **!** option. In our example, the command **:wq!** overrides the read only option and writes the file.

```
The vi history
The vi editor is an interactive text editor that is supported
by most of the UNIX operating systems.
~
~
:wq!
"myfirst" 3L, 116C written
$
```

You can override read only files that are your files or if you have access to them. Most of the system files are read only files, however, and a normal user cannot change this option for them.

Quitting a Read Only File Generally, you open a file as read only file just to look at it or read it, and your intention is not to edit and save it. To quit a read only file, you type **:q** [Return] and to force it to quit, you type **:q!** [Return].

Viewing Files You can use **view** to open the vi editor in read only mode. **view** is a version of vi that always starts in read only mode. That is, you will be protected from writing over the files. You also can start vi in read only mode by using the **-R** option, as was explained in the previous section.

The following two command lines will both open `myfirst` in read only mode:

`$ view myfirst [Return]`	Use view to read a file.
`$ vi -R myfirst [Return]`	Use vi with read only option to read a file.

The Command Option The **-c** (for command) option allows you to give specific vi commands as part of the command line. This option is useful to position the cursor or search for a pattern in a file before you begin editing. To use this option with the `myfirst` file, type **vi -c /most myfirst** and press [Return]. You are giving a search command (**/most**) as part of the command line. The vi editor copies the `myfirst` file into the temporary work buffer and places the cursor on the line with first occurrence of the word *most.*

Use the **-c** option with the `myfirst` file:

`$ vi -c /most myfirst [Return]` . . Use the **–c** command option.

-c indicates the command option is used.

/most indicates you are giving a search command as part of the command line.

The vi editor opens the `myfirst` file (copies the `myfirst` file into a temporary work buffer) and places the cursor on line two that happens to have the first occurrence of the word *most.*

```
The vi history
The vi editor is an interactive text editor that is supported
by most of the UNIX operating systems.
~
~
"myfirst" 3L, 116C
```

6.1.3 Editing Multiple Files

You can start vi and give it a list of filenames instead of one filename. Then when you finish editing one file, you can start editing the next file without reinvoking the vi editor. You press **:n** (for next) to invoke the next editing file. When you issue the **:n** command, vi replaces the contents of the working buffer with text from the next file. However, if you have modified the current text, vi displays a message like the following:

```
No write since last change (:next ! overrides)
```

You can use the **n!** command to override this protection. In this case, the changes you have made in your current editing file are lost.

To see a list of the filenames, you use the **:ar** command. vi responds by showing a list of the filenames and also indicates the name of the file being edited.

We are using `file1` and `file2` for this terminal session, and we need a file called `yourfirst` for next terminal session. You can create these files quickly using the vi editor. The following screen shows the content of each file.

$ **vi file1 [Return]** . . . . . Create the `file1` file.

```
file1
This is file 1.
~
:wq
"file1" 2L, 22C written
$
```

$ **vi file2 [Return]** . . . . . Create the `file2` file.

```
file2
This is file 2.
~
:wq
"file2" 2L, 22C written
$
```

$ **vi yourfirst [Return]** . . . Create the `yourfirst` file.

```
yourfirst
This is "yourfirst" file.
~
:wq
"yourfirst" 2L, 36C written
$
```

The following command sequences show examples of specifying more than one filename in a command line:

- Type **vi file1 file2** and press [Return] to invoke vi with two editing files, `file1` and `file2`; vi responds:

```
2 files to edit
"file1" 2 lines, 22 characters
```

```
file1
This is file 1.
~
~
"file1" 2L, 22C
```

- Type **:w** and press [Return] to save `file1.`
- Type **:ar** and press [Return] to display the names of the files. vi indicates the filename of the current file by enclosing it in brackets:

```
[file1] file2
```

```
file1
This is file 1.
~
~
:ar
[file1] file2
```

- Type **:n** and press [Return] to start `file2` for editing; vi responds:

```
"file2" 2 lines, 22 characters
```

```
file2
This is file 2.
~
~
:n
"file2" 2L, 22C
```

□ Type **:ar** and press [Return] to display the names of the files. This time the current file is `file2` and vi indicates that by enclosing the filename in brackets.

```
file1
This is file 1.
~
~
:ar
file1 [file2]
```

1. *If you type* **:n** *and there are no more files on the command line to open, the following message is displayed:*

   `No more files to edit`

2. *You can use* **:n** *to open the next file, but there is no way to go back and open the previous file.*

3. *If you type* **:n** *before saving your changes in the current file, vi displays the following message:*

   `No write since last change (:Next! Overrides)`

   In this case you can save your file (**:w**) *or type* **:n!** *and press [Return] to edit the next file without saving the current file.*

Editing Another File Another way to edit multiple files is to use the **:e** (for edit) command to switch to a new file. While in the vi editor, type **:e** followed by the name of the file and press [Return]. Usually, you save the current editing file before bringing in a new file and, unless you have made no changes, the vi editor warns you to write (save) the current file before switching to the next file.

Try the following command sequences to experiment with changing files:

□ Type **vi** and press [Return] to invoke vi without specifying the filename.

```
~
~
:e myfirst
```

□ Type **:e myfirst** and press [Return] to call in the `myfirst` file. Your current editing file is `myfirst;` vi shows the name and size of `myfirst:`

```
"myfirst" 3 lines, 103 characters
```

```
The vi history
The vi editor is an interactive text editor that is supported
by most of the UNIX operating systems.
~
~
"myfirst" 3L, 103C
```

Reading Another File The vi editor lets you read (import) a file into your current editing file. While in the vi editor command mode, type **:r** followed by the name of the file and press [Return]. The **:r** command places a copy of the specified file into the buffer after the cursor position. The specified file becomes part of your current editing file.

Use the following command sequence to import (read) another file into your current working buffer:

- Type **vi myfirst** and press [Return] to invoke the vi editor and edit `myfirst`.
- Type **:r yourfirst** and press [Return] to add the contents of `yourfirst` to the current editing file. vi shows the name and size of the imported file:

```
"Yourfirst" 2 lines, 36 characters
```

```
The vi history
The vi editor is an interactive text editor that is supported
by most of the UNIX operating systems.
~
~
:r yourfirst
```

Notice that the content of the `yourfirst` file is added right after the current line. These two lines are highlighted for your attention. We can control where a file is inserted by placing the cursor on the appropriate line. For example, you place the cursor on the last line if you want the imported file to be appended to the end of the file.

```
The vi history
yourfirst
This is "yourfirst" file.
The vi editor is an interactive text editor that is supported
by most of the UNIX operating systems.
~
~
"yourfirst" 2L, 36C
```

If you save this file (for example, using **:wq**), then the content of the `myfirst` file will be as is shown in this last screen example, with the two lines inserted after line one.

*The **:r** command adds a copy of the specified file to your current editing file, and the specified file remains intact.*

Writing to Another File The vi editor lets you write (save) a part of your current editing file into another file. You indicate the range of the lines you intend to save and use the **:w** command to write them. For example, if you want to save the text from lines 5 to 100 into a file called `temp`, type **:5,100 w temp** and press [Return]. vi saves lines 5 through 100 in a file called `temp` and shows a message similar to the following:

```
"temp" [new file] 96 lines, 670 characters
```

If the filename already exists, vi displays an error message. Then you can provide a new filename or use the **:w!** command to overwrite the existing file.

Save the first two lines in `myfirst` in a file called `temp`.

- Type **vi myfirst** and press [Return]. The vi editor is invoked with `myfirst` file.
- Type **:1,2 w temp** and press [Return]. The first two lines are saved in `temp`.

```
The vi history
The vi editor is an interactive text editor that is supported
by most of the UNIX operating systems.
~
~
:1,2 w temp
"temp" [New] 2L, 77C written
```

If the filename `temp` already exists, then you get a message such as

```
"temp" Use "w!" to write partial buffer
```

In this case, you type the **:w!** command to overwrite the existing file, or you can give it another filename.

- Type **:1,2 w! temp** and press [Return]. The first 2 lines are saved in `temp`. If `temp` already exists, it will be overwritten.
- Type **:1,2 w xyz** and press [Return]. The first two lines are saved in `xyz`, a new filename.
- Type **:e temp** and press [Return] to check the content of the `temp` file.

```
The vi history
The vi editor is an interactive text editor that is supported
by most of the
~
~
:e temp
"temp" 2L, 77C
```

6.2 REARRANGING TEXT

Deleting, copying, moving, and changing text are collectively referred to as *cut-and-paste operations.* Table 6.1 summarizes the operators or command keys that are used in combination to do cut-and-paste operations in a file. All of the commands are applicable when vi is in the command mode. With the exception of the change command, the vi editor remains in the command mode after completion of the command. The change command places the vi editor in text input mode, which means that you must press [Esc] to return vi to the command mode.

Table 6.1
The vi Editor Cut-and-Paste Keys

Key	Operation
d	Deletes a specified portion of the text and stores it in a temporary buffer. This buffer can be accessed by using the put operator (**p** or **P**).
y	Copies a specified portion of the text into a temporary buffer. This buffer can be accessed by using the put operator (**p** or **P**).
P	Places the contents of a specified buffer above the cursor position.
p	Places the contents of a specified buffer below the cursor position.
c	Deletes text and places vi in the text input mode. This is a combination of the delete and insert commands.

Assuming that you have the `myfirst` file on the screen and the cursor on *s,* the following examples show the cut-and-paste applications.

6.2.1 Moving Lines: dd and p or P

Using the delete and the put operators, you can move text from one part of a file to another:

- Press **dd**. vi deletes the current line, saves a copy of it in the temporary buffer, and moves the cursor to the *U.*
- Press **p**. vi places the deleted line below the current line.

```
The vi history
The vi editor is an interactive text editor that is supported
by most of the UNIX operating systems.
```

```
The vi history
by most of the UNIX operating systems.
```

```
The vi history
by most of the UNIX operating systems.
The vi editor is an interactive text editor that is supported
```

- Use the cursor movement keys and place the cursor on any character on the first line.
- Press **P**. vi places the deleted line above the current line.

```
The vi editor is an interactive text editor that is supported
The vi history
by most of the UNIX operating systems.
The vi editor is an interactive text editor that is supported
```

The deleted text remains in the temporary buffer, and you can move a copy of it to different places in the file.

6.2.2 Copying Lines: yy and p or P

Using the *yank* and the *put* operators, you can copy text from one part of the file to another:

- Press **yy**. vi copies the current line into a temporary buffer.
- Use the cursor movement keys to place the cursor on the first line.
- Press **p**. vi copies the contents of the temporary buffer below the current line.

```
The vi history
The vi editor is an interactive text editor that is supported
by most of the UNIX operating systems.
```

```
The vi history
The vi editor is an interactive text editor that is supported
The vi editor is an interactive text editor that is supported
by most of the UNIX operating systems.
```

- Use the cursor movement keys to move the cursor to the last line.
- Press **P**. vi copies the contents of the temporary buffer above the current line.

```
The vi history
The vi editor is an interactive text editor that is supported
The vi editor is an interactive text editor that is supported
The vi editor is an text editor which is supported
by most of the UNIX operating systems.
```

The copied text remains in the temporary buffer until the next delete or copy operation. You can copy the contents of this buffer to anywhere in the file and as many times as you wish.

6.3 SCOPE OF THE vi OPERATORS

In Chapter 4, you learned the basic vi commands; however, many vi commands operate on a block of text. A block of text can be a character, a word, a line, a sentence, or some other specified collection of characters. Using the vi commands in combination with the scope keys gives you more control over your editing task. The format for these types of commands can be represented this way:

```
command = operator + scope
```

There is no specific scope key to indicate the entire line. In order to indicate the entire line as the scope of a command, you press the operator key twice. For example, we saw that **dd** deletes a line and **yy** yanks (or copies) a line. Table 6.2 summarizes some of the common scope keys used in combination with other commands.

Table 6.2
Some of the vi Scope Keys

Scope	Operation
$	The scope is from the cursor position to the end of the current line.
0 (zero)	The scope is from just before the cursor position to the beginning of the current line.
e or **w**	The scope is from the cursor position to the end of the current word.
b	The scope is from the letter before the cursor backward to the beginning of the current word.

The following examples demonstrate the use of commands with the scope operators. To follow these examples, begin with the `myfirst` file on the screen. Using the delete, yank, and change operators on this file gives you a practical view of these commands.

6.3.1 Using the Delete Operator with Scope Keys

To delete text from the current cursor position to the end of the current line:

- ☐ Press **d$**. vi deletes text starting from the cursor position to the end of the current line and moves the cursor to the space after the word *by.*

```
The vi history
The vi editor is an interactive text editor that is supported
by most of the UNIX operating systems.
```

You can use **u** *or* **U** *to undo your most recent text changes.*

```
The vi history
The vi editor is an interactive text editor that is supported
by_
```

To delete text starting from the cursor position to the beginning of the current line:

- Press **d0**. vi deletes the text and the cursor remains on the letter *m*.

```
The vi history
The vi editor is an interactive text editor that is supported
by most of the UNIX operating systems.
```

```
The vi history
The vi editor is an interactive text editor that is supported
most of the UNIX operating systems.
```

To delete a word after the cursor position:

- Press **dw**. vi deletes the word *most* and the space after it, and moves the cursor to the letter *o*.

```
The vi history
The vi editor is an interactive text editor that is supported
by most of the UNIX operating systems.
```

```
The vi history
The vi editor is an interactive text editor that is supported
by of the operating systems.
```

To delete more than one word after the cursor position (for example, three words):

- Press **3dw**. vi deletes three words, *most*, *of*, and *the*, and the space after the, and moves the cursor to the space after the word *by*.

```
The vi history
The vi editor is an interactive text editor that is supported
by most of the UNIX operating systems.
```

```
The vi history
The vi is an interactive text editor that is supported
by_UNIX operating systems.
```

To delete to the end of a word:

- Press **de**. vi deletes the word *most* and moves the cursor to the space before the letter *o*.

```
The vi history
The vi editor is an interactive text editor that is supported
by most of the UNIX operating systems.
```

```
The vi history
The vi editor is an interactive text editor that is supported
by_of the operating systems.
```

To delete to the beginning of the previous word:

- Press **db**. vi deletes the word *by*, and the cursor remains on the space before the letter *o*.

```
The vi history
The vi editor is an interactive text editor that is supported
_of the operating systems.
```

6.3.2 Using the Yank Operator with Scope Keys

The yank operator can use the same scope keys as the delete operator. The **p** and **P** operators are used to place the yanked text in other places in the file. What portion of the text is yanked is controlled by using the scope keys.

To copy text from the current cursor position to the end of the current line, do the following:

- Press **y$**. vi copies the text starting from the cursor position to the end of the current line (is supported) to the temporary buffer, and the cursor remains on the letter *i*.
- Use the cursor movement keys to move the cursor to the end of the last line.
- Press **p**. vi copies the yanked text right after the cursor position.

```
The vi history
The vi editor is an interactive text editor that is supported
by most of the UNIX operating systems.
```

```
The vi history
The vi editor is an interactive text editor that is supported
by most of the UNIX operating systems.is supported
```

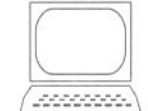

To copy text from the current cursor position to the beginning of the current line, do the following:

- Press **y0**. vi copies the text starting from the cursor position to the beginning of the current line (The vi), and the cursor remains on the letter *e*.
- Use the cursor movement keys to place the cursor at the end of the first line.
- Press **P**. vi copies the yanked text from the temporary buffer to right before the cursor position.

```
The vi history
The vi editor is an interactive text editor that is supported
by most of the UNIX operating systems.
```

```
The vi historyThe vi_
The vi editor is a text editor that is supported
by most of the UNIX operating systems.
```

6.3.3 Using the Change Operator with Scope Keys

The change operator, **c**, can use the same scope keys as the delete and yank operators. The difference in the **c** operator's function is that it changes vi from the command mode to the text input mode. After pressing **c**, you can enter text, beginning from the cursor position, and the text moves to the right. It wraps around when necessary to make room for the text you are entering. You return vi to command mode, as always, by pressing [Esc]. The change operator deletes the portion of the text indicated by the scope of the command and also places the vi editor in text input mode.

Some versions of the vi editor have a marker to mark the last character to be deleted. This marker is usually a dollar sign (**$**), and it overwrites the last character to be deleted.

The following example shows how to use the change text operator with the scope key to change a word:

- Press **cw**. vi places a marker at the end of the current word, overwrites the letter *t,* and changes to text input mode. The cursor remains on the letter *m,* the first character scheduled for change.
- Type **all** to change the word *most* to *all.*
- Press [Esc]. vi returns to command mode. Marker is removed.

```
The vi history
The vi editor is an interactive text editor that is supported
by most of the UNIX operating systems.
```

```
The vi history
The vi editor is an interactive text editor that is supported
by mos$ of the UNIX operating systems.
```

```
The vi history
The vi editor is an interactive text editor that is supported
by all of the UNIX operating systems.
```

6.4 USING BUFFERS IN vi

The vi editor has several buffers used for temporary storage. The temporary buffer (work buffer) that holds a copy of your file was discussed in Chapter 4, and when you use the write command, the contents of this buffer are copied to a permanent file. There are two categories of temporary buffers, *numbered buffers* and *named buffers* (or *alphabetic buffers*), that you can use for storing your changes and later retrieving them.

6.4.1 The Numbered Buffers

The vi editor uses nine temporary buffers numbered from 1 to 9. Each time you delete or yank text, it is placed in these temporary buffers, and you can access any of the buffers by specifying the buffer number. Each new deletion or yanking of the text replaces the previous contents of these buffers. For example, when you give the **dd** command, vi stores the deleted line in buffer 1. When you use **dd** again to delete another line, vi bumps the old contents up one buffer, in this case to buffer 2, and then stores the new material in buffer 1. This means that buffer 1 always holds the most recently changed material. The contents of the numbered buffers change each time you issue a delete or yank command. After a few text changes, of course, you lose track of what is stored in each of the numbered buffers. But keep on reading; the best is yet to come!

The contents of the numbered buffers can be recovered by using the put operator, prefixed with the buffer number. For example, to recover the material from buffer 9, you type **"9p**. The command **"9p** means copy the contents of buffer 9 to where the cursor is. The format for specifying the buffer number can be presented as follows:

double quotation mark + *n* (where *n* is the buffer number from 1 to 9) + (**p** or **P**)

The temporary numbered buffers are depicted in Figures 6.1 through 6.5. The examples explain the vi editor's sequence of events when you change text in the file.

To practice the following terminal session, you need to create a file that contains text similar to the following lines of characters and numbers. Using vi, create a file called `buffer` under the `Chapter6` directory.

```
AAAAAAAAAA
222222222222222
BBBBBBBBBB
333333333333333
CCCCCCCCCC
```

Assume that your screen looks like the screen in Figure 6.1, with five lines of text and the numbered buffers empty, because you have not yet done any editing operation.

Figure 6.1
The vi Editor's Nine Numbered Buffers

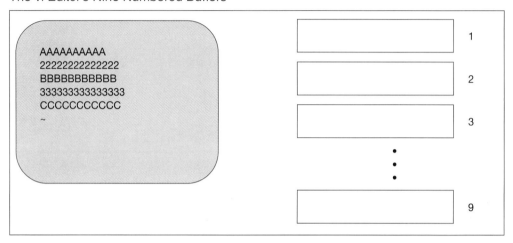

- Position the cursor on the first line and use the delete command to delete the current line. The vi editor saves the deleted line in buffer 1, and the screen and buffers look like Figure 6.2.

Figure 6.2
The Screen and Buffers After First Delete

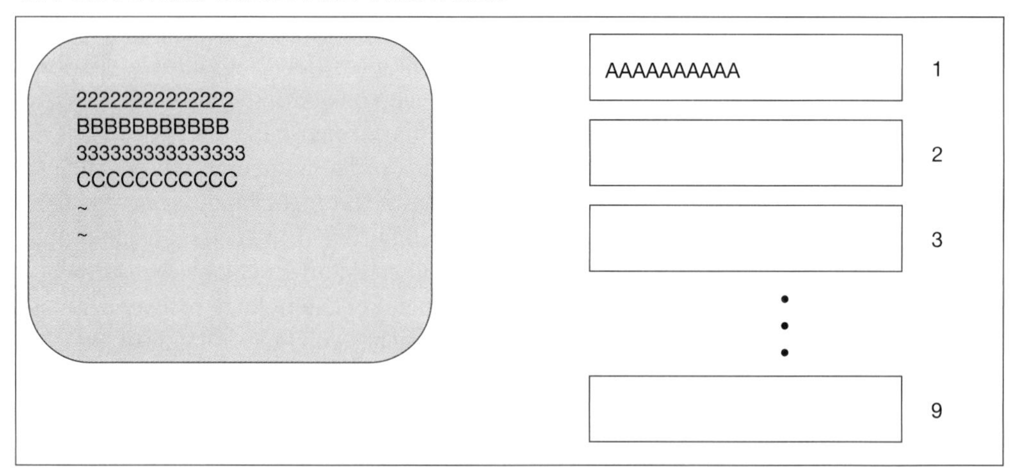

- Delete two lines using the delete command. The vi editor responds by deleting the two lines from your text and moving the contents of the temporary buffers one buffer up, to the empty buffer 1. Then it saves the deleted lines in buffer 1. The screen and buffers look like Figure 6.3.

1. The two lines deleted are saved in one buffer. The numbered buffers are not storage for just a line but for any size of text you have changed. Any amount of text, whether it is one line or 100 lines, is saved in one buffer.

2. *When all nine buffers are full, and vi needs buffer 1 for new material, the contents of buffer 9 are lost.*

Figure 6.3
The Screen and Buffers After Second Delete

□ The portion of the text that you yank is also saved in the temporary buffers. Place the cursor on the first line and then type **yy**. vi responds by copying the line you yanked into buffer 1. Remember that vi has to move all the buffers' contents to the next buffers in order to empty buffer 1, and then the new material is copied into buffer 1. Refer to Figure 6.4.

Figure 6.4
The Screen and Buffers After Yank

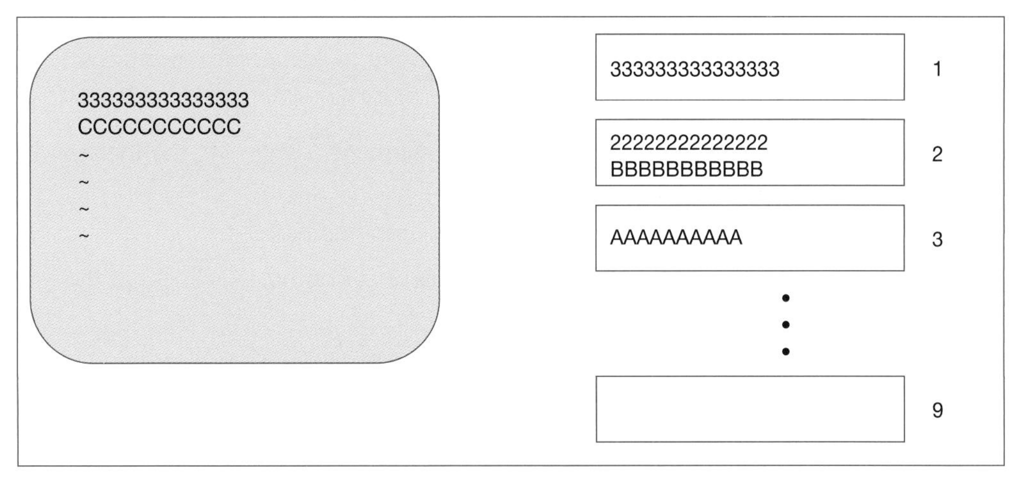

□ Now, if you remember what you have in the numbered buffers, you can access any one of them by specifying its number as part of a command. For example, to copy the contents of buffer 2 to the end of the file, type **"2p**, and vi copies the contents of buffer 2 to right after the cursor position. Figure 6.5 shows the screen and contents of the buffers after this put command.

Accessing the buffers does not change their contents.

Figure 6.5
The Screen and Buffers After the Put Command

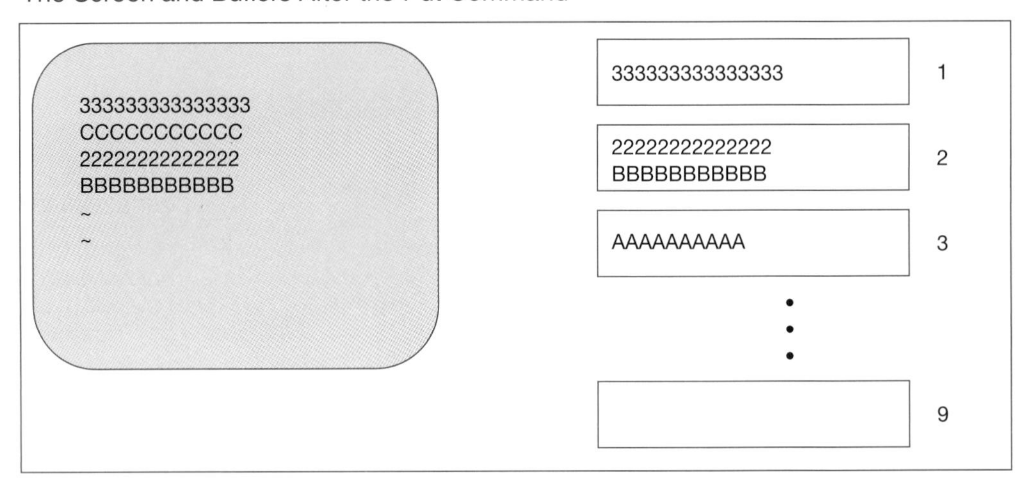

6.4.2 The Alphabetic Buffers

The vi editor also uses 26 named buffers. These buffers are named by the lowercase letters *a* through *z*, and you can refer to them by specifying their names explicitly. These buffers are similar to the numbered buffers, except the vi editor does not automatically change the contents of them every time you delete or yank from a file. It gives you more control over the operation. You can store deleted or copied text into a specified buffer, and later copy text from the named buffer to other places in your text by using the put operator. The text in a numbered buffer remains unchanged until you specify the buffer in a delete or yank operation. The format for specifying a buffer in your command is as follows:

```
double quotation mark + buffer name (a to z) + the command
```

To experiment with using commands to operate on a specific buffer, do the following things:

- Type **"wdd** to delete the current line and save a copy of it in buffer *w*.
- Type **"wp** to copy the contents of buffer *w* to the location indicated by the cursor position.
- Type **"z7yy** to copy seven lines into buffer *z*.
- Type **"zp** to copy the contents of buffer *z* (7 lines) to the location indicated by the cursor position.

1. **These commands are not displayed on the screen.**
2. **The alphabetic buffers are named in lowercase letters from *a* to *z*.**
3. **These commands do not require you to press [Return].**

6.5 THE CURSOR POSITIONING KEYS

The screen displays 24 lines of text at a time, and if your file contains more than 24 lines, you use the cursor movement keys to scroll a new line up or down into view. If your file contains 1,000 lines of text, you need about 999 keystrokes to bring line 999 into view on the screen. This is cumbersome and not practical. To overcome this problem, you use the vi editor's *paging operators*. Table 6.3 summarizes the paging operators (or command keys) and their capabilities.

Table 6.3
The vi's Paging Keys

Key	Operation
[Ctrl-d]	Scrolls the cursor down toward the end of your file, usually 12 lines at a time.
[Ctrl-u]	Scrolls the cursor up toward the beginning of your file, usually 12 lines at a time.
[Ctrl-f]	Scrolls the cursor down (forward) toward the end of your file, usually 24 lines at a time.
[Ctrl-b]	Scrolls the cursor up (backward) toward the beginning of your file, usually 24 lines at a time

[Ctrl-d] means simultaneously holding down the [Ctrl] and [d] keys; this convention also applies to the other control keys.

If you have a really large file, even the scrolling commands are not practical to position the cursor. Another way to position the cursor is to use **G** prefixed with the line number on which you want to place the cursor.

To make line number 1000 the current line, do the following:

- Type **1000G** to move the cursor to line 1000.
- Type **1G** to move the cursor to the first line.
- Type **G** to move the cursor to the end of the file.

Another useful command is [Ctrl-g], which tells you the line number of the current line. For example, if you press [Ctrl-g] while in the command mode, the vi editor responds by showing a message similar to the following:

```
"myfirst" line 30 of 90 — 30%
```

6.6 CUSTOMIZING THE vi EDITOR

The vi editor has many parameters (also called *options* or *flags*) that you can set, enable, or disable to control your working environment. These parameters have default values but are adjustable, and they include things like the tab setting, the right margin setting, and so on.

In order to see the complete list of parameters on the screen and how they are currently set in your system, enter the command mode, type **:set all**, and press [Return].

Your terminal screen shows options similar to Figure 6.6. Your system may have other options set.

Figure 6.6
Screen Showing vi's Options

```
~
~
~
:set all                noshowmode
noautoindent            number          terms5wyse50
nobeautify              readonly        noterse
hardtabs=8              report=10       window=23
noignorecase            shiftwidth=8    wrapmargin=0
magic
[Hit return to continue]
```

6.6.1 The Options Formats

The set command is used to set the options, and the options fall into three categories, each set in a different manner:

- Boolean (toggle)
- Numeric
- String

Assuming there is an option called X, the following examples show how to set up the three categories of options.

The Boolean Options

The *Boolean options* work like a toggle switch: you can turn them on or off. These options are set by typing the option name and are disabled by adding the word *no* in front of the option name. Typing **set X** enables option X, and typing **set noX** disables option X.

There is no space between the word *no* and the option's name.

The Numeric Options

The *numeric options* accept a numeric value, and depending on the option, the range of the numeric value is different. Typing **set X=12** assigns value 12 to option X.

The String Options

The *string options* are similar to the numeric options, but they accept a string value. Typing **set X=PP** assigns string PP to option X.

There is no space on either side of the equal sign.

The Set Command

The **set** command is used to set the different vi environment options, list them, or get the value of a specified option. The basic formats of the **set** command are as follows; each command is completed by pressing [Return]:

```
:set all
```

shows all the options on the screen.

```
:set
```

shows only the changed options.

```
:set X?
```

shows the value of the option X.

6.6.2 Setting the vi Environment

The behavior of the vi editor can be customized by setting the edit parameters to new values, and there are different methods you can use to set them. The direct method of changing these values is to use the vi **set** command to set the values as desired. In this case, vi must be in the command mode before you can issue a **set** command. You can set every option using this method; however, the changes are temporary, and they are in effect only for your current editing session. When you quit the vi editor, your options settings are abandoned.

This section describes some useful vi parameters (listed alphabetically), and Table 6.4 summarizes these options. Most of the option names have abbreviations; you can use the full or abbreviated names in the set command.

Table 6.4
Some of vi's Environment Options

Option	Abbreviation	Operation
autoindent	**ai**	Aligns the new lines with the beginning of the previous ones.
ignorecase	**ic**	Ignores the uppercase/lowercase difference in search options.
magic		Allow use of the special characters in search.
number	**nu**	Displays line numbers.
report		Informs you of the number of lines affected by your last command.
scroll		Sets number of lines to scroll when [Ctrl-d] command is given.
shiftwidth	**sw**	Sets number of spaces to indent. Used with autoindent option.
showmode	**smd**	Displays the vi editor modes in the right corner of the screen.
terse		Shortens the error messages.
wrapmargin	**wm**	Sets the right margin to a specified number of characters.

***autoindent* Option** The *autoindent* (**ai**) *option* aligns each new line you type in the text mode with the beginning of the previous line. This option is useful for writing computer programs in C, Ada, or other structured programming languages. You use [Ctrl-d] to backspace one level of indentation. While in text inset mode, each [Ctrl-d] backs up the number of columns specified by the *shiftwidth option*, discussed shortly. The default value for this option is set to **noai**.

***ignorecase* Option** The vi editor performs case-sensitive searches—that is, it differentiates between uppercase and lowercase letters. In order to make the vi editor ignore the letter cases, type **:set ignorecase** and press [Return].

To restore vi to the case-sensitive search, type **:set noignorecase** and press [Return].

***magic* Option** Certain characters (like bracket pairs [ ]) have a special meaning when you use them in search strings. When you toggle this option to **nomagic**, these characters no longer have special meanings. The default for this option is **magic**.

To set the *magic option*, you type **:set magic** [Return], and to unset this option you type **:set nomagic** [Return].

***number* Option** The vi editor does not ordinarily display the line numbers associated with each line. There are occasions when you want to refer to a line by its line number, and sometimes having the line numbers on the screen gives you a better feel for the size of a file and what part of the file you are editing.

To display the line numbers, type **:set** number and press [Return].

The following screen shows the `myfirst` file after the *number option* has been set.

```
1 The vi history
2 The vi editor is an interactive text editor that is supported
3 by most of the UNIX operating systems.
~
~
:set number
```

If you decide you do not want the line numbers to be displayed, type **:set nonumber** and press [Return].

The line numbers are not part of the file; they appear on the screen only while you are using the vi editor.

***report* Option** The vi editor does not give you any feedback on your editing job. For example, if you type **5dd**, vi deletes five lines starting from the current line but does not show any confirmation message on the screen. If you want to see feedback related to your editing, use the **report** parameter of the set command. This parameter is set to the number of lines that must be changed before the vi editor displays a report of the number of lines affected.

To set the *report option* to affect two-line edits, type **:set report=2** and press [Return]. Then if your editing job affects more than two lines, vi displays a report on the status line. For example, deleting two lines (**2dd**) and copying two lines (**2yy**) produces the following reports on the bottom line, respectively:

```
2 lines deleted
2 lines yanked
```

If you want to receive feedback on every change to your file, you type **:set report=0 [Return]**. Now you will receive feedback even if one character is changed in your file.

***scroll* Option** The *scroll option* is set to the number of lines that you want the screen to scroll when using [Ctrl-d] (in command mode). For example, to have the screen scroll five lines, type **:set scroll=5** and press [Return].

***shiftwidth* Option** The *shiftwidth* (**sw**) *option* sets the number of spaces used by the [Ctrl-d] key (in text input mode) when the autoindent option is in effect. The default setting for this option is **sw=8**. To change the setting to 10, for example, type **:set sw=10** and press [Return].

***showmode* Option** The vi editor does not display any visual feedback to indicate whether it is in text input mode or command mode. This can be confusing, especially for beginners. You can set the *showmode option* to provide visual feedback on the screen.

To toggle on the showmode option, type **:set showmode** and press [Return]. Then, depending on which key you use to change from command mode to text input mode, vi displays a different message at the lower right side of the screen. If you press **A** or **a** to change mode, vi displays `APPEND MODE`; if you press **I** or **i**, vi shows `INSERT MODE`; if you press **O** or **o**, vi displays `OPEN MODE`, and so on.

These messages remain on the screen until you press [Esc] to change to command mode. When there is no message on the screen, vi is in the command mode.

To turn off the showmode option, type **:set noshowmode** and press [Return].

***terse* Option** The *terse option* makes the vi editor display shorter error messages. The default for this option is **noterse**.

6.6.3 Line Length and Wraparound

Your terminal screen usually has 80 columns. When you type text and reach the end of the line (pass the 80th column), the screen starts a new line; that is what is called *wraparound.* The screen also starts a new line when you press [Return]. Thus, the length of a line on the screen could be any length from 1 to 80 characters. However, the vi editor starts a new line in your file only when you press [Return]. If you type 120 characters before pressing [Return], your text appears in two lines on the screen, but in your file it is one line of 120 characters.

Long lines can be a problem when you print a file, and it is confusing to relate the number of lines on the screen to the actual number of lines in the file. The simplest way to limit the length of a line is by pressing [Return] any time before reaching the end of the line on the screen. Another way to limit the line length is to set the wrapmargin parameter and let the vi editor insert returns automatically.

***wrapmargin* Option** The *wrapmargin option* causes the vi editor to break the text you are entering when it reaches a specified number of characters from the right margin. To set the wrapmargin to 10 (where 10 is the number of the characters from the right side of the screen), type **:set wrapmargin=10** and press [Return]. Then, when what you are typing reaches column 70 (80 minus 10), the vi editor starts a new line, just as if you had pressed [Return]. If you are typing a word as the characters pass column 70, vi moves the whole word to the next line. This means the right margin will probably be uneven. But remember, the vi editor is not a text formatter or a word processor.

The default value for the wrapmargin option is 0 (zero). To turn wrapmargin off, type **:set wrapmargin=0** and press [Return].

6.6.4 Abbreviations and Macros

The vi editor provides you with some shortcuts to make your typing faster and simpler: **:ab** and **:map** are two commands that serve this purpose.

The Abbreviation Operator The **ab** (for *abbreviation*) command lets you assign short, abbreviated words that will automatically insert any string of characters. This helps speed up your typing. Pick an easy-to-remember abbreviation for text that you often type, and, after you have set up that abbreviation in the vi editor, you can use the abbreviated word instead of typing the entire text. For example, to abbreviate the words *UNIX Operating System,* which are used often in this text, to the abbreviation *uno,* type **:ab uno UNIX Operating System** and press [Return].

In this example, *uno* is the abbreviation assigned to *UNIX Operating System*; thus, when vi is in text input mode, any time you type **uno** and then a space, vi expands *uno* to *UNIX Operating System.* If *uno* is part of another word, such as *unofficial*, no expansion occurs. It is the space before as well as after the *uno* that makes the vi editor recognize *uno* as an abbreviation and expand it.

To remove an abbreviation, you use the **unab** (for *unabbreviate*) operator. For example, to remove the *uno* abbreviation, type **:unab uno** and press [Return].

To list the abbreviations that are set, type **:ab** and press [Return].

1. *The abbreviations are assigned in the vi editor command mode and are used while you are typing in the text input mode.*
2. *The abbreviations set up are temporary; they remain in effect only during the current editing session.*

Try setting up some abbreviations as follows:

- Type **:ab ex extraordinary adventure** and press [Return] to assign *ex* to the string *extraordinary adventure.*
- Type **:ab 123 one, two, three, etc.** and press [Return] to assign *123* to the string *one, two, three, etc.*
- Type **:ab** and press [Return] to display all the abbreviations:

```
ex    extraordinary adventure
123   one, two, three, etc.
```

- Type **:unab 123** and press [Return] to remove the *123* abbreviation.

The Macro Operator The macro operator (**map**) lets you assign key sequences to a single key. Just as the abbreviation operator allows you to create a shortcut in text input mode, so **map** lets you create shortcuts in command mode. For example, to assign the command **5dd** (delete five lines) to **q**, type **:map q 5dd** and press [Return]. Then, while vi is in command mode, vi deletes five lines of text each time you press **q**.

To remove a **map** assignment, you use the **unmap** operator. Type **:unmap q** and press [Return].

To look at the list of the map keys and their assignments, type **:map** and press [Return].

The vi editor uses most of the keys on the keyboard for commands. This leaves you with a limited number of keys to assign to your key sequences. Available keys are **K**, **q**, **V**, [Ctrl-e], and [Ctrl-x].

You can also assign the function keys of your terminal with the **map** command. In this case, you type **#n** as the key name, where *n* refers to the function key number. For example, to assign **5dd** to [F2], type **:map #2 5dd** and press [Return].

Then, if you press [F2] while in command mode, vi deletes five lines of text.

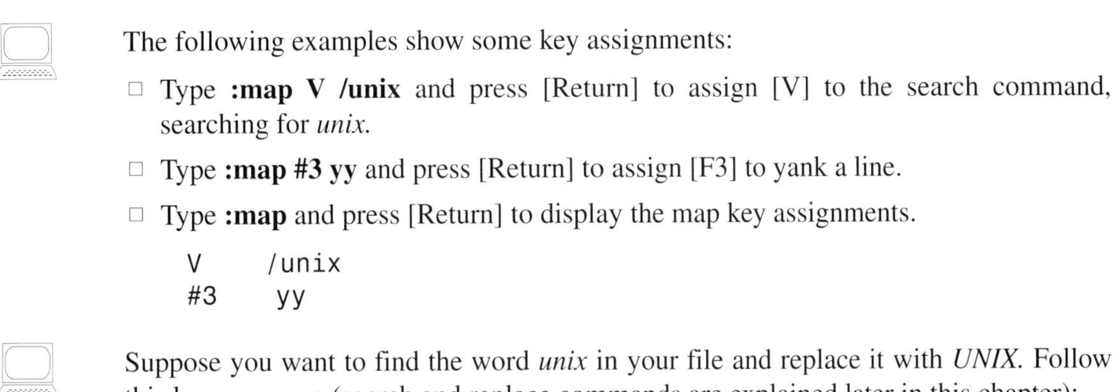

The following examples show some key assignments:

- Type **:map V /unix** and press [Return] to assign [V] to the search command, searching for *unix.*
- Type **:map #3 yy** and press [Return] to assign [F3] to yank a line.
- Type **:map** and press [Return] to display the map key assignments.

```
V    /unix
#3    yy
```

Suppose you want to find the word *unix* in your file and replace it with *UNIX.* Follow this key sequence (search and replace commands are explained later in this chapter):

- Type **:/unix** and press [Return] to search for the word unix.
- Type **cwUNIX** and press [Esc] to change the word *unix* to *UNIX* and return vi to command mode.

In map key assignment, you press [Ctrl-v][Return] to represent [Return], and [Ctrl-v][Esc] to represent the [Esc] key in the command line. Thus, to map the preceding command sequences to a single key, say **V**, type **:map V /unix** and press [Ctrl-V][Return] and type **cwUNIX** and press [Ctrl-V][Esc]. This command line uses unprintable characters ([Ctrl-V] and [Esc]), so what you see on the screen is the following:

```
:map V /unix ^McwUNIX ^[
```

1. *The map keys you create in the vi editor are temporary; they are in effect only for the current editing session.*
2. *The map keys are assigned and used while vi is in the command mode.*

You must precede [Return] and [Esc] with [Ctrl-v] if they are part of a map key assignment.

6.6.5 The `.exrc` File

All the options that you set up while you are in the vi editor are temporary; they disappear when you exit vi. To make the options permanent and save yourself the trouble of retyping them for every editing job, you can save the options settings in a file called `.exrc`.

Files starting with a . (dot) are called hidden files; you learned about these in Chapter 5.

When you start the vi editor, it automatically checks for the existence of the `.exrc` file in your current (working) directory and sets up the edit environment according to what it finds in the file. If vi does not find the .exrc file in the current directory, it checks your HOME directory and sets up options according to the .exrc file it finds there. If vi does not find a .exrc file, it assumes the default values of the options.

The way vi checks for the existence of the .exrc file gives you a powerful tool to create specialized .exrc files for your different editing needs. For example, you may create a general-purpose .exrc file under your HOME directory and a different .exrc file for your C programs under the directory in which you keep your C programs. You can use the vi editor to create a .exrc file or to modify it, if it already exists.

To create an .exrc file, type **vi .exrc** and press [Return]. Then enter the set and other commands you want to use (perhaps like those on the following screen).

```
set report=0
set showmode
set nu
set wm=10
ab uop UNIX Operating System
map q 5dd
```

Do not forget the . (dot) at the beginning of the filename.

If you create this file under your HOME directory, the edit environment will be established each time you use vi. Under the settings shown, vi displays what mode it is in, shows the line number, sets the right margin at 10 characters (line length is 70), inserts *UNIX Operating System* into your file every time you type **uop** followed by a space, and deletes five lines each time you press **q**.

1. .exrc *belongs to a group of files called startup files.*
2. *There are other utilities that use startup files similar to the* .exrc *file.*

6.7 THE LAST OF THE GREAT vi COMMANDS

Before concluding this discussion of the vi editor, we must consider one more vi operator. You know enough vi editor commands and other particulars so that you can easily and efficiently create or modify files. However, that is not all that vi can do. The vi editor has more than 100 commands and numerous variations of them that, when combined with the scope of the commands, give you detailed control over your editing job.

6.7.1 Running Shell Commands

You can run UNIX shell commands from the vi command line. This handy feature lets you temporarily put the vi editor aside and go to the shell commands. The **!** (exclamation mark) signals vi that the next command is a UNIX shell command. For example, to run the date command while in the vi editor, type **:! date** and press [Return]. The vi editor clears the screen, executes the command **date**, and you see lines similar to the following on the screen:

```
Sat Nov 27 14:00:52 EDT 2005
[Hit any key to continue]
```

Pressing a key returns the vi editor to the screen, and you can continue editing where you left off. If you want, you can also read the result of the shell commands and add it to your text. You use the **:r** (read) command followed by **!** to incorporate the command's result into your editing file.

To read the system's date and time, type **:r ! date** and press [Return]; vi responds by putting the current date and time under the current line.

The vi editor remains in the command mode.

```
The vi history
The vi editor is an interactive text editor that is supported
by most of the UNIX operating systems.
```

```
The vi history
The vi editor is an interactive text editor which is supported
Sat Nov 27 14:00:52 EDT 2005
by most of the UNIX operating systems.
```

The following command sequences show the use of **!**:

- Type **:! ls** and press [Return] to list the files in the current directory.
- Type **:! who** and press [Return] to show who is on the system.
- Type **:! date** and press [Return] to show the date and time of day.
- Type **:! pwd** and press [Return] to list the contents of the working directory.
- Type **:r ! date** and press [Return] to read the results of the **date** command and place it after the cursor position.
- Type **:r ! cal 1 2005** and press [Return] to read the calendar for January 2005 and place it after the cursor position.
- Type **:! vi mylast** and press [Return] to invoke another copy of vi to edit the `mylast` file.

6.7.2 Joining Lines

Use **J** to join two lines together. The **J** command joins the line below the current line to the current line, right after the cursor position. If the joining of the two lines results in a long line, vi wraps it around the screen.

To join two lines together, do the following:

- Use the cursor movement keys to place the cursor at the end of the first line.
- Press **J**. vi joins the line below the current line to the current line.

```
The vi history_
The vi editor is an interactive text editor that is supported
by most of the UNIX operating systems.
```

```
The vi history The vi editor is an interactive text editor that is supported
by most of the UNIX operating systems.
```

6.7.3 Searching and Replacing

There are occasions when you want to change a word throughout a file. If the file is a long one, it is cumbersome to go through the text, find each occurrence of a specific word, and change it. Additionally, the chances are good that you will miss one or two occurrences of the word. A better way is to use the vi search commands (**/** and **?**) in combination with other commands to do the job.

The following command sequence demonstrates the vi search and replacement capability:

- Type **:/UNIX** and press [Return] to search forward to find the first occurrence of the word *UNIX.*
- Type **cwunix** and press [Return] to change *UNIX* to *unix.*
- Type **n** to find the next occurrence of the word *UNIX.*
- Press **.** (dot) to repeat the last change (*UNIX* to *unix.*)
- Type **:?unix** and press [Return] to search backward from the current line to find the first occurrence of the word *unix.*
- Type **dw** to delete the word *unix.*
- Type **n** to find the next occurrence of the word *unix.*
- Press **.** (dot) to repeat the last command (**dw**) and delete the word *unix.*

6.7.4 File Recovery Option

What if the system or vi crashes while you are editing a file? Fortunately, vi provides a file recovery option that recovers the version of file that you were editing (it was in the buffer) when the system or vi crash occurred. In most cases recovery is quite easy. You start vi with the -r option on the same file you were editing when the crash happened. For example, the following command line recovers `myfirst` file:

$ **vi -r myfirst [Return]** . . Invokes vi recovery option.

If you were editing without a filename, or you don't remember the name of the file you were editing, just type the vi -r command without the filename argument.

$ **vi -r [Return]** . . . . . . Invokes vi recovery without a filename.

In this case, vi displays a list of the saved files, similar to the list in Figure 6.7.

Figure 6.7
Using the vi File Recovery Option

```
$ vi -r
/usr/preserve/david:
On Wed Aug 22 at 15:23, saved 2 lines of file "myfirst"
On Mon Oct 06 at 09:12, saved 6 lines of file "yourfirst"
/var/tmp:
No files saved.
$
```

Linux vim Editor

As was mentioned in Chapter 4, Linux provides the vim editor that is upwardly compatible to the vi editor. In fact, every time you use vi, the vim editor is invoked. Like vi, vim can be used to edit any text file. However, vim provides a lot of enhancements beyond the capablities of vi, including sophisticated file recovery and the help commands.

***help* Command** While in the vi editor, you type the following command:

:help recovery [Return] . . . Display the file recovery usage.

A full description of the file recovery commands will be displayed.

***recover* Command** When you start vi to edit a file you might get the *"ATTENTION: Found a swap file ..."* message. This means something happened the last time you were editing this file and a copy of the file was saved before the crash. In this case, you use the recover command to recover your file.

:recover [Return] . . . . . . Recover the current file.

Use the **man** command to obtain more information about the vim editor and its capabilities.

COMMAND SUMMARY

The following vi editor commands and operators discussed in this chapter. These commands complement the commands you learned in Chapter 4.

Cut-and-paste keys

These keys are used to rearrange text in your file.
They are applicable in vi's command mode.

Key	Operation
d	Deletes a specified portion of the text and stores it in a temporary buffer; this buffer can be accessed by using the put operator.
y	Copies a specified portion of the text into a temporary buffer; this buffer can be accessed by using the put operator.
P	Places the contents of a specified buffer above the cursor position.
p	Places the contents of a specified buffer after the cursor position

Paging keys

The paging keys are used to scroll a larger portion of your file.

Key	Operation
[Ctrl-d]	Scrolls the cursor down toward the end of the file, usually 12 lines at a time.
[Ctrl-u]	Scrolls the cursor up toward the beginning of the file, usually 12 lines at a time.
[Ctrl-f]	Scrolls the cursor down (forward) toward the end of the file, usually 24 lines at a time.
[Ctrl-b]	Scrolls the cursor up (backward) toward the beginning of the file, usually 24 lines at a time.

Scope keys

Using the vi commands in combination with the scope keys gives you more control in your editing tasks.

Key	Operation
$	The scope is from the cursor position to the end of the current line.
0 (zero)	The scope is from just before the cursor position to the beginning of the current line.
e or **w**	The scope is from the cursor position to the end of the current word.
b	The scope is from the letter before the cursor backward to the beginning of the current word.

Setting the vi environment

You can customize the behavior of the vi editor by setting the vi environment options. You use the **set** command to change the options' values.

Option	Abbreviation	Operation
autoindent	**ai**	Aligns the new lines with the beginning of the previous ones.
ignorecase	**ic**	Ignores the uppercase/lowercase difference in search operations.
magic		Allows the use of the special characters in a search.
number	**nu**	Displays line numbers.
report		Informs you of the number of lines affected by the last command.
scroll		Sets the number of lines to scroll when [Ctrl-d] command is given.
shiftwidth	**sw**	Sets the number of spaces to indent; used with the autoindent option.
showmode	**smd**	Displays the vi editor modes in the right corner of the screen.
terse		Shortens the error messages.
wrapmargin	**wm**	Sets the right margin to a specified number of characters.

REVIEW EXERCISES

1. What is the command line to open a file named `xyz` in read only mode?
2. What is the command line to open a file named `xyz` with cursor on the line with first occurrence of word UNIX?
3. What is **view**?
4. What are the numbered buffers?
5. What are the named buffers?
6. What is the vi command to
   a. delete a line?
   b. delete a word?
   c. copy a line?
   d. copy a word?
   e. delete to the end of the current line?

f. save two lines in a buffer called *z*?

g. copy the contents of the *z* buffer after the current line?

h. copy the contents of the buffer two after the current line?

7. What is the vi command for removing the showmode option?

8. What is the vi command to abbreviate "one two three" to "123"?

9. What is the `.exrc` file? When is the `.exrc` file is executed?

10. What is the command to execute a shell command such as **date** while in vi?

11. What is the vi command to read the current date and place it in the file under the current line?

12. What is the **set** command for vi to display confirmation messages?

13. What is the vi file recovery option?

14. What is the command to get the list of the file that were saved before a crash?

15. What is the vim editor?

16. Match the commands shown in the left column with explanations shown in the right. All the commands are applicable only in the command mode:

1. G	a. Replaces the character under the cursor with the letter *x*.
2. /most	b. Places the cursor on the last line in the file.
3. [Ctrl-g]	c. Copies four lines in buffer *x*.
4. 2dw	d. Moves cursor down one line.
5. j	e. Shows the line number of the current line.
6. "x4yy	f. Positions the cursor on line 66.
7. $	g. Deletes the character under the cursor.
8. 0 (zero)	h. Retrieves the contents of buffer 1.
9. 66G	i. Deletes two words.
10. x	j. Positions the cursor at the end of the current line.
11. rx	k. Finds the word *most*.
12. "1p	l. Positions the cursor at the beginning of the current line.

17. While using the vi editor, what command do you use to

a. set the line number option?

b. save five lines in buffer *x*?

c. read (import) the date string into your file?

d. list your current directory?

e. create an abbreviation?

f. remove an abbreviation?

g. read another file?

h. write (save) a file without quitting the vi editor?

i. delete a word?

Terminal Session

In this terminal session, create a file called `garden` and practice using the editing keys discussed in this chapter. Create this file as shown in Screen 1. Then use cut-and-paste, cursor positioning, and other commands to make it look like Screen 2. Finally, apply the following commands to your `garden` file.

1. Create an abbreviation of your name and add it to the beginning of your file.
2. Create a map key that finds a line with a specific word and deletes that line.
3. Undo the previous text changes.
4. While in vi, list your file in your current directory.
5. Read the date and time of the day, and place it after your name in the `garden` file.
6. Read another file (import it), and add it to the end of the `garden` file.
7. Save your file with another name.
8. Set the showmode option.
9. Set the line number option.
10. Move the cursor to the end of the file.
11. Move the cursor to the top of the file.
12. Move the cursor to line 10.
13. Search for the word *weeds*.
14. Set the report option to 1.
15. Remove the line number option.
16. Delete five lines; observe the vi feedback message. Use **U** to undo your deletion.
17. Copy five lines from beginning of the file to the end of the file. Observe the vi feedback message. Use **U** to undo your last editing.
18. Use the paging keys, [Ctrl-d], [Ctrl-u], and so on, and observe the results.
19. While in vi, copy five lines of your current file and save it in another file.
20. Add the current date at the beginning of your file using the **date** command.

Screen 1

```
Everywhere the trend is toward a simpler, and easy to care
garden. Few advises might help you to have less trouble with
your gardening. I am sure you have heard them before, but
listen once more. Gardening: The easy approach visit the plant
nurseries, it is good for your soul. Let me tell you that
There is no easy to care garden. Use plants that are suitable
for your climate. Native plants are good CHOICE. Before
planting, choose the right site. Use your imagination, plants
grow faster than what you think. Gardening can be made easier
and more enjoyable if you hire a gardener to do the job. Use
mulches to reduce weeds and save time in watering the plants.
Do not use too much chemicals to kill every weed insight. You
are the only one who sees the weeds, let them grow. They keep
the moisture and prevent soil erosion.
```

Screen 2

```
Gardening: The easy approach
Everywhere the trend is toward making simpler and easier-to-
care-for gardens.
However, let me tell you that there are no easy-to-care-for
gardens.
Gardening can be made easier and more enjoyable if you hire a
gardener to do the job.
Some advice might help you to have less trouble with your
gardening.
I am sure you have heard it before, but listen once more:
1. Before planting, choose the right site.
   Use your imagination. Plants grow faster than you think.
2. Visit the plant nurseries; it is good for your soul.
3. Use mulch to reduce weeds and save time in watering.
4. Use plants that are suitable for your climate.
   Native plants are a good choice.
5. Do not use too many chemicals to kill every weed in sight.
   You are probably the only one who sees the weeds.
   Let them grow.
   They keep moisture and prevent soil erosion.
```

CHAPTER 7

The UNIX File System Continued

This is the second chapter that discusses the UNIX file system and its associated commands; it complements the material discussed in Chapter 5. Chapter 7 presents more commands for manipulating files, including commands for copying files, moving files, and looking at the contents of a file. The chapter also explains the shell input/output redirection operators and file substitution metacharacters.

In This Chapter

7.1 FILE READING

Chapter 5 explained how you can use the vi editor or the **cat** command to read files. To refresh your memory, you can use vi with the read only option to read files, or you can use **cat** to view a small file. Using the **cat** command to view a large file, [Ctrl-s] to stop screen output, and [Ctrl-q] to resume screen output is very inconvenient. Try the following examples to get the feel of it. Doing so will make you appreciate the other file reading commands that let you view files one screen at a time.

Assume that your working directory is `david` and you have a file (say 20 pages long) called `large_file` in it.

- To read `large_file` using the vi editor, type **vi -R large_file** and press [Return]. This opens `large_file` for read only. Contents of `large_file` are displayed on the screen, and you can use vi commands to view the other pages.
- To read `large_file` using the **cat** command, type **cat large_file** and press [Return]. The contents of `large_file` are displayed and scroll before your eyes. You can use [Ctrl-s] and [Ctrl-q] to stop and resume scrolling.

7.1.1 The vi Editor Read Only Version: The view Command

Some UNIX systems provide the vi editor version called **view** that you can use to read files. It is a good tool for reading large files because you can use all of the vi editor commands to facilitate the reading of different parts of your file. Viewing a file using the **view** command prevents you from saving your editing or changes to the file. You can use it only to read a file. Please refer to Chapter 6 for usage examples.

7.1.2 Reading Files: The pg command

You can use the **pg** command to view files one screen at a time. A prompt sign (**:**) is produced at the bottom of the screen, and you press [Return] to continue viewing the rest of the file. The **pg** command shows EOF (end of the file) on the last line of the screen when it reaches the end of your file. You press [Return] at this point to get to the **$** prompt.

Using the **pg** command options gives you more control over the format and the way you want to view your file. Table 7.1 summarizes these options.

Unlike other commands' options, some pg options start with the plus sign (+).

*The **pg** command is not available on all systems. You can check its availability by typing **pg**, or **man pg** (to look at the manual pages about the pg commands). In each case, some sort of message is displayed announcing that **pg** is not available.*

See page x for an explanation of the icons used to highlight information in this chapter.

Table 7.1
The **pg** Command Options

Option	Operation
-n	Does not require [Return] to complete the single-letter commands.
-s	Displays messages and prompts in reverse video.
-*num*	Sets the number of lines per screen to the integer *num*. The default value is 23 lines.
-p*str*	Changes the prompt **:** (colon) to the string specified as *str*.
+*line–num*	Starts displaying the file from the line specified in *line–num*.
+/*pattern*	Starts viewing at the line containing the first occurrence of the specified *pattern*.

Assuming you have a file called `large_file` in your working directory, use the **pg** command to read it by doing the following:

- Type **pg large_file** and press [Return]. This is a simple way of looking at `large_file` one screen at a time.
- Type **pg -p Next +45 large_file** and press [Return] to view `large_file` starting from line 45 and show the prompt **Next** instead of the normal prompt **:** (colon).

Two options are used: **-p** to change the default prompt from **:** to **Next**, and +45 to start viewing from line 45.

- Type **pg -s +/hello large_file** and press [Return] to show prompt and other messages in reverse video, and start viewing from the first line that contains the word *hello.*

Two options are used: **-s** to show the prompt and other messages in reverse video, and **+/hello** to search for first occurrence of the word *hello.*

When the **pg** command displays the prompt sign **:** (or any other prompt if you have used the **-p** option), you can give commands to move forward or backward a specified number of pages or lines to view different parts of the text. Table 7.2 summarizes some of these commands.

Table 7.2
The **pg** Command Key Operators

Key	Operation
+*n*	Advances *n* screens, where *n* is an integer number.
-*n*	Backs up *n* screens, where *n* is an integer number.
+*n***l**	Advances *n* lines, where *n* is an integer number.
-*n***l**	Backs up *n* lines, where *n* is an integer number.
n	Goes to the nth screen, where *n* is an integer number.

7.1.3 Specifying Page or Line Number

You can specify a page number or line number from the beginning of the file or relative to the current page number. Use unsigned integers to indicate that the reference is to the beginning of the file. For example, type **10** to go to page 10, or **60l** (that's *six, zero, lowercase letter l*) to go to line 60 in the file.

Use signed integers to indicate that the reference is relative to the current page. For example, type **+10** to move forward 10 pages, or type **-30l** (*minus sign three, zero, lowercase letter l*) to move backward 30 lines. If you type only **+** or **-** without any numbers, the command is interpreted as **+1** or **-1**, respectively.

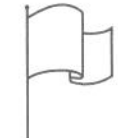

These operators are applicable only while you are viewing your file and pg is displaying the prompt sign.

If you use **pg** *with the* **-n** *option, then you do not need to press [Return] for single-letter operators.*

7.2 SHELL REDIRECTION

Among the most useful facilities that the shell provides are the *shell redirection operators.* Many UNIX commands take input from the standard input device and send the output to the standard output device. This is usually the default setting. Using the shell redirection operators, you can alter where a command gets its input and where it sends its output. The command's standard (default) input/output device is your terminal.

The shell redirection operators allow you to do the following things:

- Save the output of a process in a file.
- Use a file as input to a process.

1. *A* process *is any executable program. This could be an appropriate shell command, an application program, or a program you have written.*
2. *The redirection operators are instructions to the shell and are not part of the command syntax. Accordingly, they can appear anywhere on the command line.*
3. *Redirection is temporary and effective only with the command using it.*

7.2.1 Output Redirection

Output redirection allows you to store the output of a process in a file. Then you can edit it, print it, or use it as input to another process. The shell recognizes the greater than sign (>) and double greater than sign (>>) as output redirection operators.

The format is as follows:

```
command > filename
```

or

```
command >> filename
```

For example, to get the list of the filenames in your working directory, you use the **ls** command. Type **ls** and press [Return]. The shell default output device is your terminal screen (the standard output device). Consequently, you see the list of files on the screen. Suppose you want to save the **ls** command output (the list of filenames in the directory) in a file. One way to do that is to redirect the output of the **ls** command from the screen to a file as follows:

`$ ls > mydir.list [Return]` . . . Redirect **ls** output to `mydir.list`.

This time the **ls** command output is not sent to the terminal screen, but is saved in a file called `mydir.list`. If you open the `mydir.list` file, you see the list of files.

1. *If the specified filename already exists, then it is written over, and the contents of the existing file are lost.*
2. *If the specified filename does not exist, then the shell creates a new file.*

The double greater than sign (>>) redirection operator works the same as the greater than sign (>) operator, except for the fact that it appends the output to the specified file. If you type **ls >> mydir.list**, the shell adds the filenames in the working directory to the end of the file called `mydir.list`.

1. *If the specified file does not exist, then the shell creates a new file to save the output in it.*
2. *If the specified filename does exist, then the shell adds the output to the end of the file, and the previous contents of the file remain intact.*

The following command sequences show more examples of the output redirection operators.

To obtain a hard copy of the filenames in your home directory, do the following:

```
$ cd [Return] . . . . . . . . . . . Change to your HOME directory.
$ ls -C [Return] . . . . . . . . .  List filenames in the david
                                    directory in column format. You
                                    have two files:
myfirst     yourlast
$ ls -C > mydir.list [Return] . .   Save the output in mydir.list.
$_ . . . . . . . . . . . . . . . .  Done. The prompt is back.
$ cat mydir.list [Return] . . . .   Check what you have in
                                    mydir.list.
myfirst     yourlast
$ lp mydir.list [Return] . . . . .  Print the list.
request id is lp1-8056 (1 file)
$_ . . . . . . . . . . . . . . . .  Ready for the next command.
```

To append the list of the users on the system to `mydir.list`, do the following:

```
$ who >> mydir.list [Return] . .    Append the list of the users on the
                                    system to mydir.list. Now
                                    mydir.list contains the list of
                                    filenames and the list of users
                                    currently logged in.
```

```
$ date > mydir.list [Return]  . .  Save the output of the command
                                   date in mydir.list. This time
                                   the previous contents of the
                                   mydir.list are lost, and all you
                                   have in it are the results of the last
                                   command.
$_ . . . . . . . . . . . . . . . . Ready for the next command.
```

To save the current year's calendar in this_year, print it, and then remove it from the file, do the following:

```
$ cal > this_year [Return]  . . .  Save output of cal in this_year.
$ lp this_year [Return] . . . . .  Print it.
request id lp1-6889 (1 file)
$ rm this_year [Return]   . . . .  Remove it.
$_ . . . . . . . . . . . . . . . . Ready for the next command.
```

7.2.2 Input Redirection

The input redirection operator allows you to issue commands or run programs that get their input from a specified file. The shell recognizes the less than sign (<) as the input redirection operator. The format is as follows:

```
command < filename
```

or

```
command << word
```

For example, to send mail to another user, you use the **mailx** command (**mailx** is discussed in Chapter 9) and type **mailx daniel < memo**. This command tells the computer to send mail to the user called *daniel* (his user id). The input to **mailx** is not coming from the standard input device, your terminal, but from the file called memo. Thus, the input redirection operator (<) is used to indicate where the input comes from.

Try using the **cat** command with the input redirection operator.

- ☐ Type **cat < mydir.list** and press [Return] to display the contents of mydir.list, UNIX responds:

```
myfirst     yourlast
```

Using the **cat** command with the input redirection operator gives you the same result as the **cat** command with the filename as an argument (cat mydir.list). There are other commands that work in the same manner. If you specify a filename on the command line, the command takes its input from the specified filename. If you do not specify any argument, the command takes its input from the default input device (your keyboard), and if you use the input redirection operator to specify where the input comes from, then the command takes its input from the file you specify.

The redirection operator (<<) is used mostly in script files (shell programs) to provide standard input to other commands.

7.2.3 The cat Command Revisited

Now that you know about the shell redirection capabilities, we can explore the **cat** command in more detail. The **cat** command was introduced in Chapter 5, where it was used to show the contents of a small files on the screen. However, the **cat** command can be used for many things other than displaying files.

To refresh your memory, let's look at the following command lines:

```
$ cat myfirst [Return]  . . . . Display myfirst.
$ cat -n myfirst [Return] . . . Display myfirst with line numbers.
```

```
$ cat -n myfirst
     1  The vi history
     2  The vi editor is an interactive test editor that is supported
     3  by most of the UNIX operating systems.
$_
```

The **cat -n** command is used throughout this book to display content of the files. It is convenient to have the line numbers when one or more lines are referenced in the text.

Creating Files

By using the **cat** command with the output redirection operator (>), you can create a file. For example, if you wanted to create a file called `myfirst`, you would type **cat > myfirst**. This command means that the output of the **cat** command is to be redirected from the standard output device (your terminal) to a file called `myfirst`. The input comes from the standard input device—your keyboard. In other words, you type the text, and **cat** saves it in the `myfirst` file. You signal the end of the file by pressing [Ctrl-d].

This feature of the **cat** command is useful for creating small files quickly. Of course, you can also use it to create long files, but you must be a very accurate typist because after you press [Return] you cannot edit the text you have typed. The following command sequences show how to create a file using the **cat** command.

Try using the **cat** command with the output redirection operator to create a file.

```
$ cat > myfirst [Return]  . . . Create a file called myfirst.
_ . . . . . . . . . . . . . . . Cursor ready for your input. Let's say
                                you type the following:
I wish there were a better way to learn
UNIX. Something like having a daily UNIX pill.
[Ctrl-d] . . . . . . . . . . .  End your typing.
$_ . . . . . . . . . . . . . .  Ready for the next command.
$ cat myfirst [Return] . . . .  Check to see if myfirst has been
                                created; display it on the screen.
I wish there were a better way to learn
UNIX. Something like having a daily UNIX pill.

$_ . . . . . . . . . . . . . .  Ready for the next command.
```

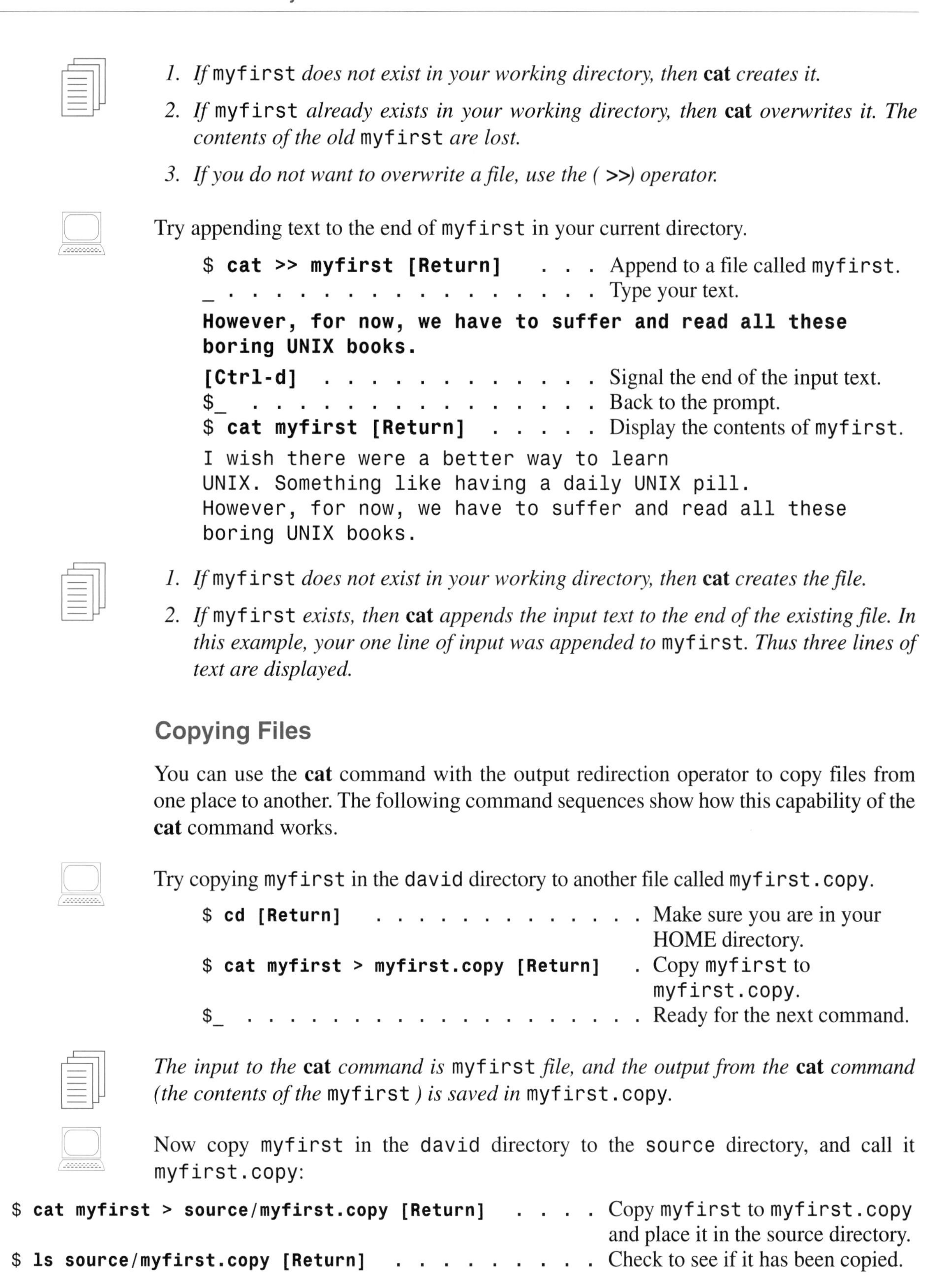

1. *If* `myfirst` *does not exist in your working directory, then* **cat** *creates it.*
2. *If* `myfirst` *already exists in your working directory, then* **cat** *overwrites it. The contents of the old* `myfirst` *are lost.*
3. *If you do not want to overwrite a file, use the (>>) operator.*

Try appending text to the end of `myfirst` in your current directory.

```
$ cat >> myfirst [Return]   . . . Append to a file called myfirst.
_ . . . . . . . . . . . . . . . . . Type your text.
However, for now, we have to suffer and read all these
boring UNIX books.
[Ctrl-d] . . . . . . . . . . . . . Signal the end of the input text.
$_  . . . . . . . . . . . . . . . . Back to the prompt.
$ cat myfirst [Return] . . . . . Display the contents of myfirst.
I wish there were a better way to learn
UNIX. Something like having a daily UNIX pill.
However, for now, we have to suffer and read all these
boring UNIX books.
```

1. *If* `myfirst` *does not exist in your working directory, then* **cat** *creates the file.*
2. *If* `myfirst` *exists, then* **cat** *appends the input text to the end of the existing file. In this example, your one line of input was appended to* `myfirst`*. Thus three lines of text are displayed.*

Copying Files

You can use the **cat** command with the output redirection operator to copy files from one place to another. The following command sequences show how this capability of the **cat** command works.

Try copying `myfirst` in the `david` directory to another file called `myfirst.copy`.

```
$ cd [Return]  . . . . . . . . . . . . . Make sure you are in your
                                          HOME directory.
$ cat myfirst > myfirst.copy [Return]  . Copy myfirst to
                                          myfirst.copy.
$_  . . . . . . . . . . . . . . . . . . . Ready for the next command.
```

The input to the **cat** *command is* `myfirst` *file, and the output from the* **cat** *command (the contents of the* `myfirst` *) is saved in* `myfirst.copy`.

Now copy `myfirst` in the `david` directory to the `source` directory, and call it `myfirst.copy`:

```
$ cat myfirst > source/myfirst.copy [Return]  . . . . Copy myfirst to myfirst.copy
                                                        and place it in the source directory.
$ ls source/myfirst.copy [Return]  . . . . . . . . . Check to see if it has been copied.

myfirst.copy
$_  . . . . . . . . . . . . . . . . . . . . . . . . . Yes, myfirst.copy is in the
                                                        source directory.
```

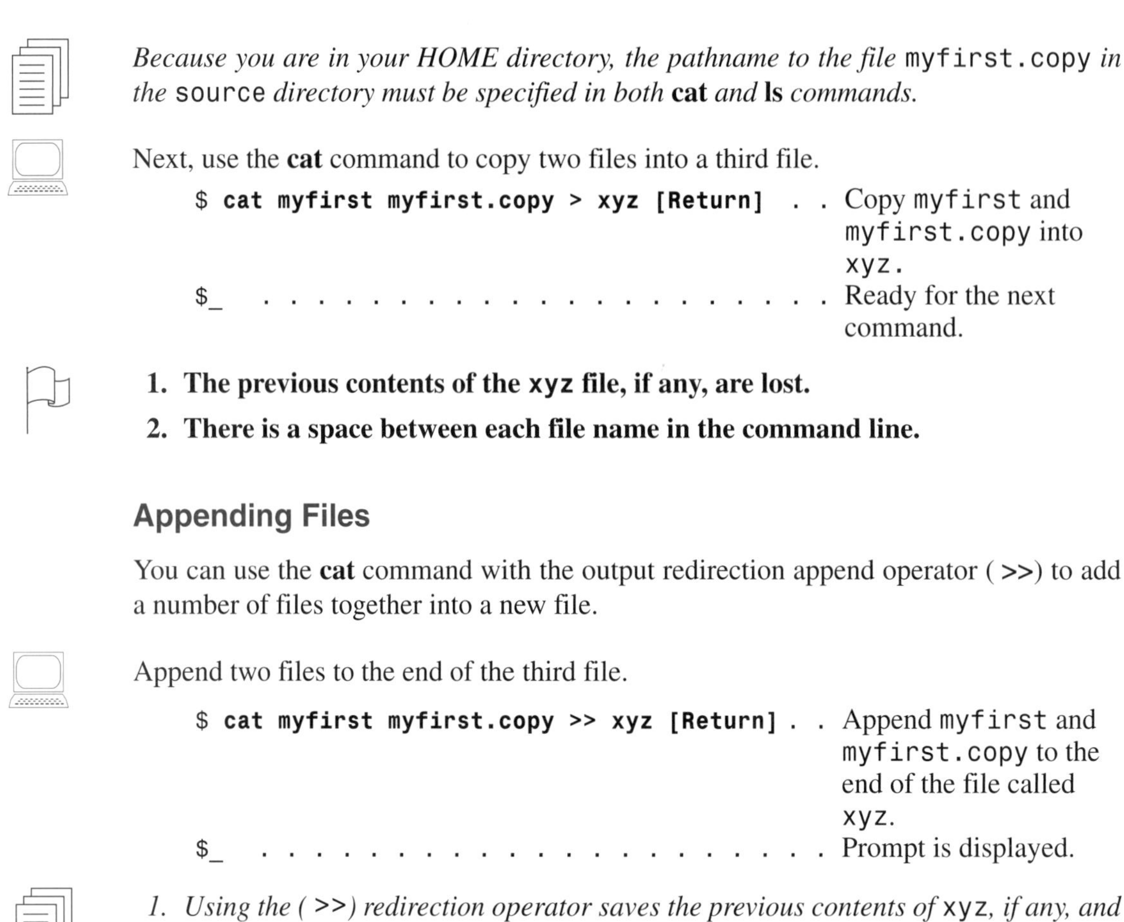

Because you are in your HOME directory, the pathname to the file `myfirst.copy` *in the* `source` *directory must be specified in both* **cat** *and* **ls** *commands.*

Next, use the **cat** command to copy two files into a third file.

```
$ cat myfirst myfirst.copy > xyz [Return] . . Copy myfirst and
                                               myfirst.copy into
                                               xyz.
$_ . . . . . . . . . . . . . . . . . . . . . . Ready for the next
                                               command.
```

1. **The previous contents of the xyz file, if any, are lost.**
2. **There is a space between each file name in the command line.**

Appending Files

You can use the **cat** command with the output redirection append operator (>>) to add a number of files together into a new file.

Append two files to the end of the third file.

```
$ cat myfirst myfirst.copy >> xyz [Return] . . Append myfirst and
                                               myfirst.copy to the
                                               end of the file called
                                               xyz.
$_ . . . . . . . . . . . . . . . . . . . . . . Prompt is displayed.
```

1. *Using the (>>) redirection operator saves the previous contents of* `xyz`*, if any, and the two files are added to the end of* `xyz`*.*
2. *You may have more than two filenames in the command line, but they must be separated by a space.*
3. *The files are appended to the specified output file in the same sequence in which the input files are specified.*

7.3 ENHANCED FILE PRINTING

The **lp** command sends your file to the printer as it is; it does not change the appearance or format of your file. You can, however, improve the appearance of the output by formatting it—for example, adding page numbers, page headings, and double-spaced lines to a document before sending it to the printer or viewing it on the screen.

You use the **pr** command to format a file before printing or viewing it. The **pr** command with no options formats the specified file into pages 66 lines long. It puts a five-line heading at the top of the page that consists of two blank lines, one line of information about the specified file, and two more blank lines. The information line includes the current date and time, the name of the specified file, and the page number. The **pr** command also produces five blank lines at the end of each page.

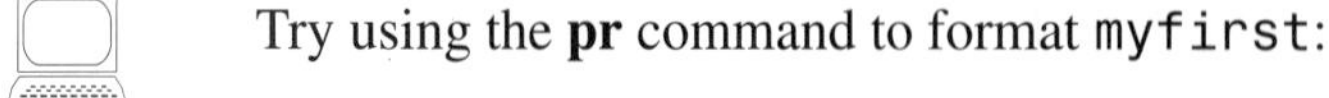

Try using the **pr** command to format `myfirst`:

- ☐ Type **pr myfirst** and press [Return] to format a file called `myfirst`. (See Figure 7.1.)

Figure 7.1
File Printed with Page Formatting: Five-Line Heading and Five-Line Blank Footing

```
                                [2 blank lines]
Nov 28  16:30  2001  myfirst  Page 1

                                [2 blank lines]
The vi history
The vi editor is an interactive text editor which is supported
by most of the UNIX operating systems. However, . . .

rest of the page . . .


                                [5 blank lines at bottom of page]
```

The **pr** command output is displayed on your terminal, the standard output device. But most of the time you will want to format files for a hard copy printout. One way to do that is to use the output redirection operator. There are other ways to send formatted files to the printer, such as using the pipe operator (¦) (which is explained in Chapter 8). Let's save the formatted version of `myfirst` in another file and then print it.

`$` **`pr myfirst > pout [Return]`** . . .	Save the formatted copy of `myfirst` in a file called `pout`.
`$` **`lp pout [Return]`**	Print `pout`.
`requested id is lp1-8045 (1file)`	
`$` **`rm pout [Return]`**	Delete `pout` if you do not need it.
`$_`	Ready for the next command.

pr Options

It is not enough to place five lines at the top and five lines at the bottom of each page and call it formatting. The **pr** options allow you to format a file's appearance with a little more sophistication. Table 7.3 summarizes the **pr** command options.

1. *The* **-m** *or* **-columns** *is used to produce multicolumn output.*
2. *The* **-a** *option can only be used with the* **-columns** *option and not* **-m**.

The following command sequences show the output of the **pr** command using different options.

The examples assume you have two files in your working directory. Let's use the **cat** command to create them.

`$` **`cat > names [Return]`**	Create a file called `names`.
`David [Return]`	
`Daniel [Return]`	
`Gabriel [Return]`	
`Emma [Return]`	
`[Ctrl-d]`	

Table 7.3
The **pr** Command Options

Option		Operation
UNIX	Linux Alternative	
+*page*	**--pages=***page*	Starts displaying from the specified *page*. The default is page 1.
-columns	**--columns=***columns*	Displays output in the specified number of *columns*. The default is one column.
-a	**--across**	Displays output in columns across (rather than down) the page, one line per column.
-d	**--double-space**	Displays output double spaced.
-h *string*	**--header=***string*	Replaces the filename in the header with the specified *string*.
-l *number*	**--length=***number*	Sets the page length to the specified *number* of lines. The default is 66 lines.
-m	**--merge**	Displays all the specified files in multiple columns.
-p		Pauses at the end of each page and sounds the bell.
-s *character*	**--separator=***character*	Separates columns with a single specified *character*. If character is not specified, then [Tab] is used.
-t	**--omit-header**	Suppresses the five-line header and five-line trailer.
-w *number*	**--width=***number*	Sets line width to the specified number of characters. The default is 72.
	--help	Displays help page and exits.
	--version	Displays version information and exits.

```
$ cat > scores [Return] . . . . . Create a file called scores.
90 [Return]
100 [Return]
70 [Return]
85 [Return]
[Ctrl-d]
$_ . . . . . . . . . . . . . . . . Ready for the next command.
```

Notice that most of the options for the **pr** command under UNIX can be used also for Linux. However, Linux provides some alternative options that do the same job and some additional options that can be used only under Linux.

For example, look at the following commands:

```
$ pr -a myfirst [Return]. . . . . Works for UNIX and Linux.
$ pr --across myfirst [return]. . Works only for Linux.
$ pr --help [return]. . . . . . . Works only for Linux.
```

To show the names in column format, and change the heading to *STUDENT LIST*, type **pr -2 -h "STUDENT LIST" names** and press [Return].

```
                                        [2 blank lines]
Nov 28 2005 14:30 STUDENT LIST [Page 1]
                                        [2 blank lines]
David                                   Gabriel
Daniel                                  Emma



                                        [5 blank lines at bottom of page]
```

The **-h** option changes the heading, but if the specified string has embedded white space, then you must put the string in quotation marks.

However, your screen will scroll up and you will not see the entire file. One way to observe the output on the screen is to redirect the **pr** output to a file, and then use the vi editor or the **view** command (the read only version of vi) to see the formatted output.

- Type **pr myfirst > outfile** and press [Return]. You have saved the output of the **pr** command in `outfile`.
- Type **view outfile** and press [Return] to see the output generated by the **pr** command.

If the **view** *command is not available on your system, type* **vi -R** *(for read only) instead.*

Display `names` in two columns across the page and suppress the header:

- Type **pr -2 -a -t names** and press [Return].

```
David          Daniel
Gabriel        Emma
```

The difference between options **-2** and **-2 -a** is the order in which the columns are arranged.

Show the files `names` and `scores` side by side:

- Type **pr -m -t names scores** and press [Return]

```
David          90
Daniel         100
Gabriel        70
Emma           85
```

The **-m** option shows the specified files side by side in the same order in which the filenames are specified in the command line.

Display names in 2 columns, separated by @ character, and omit header:

- ☐ Type **pr -2 -s@ -a names** and press [Return].

```
David@Daniel
Gabriel@Emma
```

Notice the **-s@** option caused the names to be separated by @.

7.3.1 Practicing Linux Alternative Command Options

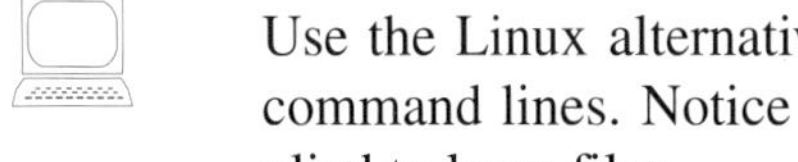

Use the Linux alternative options for the **pr** command, as suggested in the following command lines. Notice some of the options, such as *pages*, are meaningful when applied to large files.

$ **pr --pages=2 large_file [Return]**	Display large_file starting from page 2.
$ **pr --columns=2 myfirst [Return]**	Display myfirst in two columns.
$ **pr --double-space myfirst [Return]**	Display myfirst in double space.
$ **pr --omit-header --columns=2 --across names [Return]**	Display names across two columns with no header.
$ **pr --omit-header --columns=2 --separator=@ names [Return]**	Display names in two columns separated by @ sign.
$ **pr --help [Return]**	Display the help page.

Read the help page and familiarize yourself with other options available for the **pr** *command.*

7.4 FILE MANIPULATION COMMANDS

Some of the file manipulation commands were discussed in Chapter 5. From that discussion, you know how to create directories (using the **mkdir** command), create files (using the **vi** and **cat** commands), and delete files and directories (using the **rm** and **rmdir** commands). Now we will look at a few more commands to increase your knowledge of file manipulation in UNIX. These commands are used to copy (**cp**), link (**ln**), and move (**mv**) files. The general format of these commands is as follows:

```
command source target
```

where *command* is any of the three commands, *source* is the name of the original file, and *target* is the name of the destination file.

7.4.1 Copying Files: The cp Command

The **cp** (copy) command is used to create a copy (duplicate) of a file. You can copy files from one directory to another, make a backup copy of a file, or just copy files for the fun of it!

Suppose you have a file called REPORT in your current directory and want to create a copy of it. To do so, you type **cp REPORT REPORT.COPY** and press [Return].

REPORT is the source file and REPORT.COPY is the target file. If you do not provide the correct pathname/filename for the source or target file, **cp** complains by showing a message similar to the following:

```
File cannot be copied onto itself
0 file(s) copied
```

Figure 7.2 shows your directory structure before and after application of the **cp** command.

If the target file already exists, then its contents are destroyed.

Figure 7.2
An Example Application of the **cp** Command

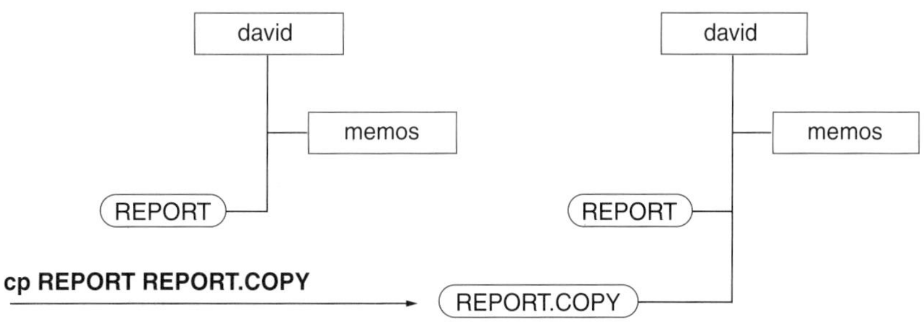

The following command sequences show how the **cp** command works.

```
$ ls [Return]  . . . . . . . . . . . List the current directory files.
memos        REPORT
$ cp REPORT REPORT.COPY [Return]  . Copy REPORT to REPORT.COPY.
$ ls [Return]. . . . . . . . . . . . See the list of files.
                                     REPORT.COPY is in the list, as
                                     you expected.
memos        REPORT REPORT.COPY
$ cp REPORT REPORT [Return]  . . . . Source and target filenames are
                                     the same.
File cannot be copied onto itself
0 file(s) copied
$_ . . . . . . . . . . . . . . . . . Ready for the next command.
```

Copy a file from your current directory to another directory. Figure 7.3 shows your directory structure before and after the **cp** command.

```
$ cp REPORT memos [Return]  . . . . Create a copy of REPORT in
                                    memos.
$ ls memos [Return]  . . . . . . .  List files in memos directory,
                                    and REPORT is there, as you
                                    expected.
REPORT
```

Figure 7.3
Another Example Application of the **cp** Command

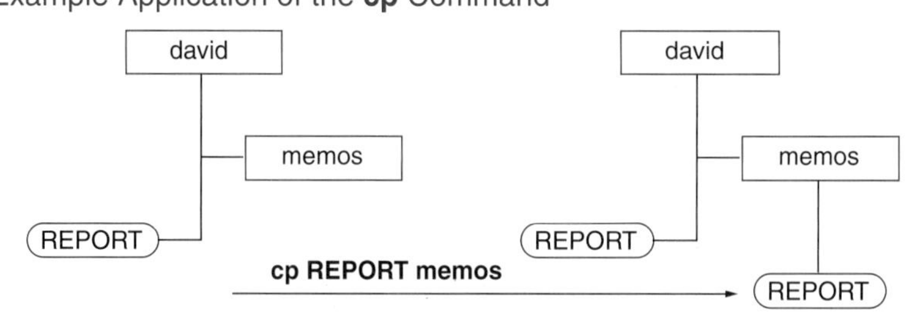

When the target file is a directory name, then the source file is copied in the specified directory with the same filename as the source filename.

You can also copy multiple files to another directory. Suppose you have files called `names` and `scores` under your current directory (david) and you want to copy them to the `memos` directory. Figure 7.4 shows your directory structure before and after the **cp** command.

```
$ cp names scores memos [Return] . Copy names, scores in current
                                   directory to the memos directory.
$_ . . . . . . . . . . . . . . . . Ready for the next command.
```

1. *The files called* `names` *and* `scores` *are in your current directory.*
2. *The filenames on the command line are separated by at least one space.*
3. *The last filename must be a directory name. In this case,* `memos` *is a directory name.*

Figure 7.4
Using the **cp** Command to Copy Multiple Files to Another Directory

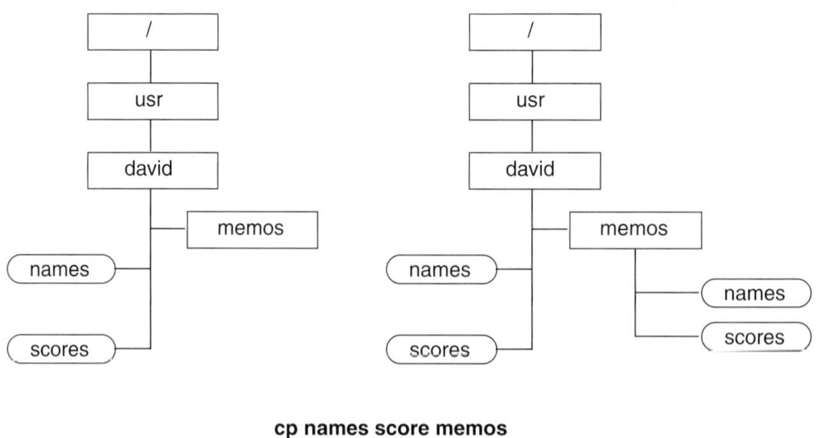

cp Options

Table 7.4 summarizes the **cp** command options.

-b Option The **-b** (backup) option creates a backup copy of the file if the file you want to copy already exits in the target directory. This protects you from overwriting an existing file.

Table 7.4
This **cp** Command Options

Option		Operation
UNIX	Linux Alternative	
-b	**--backup**	Makes a backup of the specified file if file already exists.
-i	**--interactive**	Asks for confirmation if the target file already exists.
-r	**--recursive**	Copies directories to a new directory.
	--verbose	Explain what is being done.
	--help	Displays the help page and exits.

Use the **cp** command with the **-b** (backup) option. Assuming you already have a file called REPORT under the memos directory, a backup copy of REPORT will be created. This means the REPORT file under memos is not overwritten. Figure 7.5 shows your directory structure before and after the cp -b command.

```
$ cp -b  REPORT memos [Return]  . . Move REPORT to memos and
                                    create a backup of REPORT if it
                                    already exits in memos.
$ ls memos [Return] . . . . . . . . List files under memos.
      REPORT   REPORT~
$_  . . . . . . . . . . . . . . . . Ready for the next command.
```

Notice that REPORT is backed up and the backup filename is REPORT~.

Figure 7.5
Using the **cp** Command with the **-b** Option

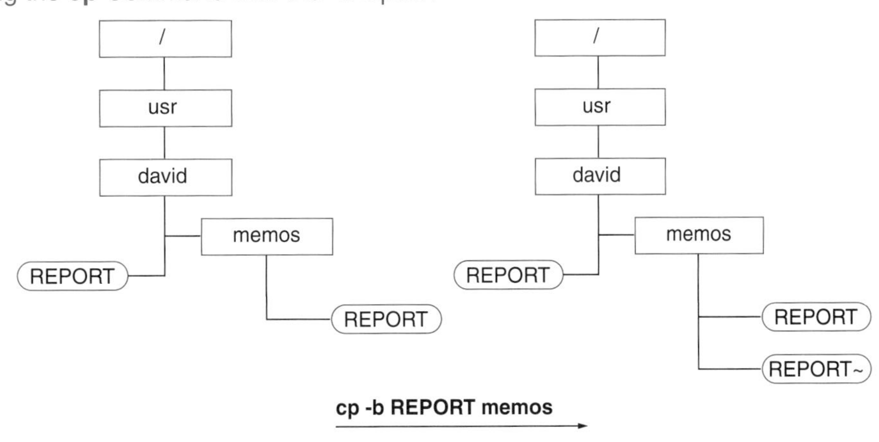

-i Option The **-i** option protects you from overwriting an existing file. It asks for confirmation if the target file already exists. If your reply is *yes*, it copies the source file, overwriting the existing file. If your answer is *no*, then it quits, and your existing file remains intact.

-r Option It takes a long time and is a tedious job to copy files one by one if you have a long list of files to copy. You can use **cp** with the **-r** option to copy directories and all their contents into a new directory.

Try using the **cp** command with the **-i** option.

```
$ cp -i REPORT memos [Return] . . . . . Make a copy of REPORT under memos.
Target file already exists overwrite? . Shows confirmation prompt; press [y] [Return] for yes or [n] [Return] for no.
$ . . . . . . . . . . . . . . . . . . . Ready for the next command.
```

Suppose you have a directory called `memos` in your current directory and you want to copy all the files and subdirectories under `memos` to the `david.bak` directory. Figure 7.6 shows your directory structure before and after the **cp -r** command. Notice there is a sub directory called `important` under `memos` that contains a file called `resume`. Copy files and subdirectories in `david` to another directory called `david.bak` using the **-r** option.

```
$ cp -r ./memos ./david.bak [Return] . . Copy memos directory and all the files in it to david.bak.
$ . . . . . . . . . . . . . . . . . . . Ready for the next command.
```

1. *If* `david.bak` *exists in your current directory, then files and directories in* `memos` *are copied into* `david.bak`.
2. *If* `david.bak` *does not exist in your current directory, then it is created and all the files and directories including* `memos` *itself are copied into* `david.bak`. *Now the pathname for files in* `memos` *under* `david.bak` *is* `./david.bak/memos`.

Figure 7.6
Using the **cp** Command with the **-r** Option

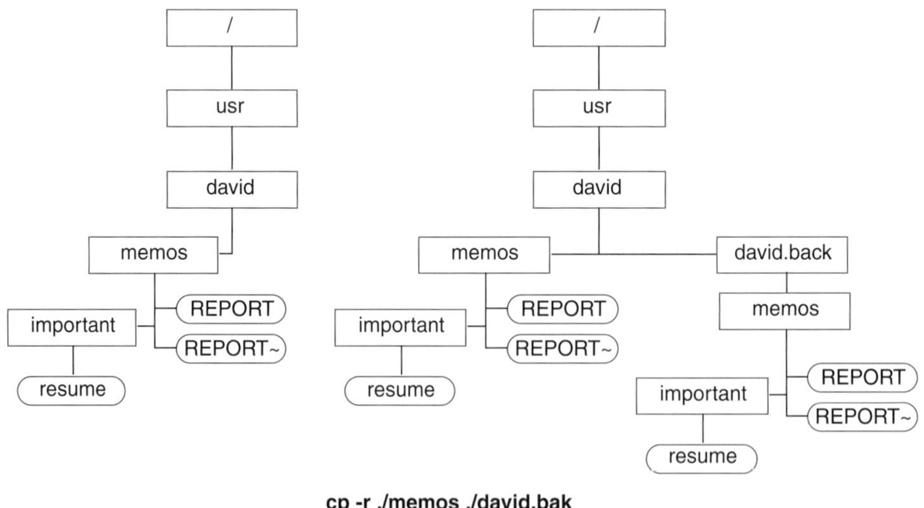

Practicing Linux Alternative cp Options

Use the Linux alternative options for the **cp** command, as suggested in the following command lines:

```
$ cp --interactive REPORT memos [Return] . . . . . Same as cp -i REPORT memos.
$ cp --recursive ./memos ./david.bak [Return] . . Same as cp -r /memos/david.bak.
$ cp --help [return] . . . . . . . . . . . . . . . Display the help page.
```

Read the help display and familiarize yourself with other options available for the **cp** *command in Linux.*

Copy files using the **-verbose** or **-v** option.

```
$ cp -v names scores memos [Return] . . . . . Copy names and scores in the current
                                              directory to the memos directory.
names -> memos/names
scores -> memos/scores
$_ . . . . . . . . . . . . . . . . . . . . . . Ready for the next command.
```

Notice the feedback from the **-v** *option that shows what file is copied and where it is copied.*

7.4.2 Moving Files: The mv Command

You use the **mv** command to move a file from one place to another or to change the name of a file or a directory. For example, if you have a file called `REPORT` in your current directory and you want to change its name to `REPORT.OLD`, you type **mv REPORT REPORT.OLD** and press [Return].

Figure 7.7 shows your directory structure before and after application of the **mv** command to rename `REPORT`.

Move `REPORT` to the `memos` directory.

```
$ mv REPORT memos [Return] . . . . Move REPORT to memos.
$_ . . . . . . . . . . . . . . . . Ready for the next command.
```

Figure 7.8 shows your directory structure after application of the **mv** command to move `REPORT`.

Figure 7.7
Using **mv** to Rename a File

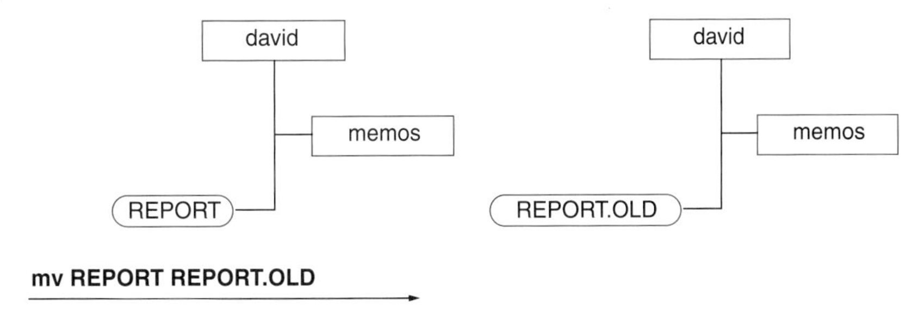

Figure 7.8
Using **mv** to Move a File

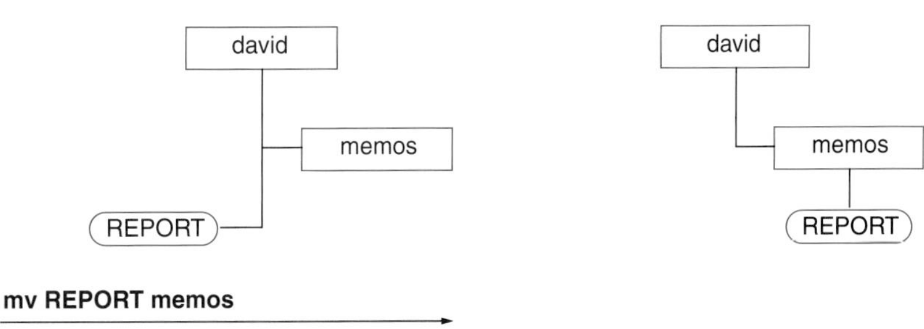

Both the **cp** and **mv** command accept more than two arguments, but the last argument must be a directory. For example, the command

```
cp xfile yfile zfile backup
```

copies `xfile`, `yfile`, and `zfile` to the directory called `backup`, provided that the `backup` directory exists.

mv Options

Table 7.5 shows some of the **mv** command options. The following command lines show the usage of the **mv** command options.

Table 7.5
The **mv** Command Options

Option		Operation
UNIX	Linux Alternative	
-b	**--backup**	Makes backup of the specified file if file already exists.
-i	**--interactive**	Asks for confirmation if the target file already exists.
-f	**--force**	Removes target file if file already exists and does not ask for confirmation.
-v	**--verbose**	Explains what is being done.
	--help	Displays the help page and exits.
	--version	Displays version information and exits.

Use the **mv** command with the **-b** (backup) option. Assuming you already have a file called `REPORT` under the `memos` directory, a backup of `REPORT` will be created. This means the `REPORT` file under `memos` is not overwritten.

```
$ mv -b  REPORT memos [Return] . . Move REPORT to memos and
                                   create a backup of REPORT if it
                                   already exits in memos.
$ ls memos [Return] . . . . . . . Check files under memos.
       REPORT   REPORT~
$_ . . . . . . . . . . . . . . . . Ready for the next command.
```

Notice that `REPORT` is backed up and the file name is `REPORT~`.

Use the **mv** command with the **-i** (confirmation) option. Assuming you already have a file called `REPORT` under the `memos` directory, a confirmation prompt is displayed and you press [y]es [Return] or [n]o [Return] to signal your intention.

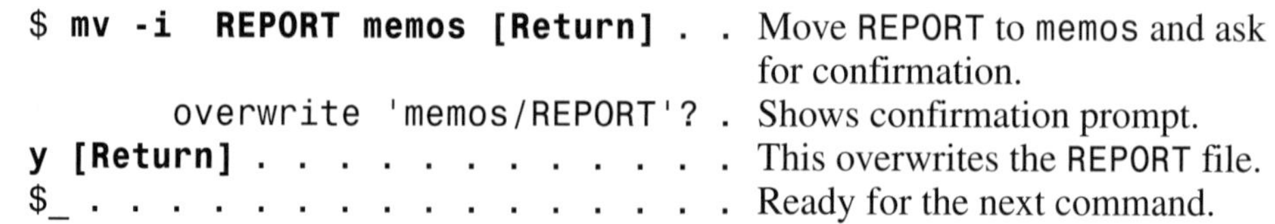

```
$ mv -i  REPORT memos [Return] . . Move REPORT to memos and ask
                                   for confirmation.
       overwrite 'memos/REPORT'? . Shows confirmation prompt.
y [Return] . . . . . . . . . . . . This overwrites the REPORT file.
$_ . . . . . . . . . . . . . . . . Ready for the next command.
```

Use the **mv** command with the **-v** (verbose) option.

```
$ mv -v  REPORT memos [Return] . .  Move REPORT to memos.
        REPORT -> memos/REPORT
$_ . . . . . . . . . . . . . . . .  Ready for the next command.
```

Practicing Linux Alternative mv Options

Use the Linux alternative options for the **mv** command, as suggested in the following command lines:

```
$ mv --backup  REPORT memos [Return] . . . . . .  Same as mv -b REPORT memos.
$ mv --interactive  REPORT memos [Return] . . .  Same as mv -i REPORT memos.
$ mv -verbose  REPORT memos [Return] . . . . . .  Same as mv -v REPORT memos.
$ mv --version [Return] . . . . . . . . . . . .  Display version information.
$ cp --help [Return] . . . . . . . . . . . . . .  Display the help page.
```

Read the help page and familiarize yourself with other options available for the **mv** *command.*

7.4.3 Linking Files: The ln Command

You can use the **ln** command to create new links (names) between an existing file and a new filename. This means you can create additional names for an existing file and refer to the same file with different names. For example, suppose you have a file called REPORT in your current directory, and you type **ln REPORT RP** and press [Return]. This creates a file named RP in your current directory and links that name to REPORT. Now, REPORT and RP are two names for a single file. Figure 7.9 shows your directory structure before and after the **ln** command application.

Figure 7.9
Using the **ln** Command to Link Filenames

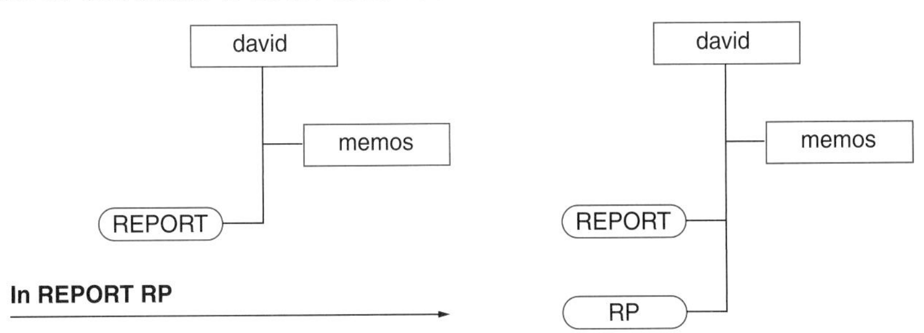

At first glance, this looks like the **cp** command, but it is not! The **cp** command physically copies the file into another place, and you have two separate files. Whatever changes you make in one are not reflected in the other. The **ln** command, however, just creates another filename for the same file; no new file is created. If you change anything in any of the linked files, the changes are there in the file regardless of the name you use to refer to it.

To experiment with the **ln** command, try the following command sequence:

$ **cat > xxx [Return]** . .	Create a file called xxx. Now type the following line in the file: Line 1:aaaaaa
$ **[Ctrl-d]**	Signal end of the input.
$ **ln xxx yyy [Return]** .	Link yyy to xxx.
$ **cat yyy [Return]** . . .	Display the contents of the xxx file, but use the new filename yyy; the output is, as you expected, the contents of xxx:

```
Line 1:aaaaaa
```

$ **cat >> yyy [Return]** .	Append a line to the end of yyy, and type the following line: Line 2:bbbbbb
[Ctrl-d]	Signal the end of your input.
$ **cat yyy [Return]** . . .	Display the contents of yyy. You have two lines in yyy, as you expected:

```
Line 1:aaaaaa
Line 2:bbbbbb
```

$ **cat xxx [Return]** . . .	Display the contents of xxx. You have two lines in it because xxx and yyy are actually the same file:

```
Line 1:aaaaaa
Line 2:bbbbbb
```

$_	Ready for the next command.

If you specify an existing directory name as the new filename, you can access the file in the specified directory without typing its pathname. For example, suppose you have a file called REPORT and a subdirectory called memos in your working directory, and you type **ln REPORT memos** and press [Return]. Now you can access REPORT from the memos directory without specifying its pathname, in this case ../REPORT.

If you want to specify a different name for the file in a different directory, type **ln REPORT memos/RP** and press [Return]. Now RP in the memos directory is linked to REPORT, and from memos you can use the filename RP to refer to REPORT.

Figure 7.10 shows your directory structure before and after application of the **ln** command.

Chapter 5 explained that the **ls -l** command lists the filenames in the current directory in a long format and that the second column in the long-format output shows the number of the links.

Figure 7.10
Using the **ln** Command to Link Filenames to a Directory

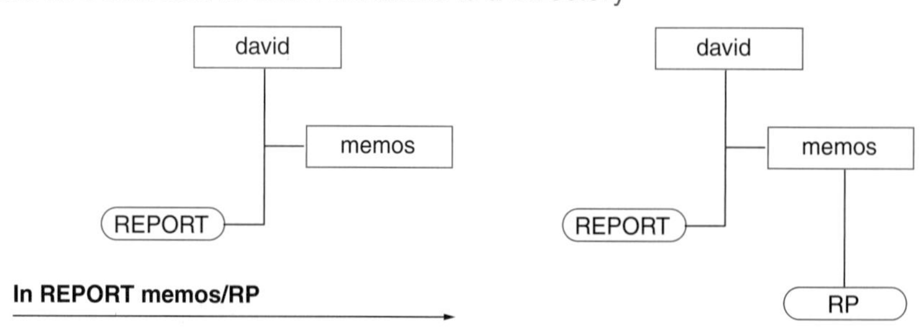

To list the files in `david` in long format, type **ls -l** and press [Return].

```
$ ls -l
total 2
drwx rw- --- 1      david student      32 Nov 28 12:30 memo
-rwx rw- --- 1      david student     155 Nov 18 11:30 REPORT
```

To link `REPORT` to `RP` and list the files using the **ls** command with the **-l** option to see the number of links, type **ln REPORT RP** and press [Return]. Then type **ls -l** and press [Return].

```
$ ln REPORT RP
$ ls -l
total 3
drwx rw- ---   1    david  student  32    Nov 28 12:30   memo
-rwx rw- ---   2    david  student 155    Nov 18 11:30   REPORT
-rwx rw- ---   2    david  student 155    Nov 18 11:30   RP
```

When you create a file, you also establish a link between the directory and the file. Thus, the link count for every file is at least one, and subsequent use of **ln** *adds to the number of links.*

Linux Alternative Options for the ln Command

Use the Linux alternative options for the **ln** command, as suggested in the following command lines:

$ ln --version [Return] . . Display version information.
$ ln --help [Return] . . . Display the help page.

Read the help page and familiarize yourself with other options available for the **ln** *command.*

Some Final Words

The three commands **cp**, **mv**, and **ln** all affect the filenames and work in a similar manner, but they are different commands and are used for different purposes:

- **cp** creates a new file.
- **mv** changes the filename or moves files from one place to another.
- **ln** creates additional names (links) for an existing file.

7.4.4 Counting Words: The wc Command

You can use the **wc** command to find out the number of lines, words, or characters in a file or a list of specified files.

The following command sequences show the output of the **wc** command, assuming you have the `myfirst` file in your current directory.

First display the contents of `myfirst`. Then show the number of lines, words, and characters in it:

```
$ cat myfirst [Return] . . . . . . . . . . Display the contents of myfirst.
I wish there were a better way to learn
UNIX. Something like having a daily UNIX pill.
However, for now, we have to suffer and read all these boring UNIX books.
$ wc myfirst [Return] . . . . . . . . . . Count the number of lines, words, and
                                          characters in myfirst.
4 30 155 myfirst
$_              Ready for the next command.
```

The first column shows the number of lines, the second column shows the number of words, and the third column shows the number of characters.

A word is considered a sequence of characters with no white space (space or tab character). Therefore, what? *is one word, and* what ? *counts as two words.*

If no filename is specified, then **wc** gets its input from the standard input device (keyboard). You signal the end of the input by pressing [Ctrl-d], and **wc** shows the results on the screen.

Use the **wc** command to get the count of input from the keyboard:

```
$ wc [Return]   . . . . . .  Invoke the wc command with no filename.
_   . . . . . . . . . . . .  The sign is the prompt indicating that the
                             shell is waiting for the rest of the
                             command. Type the following text:
The wc command is useful to find out how large your file is.
[Ctrl-d]  . . . . . . . . .  End your input; wc displays the output.
2 13 48
$_  . . . . . . . . . . . .  Ready for the next command.
```

You can specify more than one filename as the argument. In this case, the output shows one line of information for each file, and the last line shows the total counts.

Suppose you have two files in your current directory. The following command sequences show the output of the **wc** command specifying the two filenames as arguments.

Count the number of lines, words, and characters in the specified files:

```
$ wc myfirst yourfirst [Return] . Show the counts of myfirst and
                                  yourfirst.
24      10     400 myfirst
 3     100     400 yourfirst
27     110     800 total
$_  . . . . . . . . . . . . . . . Ready for the next command.
```

Table 7.6
This **wc** Command Options

Option		Operation
UNIX	**Linux Alternative**	
-l	**--lines**	Reports the number of lines.
-w	**--words**	Reports the number of words.
-c	**--chars**	Reports the number of characters.
	--help	Displays the help page and exits.
	--version	Displays version information and exits.

wc Options

You can use the wc command with options to get the number of lines, words, or characters only, or in any combination. Table 7.6 summarizes the wc command options.

1. *When no option is specified, the default is all options (***-lwc***).*
2. *You can use any combination of the options.*

The following command sequences show the use of the **wc** options.

Count the number of lines in `myfirst`:

```
$ wc -l myfirst [Return]  . Report only the number of lines:
4 myfirst
$_ . . . . . . . . . . . . . Ready for the next command.
$ wc -lc myfirst [Return] . Report the number of lines and characters.
155 4 myfirst
$_ . . . . . . . . . . . . . The prompt is back.
```

Save the output of the **wc** command in a file and print it.

```
$ wc myfirst > myfirst.count [Return]  . . Use the output redirection operator to save
                                           the wc report in myfirst.count.
$ lp -m myfirst.count [Return] . . . . . . Print myfirst.count and report when the
                                           job is done.
$_ . . . . . . . . . . . . . . . . . . . . Ready for the next command.
```

Linux Alternative Options for the wc Command

Use the Linux alternative options for the **wc** command, as suggested in the following command lines.:

```
$ wc --lines  myfirst [Return] . . . . . . Report only the number of lines.
$ wc --chars --lines myfirst [Return] . . . Report the number of lines and characters
                                           in myfirst.
$ cp --help [return] . . . . . . . . . . . Display the help page.
```

Read the help page and familiarize yourself with other options available for the **wc** *command in Linux.*

7.5 FILENAME SUBSTITUTION

Most file manipulation commands require filenames as arguments. When you want to manipulate a number of files, for example, to transfer all files with a filename starting with letter *a* to another directory, typing all the filenames one by one is tiring and boring. The shell supports *file substitution*, which allows you to select those filenames that match a specified pattern. These patterns are created by specifying filenames that contain certain characters that have a special meaning to the shell. These special characters are called *metacharacters* (or *wild cards*). Table 7.7 summarizes the wild cards that can stand for one or more characters in a file name.

Table 7.7
The Shell File Substitution Metacharacters

Character	Operation
?	Matches any single character.
*	Matches any string, including the empty string.
[list]	Matches any one of the characters specified in the *list*.
[!list]	Matches any one of the characters not specified in the *list*.

File substitution metacharacters (wild cards) can be used in any part—at the beginning, middle, or end—of the filename to create a search pattern.

For the terminal sessions in this section you need to create new files and directories. It is a good idea to create a directory for each chapter to be used for the terminal sessions and exercises of that chapter. Under the `Chapter7` directory, create a subdirectory called `7.5`. If you have not created a `Chapter7` directory in your HOME directory, here is one way to do this:

```
$ cd [Return] . . . . . . .  Change to the HOME directory.
$ mkdir Chapter7 [Return]  . Create Chapter7 directory.
$ cd Chapter7 [Return] . . . Change current directory to Chapter7.
$ mkdir 7.5 [Return] . . . . Create a directory called 7.5.
```

Now create the following files using **cat** command or vi editor. There is no need to type anything in the files, or just type in few words if you want. We only need these files for their names. Create the following six filenames:

```
report     report1     report2     areport     breport     report32
```

7.5.1 The ? Metacharacter

The question mark (**?**) is a special character that the shell interprets as a single character substitution and expands the filename accordingly.

Try the following command sequences to see how the **?** metacharacter works:

```
$ ls -C [Return] . . . . . Check the filenames in your working
                           directory. You have the following files:
report    report1    report2    areport    breport
report32
$ ls -C report? [Return] . Use a single question mark in the filename.
report1    report2
$_ . . . . . . . . . . . . Ready for the next command.
```

The shell expands the filename `report?` to the filename `report` followed by exactly one character, any character. Thus, the filenames `report1` and `report2` are the only two files that match the pattern.

```
$ ls report?? [Return] . . Use two question marks as special characters.
report32
$_ . . . . . . . . . . . . Ready for the next command.
```

The shell expands the filename `report??` to the filename `report` followed by exactly two characters, any characters. Thus, the filename `report32` is the only file that matches the pattern.

```
$ ls -C ?report [Return] . Put ? at the beginning of the filename.
areport    breport
$_ . . . . . . . . . . . . Ready for the next command.
```

The shell expands the filename `?report` to the filename `report` preceded by exactly one character, any character. Thus the filenames `areport` and `breport` are the only two files that match the pattern.

7.5.2 The * Metacharacter

The asterisk (*) is a special character that the shell interprets as any number of characters (including zero characters) of substitution in a filename, and expands the filename accordingly.

Try the following command sequences to see how the * metacharacter works:

```
$ ls -C [Return] . . . . . . Check the filenames in your working
                             directory. You have the following files:
report    report1    report2    areport    breport
report32
$ ls -C report* [Return] . . List all filenames that begin with the
                             word report:
report report1 report2 report32
```

The shell expands the filename `report*` to the filenames `report` followed by any number of characters, any characters. Thus the file names `areport` and `breport` are the only two filenames that do not match the pattern. The * wild card includes zero characters following the specified pattern. Thus, the file name `report` followed by no character matches the pattern and is displayed.

```
$ ls -C *report [Return] . . . . . List all the filenames that end with the word report:
report     areport     breport
$_ . . . . . . . . . . . . . . . . . . Ready for the next command.
```

The shell expands the filename `*report` to the filename `report` preceded by any number of characters, any characters. Thus the filenames `report`, `areport`, and `breport` are the only files that match the pattern. Remember, the * wild card includes zero characters following the specified pattern. Therefore, the filename `report` preceded by no character matches the pattern and is displayed.

```
$ Is -C r*2 [Return] . . . . . . . . List all files that start with r and end with 2:
report2     report32
$_ . . . . . . . . . . . . . . . . . . Ready for the next command.
```

The shell expands the file name **r*2** to the filename **r** followed any number of characters, but the last character of the filename must be the character 2, referring to any file that starts with *r* and ends with *2*.

7.5.3 The [ ] Metacharacters

The open and close brackets are special characters that surround a string of characters. The shell interprets this string of characters as filenames that contain the specified characters, and expands the filenames accordingly.

Using the **!** before the specified string of characters (but within the brackets) causes the shell to display the filenames that do *not* contain the characters in the string at the specified position.

Experiment with the bracket metacharacters by doing the following:

- To list all the filenames that start with *a* or *b*, type **ls -C [ab]*** and press [Return], UNIX responds:

```
areport     breport
$_ . . . . . . . . . . . . . Ready for the next command.
```

The shell expands the filename **[ab]*** to the filename `a` or `b` followed by any number of characters, any characters. Thus `areport` and `breport` are the only two filenames that match the specified pattern.

- To list all the filenames that do *not* start with *a* or *b*, type **ls -C [!ab]*** and press [Return], UNIX responds:

```
report     report1     report2     report32
```

You can use the [ ] metacharacters to specify a range of characters or digits. For example, **[5-9]** *means the digits 5, 6, 7, 8, or 9; and* **[a-z]** *means all the lowercase letters of the alphabet.*

The following command sequence shows the use of brackets with a specified alphabet or digits range:

- Type **ls *[1-32]** and press [Return] to list all the filenames that end with the digits 1 to 32, UNIX responds:

```
report1     report2     report32
```

```
$ ls -C [A-Z] [Return] . . . . .   Show all the single-capital-letter
                                   file names, assuming you have
                                   some single-letter file names.
A          B          D     W
$_ . . . . . . . . . . . . . . .   The prompt appears.
```

7.5.4 Metacharacters and Hidden Files

To use the metacharacters for displaying hidden files—filenames starting with **.** (dot)—you must explicitly have the **.** (dot) as part of the specified pattern.

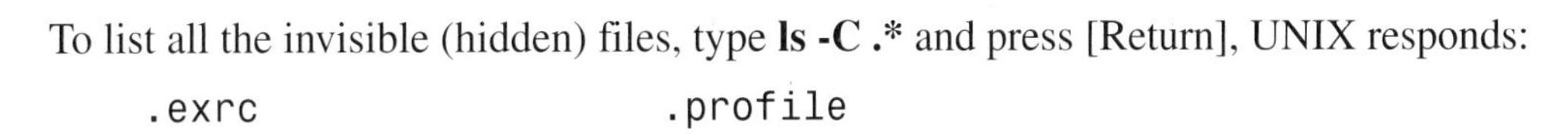

To list all the invisible (hidden) files, type **ls -C .*** and press [Return], UNIX responds:

```
.exrc                  .profile
```

The shell expands the filename `.*` to the filename `.` (dot) followed by any number of characters, any characters. Thus, only the hidden files are displayed.

The pattern is `.*` and there is no space between the dot and the asterisk.

Wild cards are not limited in use to only the **ls** *command. You can use wild cards with other commands that need filename arguments.*

To experiment with some more examples of filename substitution, try the following:

```
$ rm *.* [Return] . . . . . . . .       Delete all the files with filenames that contain at least
                                        one dot.
$ rm report? [Return] . . . . . .       Delete all filenames that begin with report and end with
                                        only one character, any character.
$ cp * backup [Return]  . . . . .       Copy all files from the current directory to the backup
                                        directory.
$ mv file[1-4] memos [Return] . .       Move file1, file2, file3, and file4, indicated by
                                        the range [1–4], to the memos directory.
$ rm report* [Return] . . . . . .       Delete all the filenames that start with the string report.
$_  . . . . . . . . . . . . . . .       Ready for the next command.
```

In the last example, there is no space between the string **report** *and the asterisk wild card. The* **rm report*** *command deletes all the files with filenames that begin with* **report**.

```
$ rm report * [Return] . . . . .   Delete all files.
$_ . . . . . . . . . . . . . . .   Ready for the next command.
```

In the preceding example, there is a space between the string *report* and the asterisk wild card. This space can have disastrous consequences. The command `rm report *` is interpreted as "delete a file called report, and then delete all the other files." In other words, all files in the current directory are deleted.

7.6 MORE FILE MANIPULATION COMMANDS

The following commands facilitate the search for a specific file in a crowded hierarchy of directories, and display a specified portion of a file for a quick look.

7.6.1 Finding Files: The find Command

You can use the **find** command to locate files that match a given set of criteria in a hierarchy of directories. The criterion may be a filename or a specified property of a file (such as its modification date, size, or type). You can also direct the command to remove, print, or otherwise act on the file. The **find** command is a very useful and important command that is not used to its full potential. Maybe its unusual command format is discouraging.

The format of the **find** command is different from that of the other UNIX commands. Let's look at its syntax:

```
find pathname  search options  action option
```

pathname indicates the directory name from which **find** begins its search, and it then continues down to the subdirectories and their subdirectories and so on. This process of branching search is called a *recursive search.* The *search options* part identifies which file you are interested in, and the *action option* part tells what to do with the file once it is found.

Let's look at a simple example:

```
$ find . -print [Return]
```

This command displays the names of all the files in the specified directory and all its subdirectories.

1. *The specified directory is indicated by a dot, meaning the current directory.*
2. *The options after the pathname always start with a hyphen (-). The action option part indicates what to do with the files. In this case,* **-print** *means to display them.*

Don't forget the -print action key. Without it, find does not display any pathnames.

find Options

Table 7.8 shows a partial list of the find options and explanations of the options.

-name Option You use the search option to find a file by its name. You type **-name** followed by the desired filename. The file name can be a simple filename or you can use the shell wild card substitution: [ ], ?, and *. If you use these special characters, place the filename in single quotation marks. Let's look at some examples. The ±*n* notation in

Table 7.8
The **find** Command Search Options

Operator	Description
-name *filename*	Finds files with the given *filename.*
-size ±*n*	Finds files of the size *n*.
-type *file type*	Finds files with the specified access mode.
-atime ±*n*	Finds files that were accessed *n* days ago.
-mtime ±*n*	Finds files that were modified *n* days ago.
-newer *filename*	Finds files that were modified more recently than *filename.*

Table 7.8 is a decimal number that can be specified as ±*n* (meaning more than n) or -*n* (meaning less than *n*) or *n* (meaning exactly *n*).

Find some files by name.

`$ find . -name first.c -print [Return]` . .	Find files named `first.c`.
`$ find . -name "*.c" -print [Return]` . . .	Find all files whose names end in `.c`.
`$ find . -name "*.?" -print [Return]` . . .	Find all files whose names end with a single character preceded by a period.

1. *In all of the preceding commands, the directory name is the current directory. The* **.** *(dot) represents the current directory.*
2. *The* **-name** *option identifies the filename. The wild card may be used to generate the filename.*
3. *The action part* **-print** *is used to display the name of the file found.*

-size ±*n* Option You use this search option to find a file by its size in blocks. You type **-size** followed by the number of blocks indicating the size of the file to be checked. The plus or minus sign before the number of blocks indicates greater than or less than, respectively. Let's look at some examples.

Find some files by size:

`$ find . -name "*.c" -size 20 -print [Return]`	Find files that are exactly 20 blocks large.

This command finds all files that have filenames that end with *.c* and that are exactly 20 blocks large.

`$ find . -name "*.c" -size +20 -print [Return]` . . .	Find files that are larger than 20 blocks.
`$ find . -name "*.c" -size -20 -print [Return]` . . .	Find files that are smaller than 20 blocks.

-type Option You use this search option to find a file by its type. You type *-type* followed by a letter specifying the file type. The *file types* are as follows:

- *b*: a block special file (such as your disk)
- *c*: a character special file (such as your terminal)
- *d*: a directory file (such as your directories)
- *f*: an ordinary file (such as your files)

Find files using the *file type* option:

`$ find $HOME -type f -print [Return]` . .	Use the **-type** option.

This command finds all the ordinary files in your HOME directory and displays their pathnames. Recall, $HOME holds the absolute pathname to your HOME directory.

-atime Option You use this search option to find a file by its last access date. You type **-atime** followed by the number of days since the file was last accessed. The plus or minus sign before the number of days indicates greater than or less than, respectively. Let's look at some examples.

Find files by their last access times:

`$ find . -atime 10 -print [Return]` . . Find and display files last accessed exactly 10 days ago.

The previous command displays the names of the files that have not been read for exactly 10 days.

`$ find . -atime -10 -print [Return]` . . Find and display files last accessed less than 10 days ago.

`$ find . -atime +10 -print [Return]` . . Find and display files last accessed more than 10 days ago.

-mtime Option You use this search option to find a file by its last modification date. You type **-mtime** followed by the number of days since the file was last modified. The plus or minus sign before the number of days indicates greater than or less than, respectively. Let's look at some examples.

Find files by their last modification times:

`$ find . -mtime 10 -print [Return]` . . Find and display files last modified exactly 10 days ago.

The previous command displays the names of the files that are exactly 10 days old.

`$ find . -mtime -10 -print [Return]` . . Find and display files last modified less than 10 days ago.

`$ find . -mtime +10 -print [Return]` . . Find and display files last modfied more than 10 days ago.

-newer Option You use this search option to find a file modified more recently than a file with a specified filename.

Let's look at an example:

`$ find . -newer first.c -print [Return]` . Find and display files modified more recently than `first.c` was.

Action Options

The action options tell **find** what to do with files once they are found. Table 7.9 summarizes the three action options.

Table 7.9
The **find** Command Action Options

Operator	Description
-print	Prints the pathname for every file found.
-exec *command* \;	Lets you give *commands* to be applied to the files.
-ok *command* \;	Asks for confirmation before applying the command.

-print Option The **-print** action option displays the pathnames of the files found that match the specified criterion.

Find pathnames to files called `first.c`, starting from the home directory.

```
$ find $HOME -name first.c -print [Return]  . Find first.c and
                                              display the pathnames.
/usr/david/first.c
/usr/david/source/first.c
/usr/david/source/c/first.c
$_ . . . . . . . . . . . . . . . . . . . . . . Receive prompt.
```

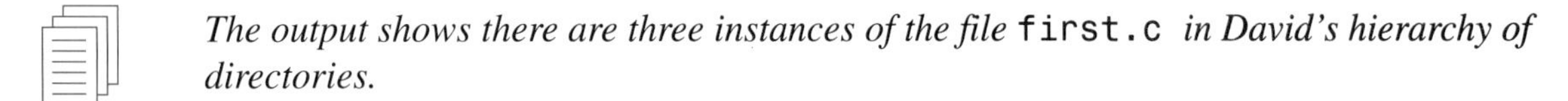

The output shows there are three instances of the file `first.c` *in David's hierarchy of directories.*

-exec Option The **-exec** action option lets you give a command to be applied to the found files. You type **-exec** followed by the specified command, a space, backslash, and then a semicolon. You can use a set of braces ({}) to represent the name of the files found. An example will help to clarify the sequence.

Find and delete all instances of the `first.c` file that are 90 days old:

```
$ find . -name first.c -mtime +90 -exec rm { }\; [Return]
$_ . . . . . . . . . . . . . . . . . . . . . . Receive prompt.
```

The search starts from your current directory (represented by **.**) and is continued through the hierarchy of the directories. The **find** command and the **-exec rm** command locate and the remove instances of `first.c` that are 90 days old.

Two search options are in effect: the **-name** and the **-mtime** options. That means **find** is seeking files that simultaneously satisfy these two search categories.

The command is comprised of many parts, and its syntax is peculiar:

1. **The -exec option followed by the command (in this case rm)**
2. **A set of braces *({ })* followed by a space**
3. **A backslash *(\)* followed by a semicolon**

All instances of `first.c` are deleted. No warning or feedback message is displayed. When you see the **$** prompt, the job is done.

-ok Option The **-ok** action option is just like the **-exec** option, except that it asks for your confirmation before applying the command to the file.

Find and delete all instances of the `first.c` file, but ask for confirmation before deleting any file:

```
$ find . -name first.c +90 -ok rm { }\; [Return]
$_ . . . . . . . . . . . . . . . . . . . . . . Receive prompt.
```

If a file, for example `first.c`, satisfies the criterion, then the following prompt is displayed:

```
<rm ... ./source/first.c> ?
```

If you reply with [Y] or [y], the command is executed (in this case, `file.c` is deleted); otherwise, your file remains intact.

You can also use the logical operators **or**, **and**, *and* **not** *to combine the search options. The search starts from your current directory, and it is continued through the hierarchy of the directories.*

It is highly recommended to use the **man** command to display the manual pages for the **find** command and its many options.

$ man find [Return] . . . . . . Display **man** pages for **find**.

Under Linux, you can use the **--help** option to display the help page:

$ find --help [Return] . . . . Display the help page for find.

7.6.2 Displaying the Beginning of a File: The head Command

You can use the head command to display the beginning of a specified file. The head command gives you a quick way to check the first few lines of a file. For example, to display the first part of the file called `MEMO` in your current directory, you type the following:

$ head MEMO [Return]

By default, head shows the first 10 lines of the specified file. You can override the default value by specifying the number of lines. For example, to display the first five lines of the file `MEMO`, you type the following:

$ head -5 MEMO [Return]

The specified number of line must be a positive integer number.

You can specify more than one filename on the command line. For example, to show the first five lines of the files called `MyFile`, `YourFile`, and `OurFile`, you type the following:

$ head MyFile YourFile OurFile [Return]

When more than one file is specified, the start of each file will look like:

```
==> filename <==
```

The following command sequences demonstrate the use of the **head** command.

$ head -5 *File [Return] . . . Display the first five lines of all files in the current directory ending with *File*.

$ head * [Return] . . . . . . Display the first 10 lines of all files in the current directory.

$ head -15 MEMO [Return] . . . Display the first 15 lines of the `MEMO` file.

head Options

Table 7.10 lists some of the **head** command options. For more detail, use the **man** command to read the man pages.

Table 7.10
The **head** Command Options

Option		Operation
UNIX	Linux Alternative	
-l	--lines	Counts by lines. This is the default option.
-c	--chars=num	Counts by characters.
	--help	Displays the help page and exits.
	--version	Displays the version information and exits.

7.6.3 Displaying the End of a File: The tail Command

You can use the **tail** command to display the last part (the tail end) of a specified file. The **tail** command gives you a quick way to check the contents of a file. For example, to display the last part of the file called MEMO in your current directory, you type the following:

```
$ tail MEMO [Return]
```

By default, **tail** shows the last 10 lines of the specified file. You can override the default value by using one of the available options.

tail Options

Table 7.11 lists some of the **tail** command options,which are similar to the options for the **head** command. When more than one filename is specified on the command line, each file starts with a header line similar to the following line:

```
==> filename <==
```

For more detail, use the **man** command and read the manual pages.

1. *If a plus sign precedes the option,* **tail** *counts from the beginning of the file.*
2. *If a hyphen precedes the option,* **tail** *counts from the end of the file.*
3. *If a number precedes the option,* **tail** *uses that number instead of the default 10 lines count.*

Table 7.11
The **tail** Command Options

Option		Operation
UNIX	Linux Alternative	
-l	--lines	Counts by lines. This is the default option.
-c	--chars=num	Counts by characters.
	--help	Displays the help page and exits.
	--version	Displays the version information and exits.

The following command sequences demonstrate the use of the **tail** command:

$ tail MEMO [Return] . . . . . . Display the last 10 lines (no option).
$ tail 11 MEMO [Return] . . . . Display the last 11 lines (11).
$ tail -4 MEMO [Return] . . . . Display the last 4 lines (-4).
$ tail -10c MEMO [Return] . . . Display the last 10 characters (-10c).
$ tail +50 MEMO [Return] . . . . Skip 50 lines from the beginning of the file MEMO and display the rest of the file.

You can specify only one filename as an argument with the **tail** *command.*

Practicing Linux Alternative tail Options

Use the Linux alternative options for the **tail** command, as suggested in the following command lines:

$ tail --chars=10 MEMO [Return] . . Same as `tail -10c MEMO`.
$ tail --lines=5 MEMO [Return] . . Same as `tail -10l MEMO`.
$ tail --version [Return] . . . . . Display version information.
$ tail --help [Return] . . . . . . Display the help page.

7.6.4 Selecting Portions of a File: The cut Command

You can use the **cut** command to "cut out" specific columns or fields from files. Many files are collections of records, with each record consisting of several fields. You might be interested in some of the fields or columns contained in a file. Figure 7.11 shows an example of such a file called phones. Each record in this file consists of five fields, and each field is separated by a space or a tab.

Figure 7.11
A File Called phones

```
$ cat phones
David Back     (909) 999999      dave@xyz.edu
Daniel Knee    (808) 888888      dan@xyz.edu
Gabe Smart     (707) 777777      gabe@xyz.edu
$
```

cut Options

Let's suppose you have a file called phones that contains names, phone numbers, and E-mail addresses, similar to the file shown in Figure 7.11. Table 7.12 summarizes the **cut** command options. The following command sequences show the usage of **cut** and some of its options.

For the terminal sessions in this section, you need to create the phones file. Under the Chapter7 directory, create a subdirectory called 7.6. This is only a suggestion and you can create this file under any directory you wish. Now, use the **vi** or **cat** command to create the phones file similar to the file shown in Figure 7.11.

Table 7.12
The **cut** Command Options

Option		Operation
UNIX	Linux Alternative	
-f	**--fields**	Specifies the field position.
-c	**--characters**	Specifies the character position.
-d	**--delimiter**	Specifies the field separator (delimiter) character.
	--help	Displays the help page and exits.
	--version	Displays the version information and exits.

-f Option Following the **-f** option, a list of the fields is specified. Fields are assumed to be separated in the file by a delimiter character (the default is the tab character.) For example, **-f 1** indicates field one and **-f 1,7** indicates fields one and seven.

Use the **cut** command with the **-f** option to display the *first* field from the phones file.

```
$ cut -f 1 phones [Return] . . . Show the first field in the phones file.
David Back
Daniel Knee
Gabe Smart
$_ . . . . . . . . . . . . . . . And the prompt appears.
```

Remember, the default field delimiter is the tab character.

Use the **cut** command with the **-f** option to display the *first* and *third* fields from the phones file.

```
$ cut -f 1,3 phones [Return] . . Show the first and third fields in the
                                 phones file.
David Back   dave@xyz.edu
Daniel Knee  dan@xyz.edu
Gabe Smart   gabe@xyz.edu
$_ . . . . . . . . . . . . . . . Prompt.
```

-c Option Following the **-c** option, the character positions are specified. For example, **-c 1-10** indicates the first 10 characters of each line.

Display the first four characters of each line from the phones file.

```
$ cut -c 1-4 phones [Return] . . Show columns 1 to 4.
Davi
Dani
Gabe
$
```

Use the **cut** command with no filename:

```
$ date | cut -c 12-13 [Return] . . . . No filename is specified.
18
$_ . . . . . . . . . . . . . . . . . . . Prompt.
```

1. *The ¦ is a pipe operator that is used to pass the* **date** *output as input to the* **cut** *command. The pipe operator is explained in Chapter 8.*
2. *The output of the* **date** *command (date string) is passed to the* **cut** *command. The* **cut** *command displays the characters in positions 12 and 13 of the date string, which happen to be the hour field.*

-d Option The character following **-d** is the field delimiter. The default character is tab. The space character or other characters with special meaning must be enclosed in double quotation marks. The delimiter is the character that separates the fields in your file.

Show one field using a space character as the field separator (delimiter):

```
$ cut -d " " -f 1 phones [Return]   . . Show the first field in the
                                         phones file.
David
Daniel
Gabe
$_ . . . . . . . . . . . . . . . . . . . And the prompt appears.
```

If the field separator is a space character, make sure it is enclosed in quotation marks.

Practicing Linux Alternative cut Options

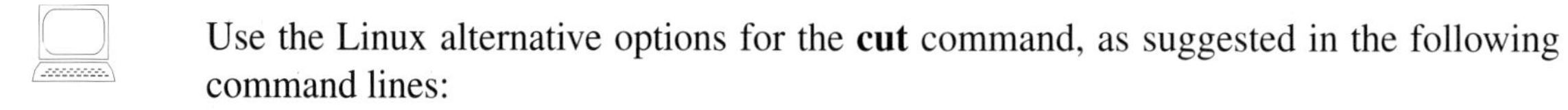

Use the Linux alternative options for the **cut** command, as suggested in the following command lines:

```
$ cut --characters 1-4 phones [Return]. . . . . . . Same as cut -c 1-4 phones.
$ cut --delimiter " " --fields 1 phones [Return]. . Same as cut -d " " -f 1
                                                     phones.
$ cut --version [Return]. . . . . . . . . . . . . . Display version information.
$ cut --help [Return] . . . . . . . . . . . . . . . Display the help page.
```

Read the help page and familiarize yourself with other options available for the **cut** *command.*

7.6.5 Joining Files: The paste Command

You can use the **paste** command to join files together line by line, or you can create new files by pasting together fields from two or more files. Figure 7.12 shows the output of the **paste** command if you have the two files called `first` and `last` in your current directory.

paste Options

Table 7.13 lists the **paste** command options. For the terminal sessions in this section, you need to create two files called `first` and `last`. Use the vi editor or the **cat** command to create these files, similar to the files in Figure 7.12. Remember, you only type in the names in the files, not the command line and the prompt sign.

Figure 7.12
The **paste** Command

```
$ cat first
David
Daniel
Gabriel
$ cat last
Back
Knee
Smart
$ paste first last
David          Back
Daniel         Knee
Gabriel        Smart
$_
```

Table 7.13
The **paste** Command Options

Option		Operation
UNIX	Linux Alternative	
-d	**--delimiters**	Specifies the field separator (delimiter) character.
	--help	Displays the help page and exits.
	--version	Displays the version information and exits.

-d Option The **-d** (delimiter) option specifies a specific delimiter character. The default is the tab character.

Use the **paste** command, indicating the **:** (colon) as the field separator.

```
$ paste -d : first last [Return]  . .  Use : as the field separator.
David:Back
Daniel:Knee
Gabriel:Smart
$_ . . . . . . . . . . . . . . . . . . The prompt is back.
```

Use the **paste** command, indicating the space character as the field separator.

```
$ paste -d " " first last [Return]  .  Use the space character as the
                                       field separator.
David Back
Daniel Knee
Gabriel Smart
$_ . . . . . . . . . . . . . . . . . . Prompt.
```

If the delimiter character is the space character, make sure to enclose it in quotation marks.

Practicing Linux Alternative paste Options

Use the Linux alternative options for the **paste** command, as suggested in the following command lines:

```
$ paste --delimiters : first last [Return]. . . . Same as paste -d : first last.
$ paste --version [Return]. . . . . . . . . . . . Display version information.
$ paste --help [Return] . . . . . . . . . . . . . Display the help page.
```

7.6.6 Another Pager: The more Command

The **more** command is another pager provided for your convenience. Like **pg**, you can use **more** to browse or page through a text file. It pauses after each page (screen), and then it displays the word *More* and the percentage of the characters displayed so far at the bottom of the screen.

```
More-(11%)
```

For screen continuity, **more** *provides two lines of overlap between screens.*

more Options

[Spacebar] Pressing the [Spacebar] moves the cursor ahead one screen.

[Return] Key Pressing the [Return] key scrolls ahead one line.

Q or q Key Pressing the [q] or [Q] key exits the **more** command.

Table 7.14 lists other options you can use when you issue the **more** command.

For the following terminal sessions, you need a large file to be able to practice all the options available for the **more** command. An easy way to create a large file is to use the **man** and redirection commands, instead of using vi to create a large file by typing pages and pages of text. Following is the command line to create a large file:

```
$ man who > who [Return] . . Save who description in a file called who
```

Table 7.14
The **more** Command Options

Option		Operation
UNIX	**Linux Alternative**	
-*lines*	**-num** *lines*	Displays the indicated number of *lines* per screen.
+*line-number*		Starts up at *line-number*.
+/*pattern*		Starts two lines above the line containing the *pattern*.
-c	**-p**	Clears the screen before displaying each page instead of scrolling. This is sometimes faster.
-d		Displays the prompt **[Hit space to continue, Del to abort]**.
	-- help	Displays the help page and exits.

In this case, the description of the **who** command is redirected to a file called who. The choice of the **who** command and the name of the file are both arbitrary, and you can choose any command from **man** and redirect it to a file of any name you choose. But by using the example command line, you now have a file called who that is few pages long.

The following examples show the use of the **more** command line options:

$ more -10 who [Return] . . . . Display the who file with 10 lines per screen.

The previous command sets the number of lines per screen to 10 rather than the default that is number of lines in the terminal screen less two lines, usually 22 lines.

$ more +100 who [Return] . . . . Start at line 100.

The previous command displays the who file one screenful at a time, starting at line number 100.

$ more +/User who [Return] . . . Start with the word *User*.

The previous command displays the who file one screenful at a time, starting two lines above the line containing the word User.

$ more -cd who [Return] . . . . Display who in clear mode and display prompt.

As with other commands, more than one option can be used on a command line. In the previous example, the -c (clear screen) and -d (display prompt) options are used to display the who file.

While using the **more** command, the following keys can be used to further control the paging process.

Assuming you have issued the following command line:

$ more +100 -cd2 who [Return]

your screen would be similar to the following screen:

```
      field is the name of the program executed by  init
      as  found  in /sbin/inittab.  The state, line, and
--More--(42%)[Hit space to continue, Del to abort]
```

The prompt line is the result of using the **-d** option, and the screen size is two lines, as was set by the **-2** option.

□ Press the = key to see the current line.

At this point, the number of current line (101) is displayed, and your screen will be similar to the following:

```
      field is the name of the program executed by  init
      as  found  in /sbin/inittab.  The state, line, and
101
```

□ Press **h** to obtain a list of the available options.

At this point, the help screen is displayed and your screen will be similar to the following screen. To save space, a partial list of the commands is shown.

```
 Star (*) indicates argument becomes new default.
-------------------------------------------------------------
<space>                 Display next k lines of text [current screen size]
z                       Display next k lines of text [current screen size]*
<return>                Display next k lines of text [1]*
d or ctrl-D             Scroll k lines [current scroll size, initially 11]*
q or Q or <interrupt>   Exit from more
~
~
-------------------------------------------------------------
--More--(84%)[Hit space to continue, Del to abort]
```

At this point, you can press the [Spacebar] to display the next screen of text, or press the [Return] key to display the next line of text. You use the [q] or [Q] key to exit from **more**.

7.6.7 Linux Pager: The less Command

The **less** command is yet another pager that is provided under Linux. This command is a program similar to **more**, but it allows backward movement in the file as well as forward movement. Also, **less** starts up faster with large input files than text editors like vi do. The **less** commands are based on both **more** and vi. The **less** command is much more sophisticated pager than **more**. Please use the **man** command or the **-?** and **--help** options to view the many available options for the **less** command.

$ **less -? [Return]** . . . . . Display usage and list of options.
$ **less --help [Return]** . . . Display usage and list of options.
$ **man less [Return]** . . . . . Display manual pages.

7.7 UNIX INTERNALS: THE FILE SYSTEM

How does the UNIX file system keep track of your files? How does it know the location of your files on the disk? From your point of view, you create directories to organize your disk space, and directories and files have filenames to identify them. This hierarchical structure of directories and files is a logical view of the file system. Internally, however, UNIX organizes the disk and keeps track of files in a different manner.

The UNIX file system associates every filename with a number called its *i-node number* and identifies each file by its i-node number. UNIX keeps all these i-node numbers in a list, appropriately called an *i-node list.* The list is saved on the UNIX disk.

7.7.1 UNIX Disk Structure

Under UNIX, a disk is a standard block device, and a UNIX disk is divided into four blocks (regions):

- Boot block
- Super block
- i-node list block
- Files and directories block

The Boot Block The *boot block* holds the *boot program*, a special program that is activated at the system boot time.

The Super Block The *super block* contains information about the disk itself. This information includes the following:

- Total number of disk blocks
- Number of free blocks
- Block size in bytes
- Number of used blocks

The i-list Block The *i-list block* keeps the list of i-nodes. Each entry in this list is an i-node, a 64-byte storage area. The i-node of a regular file or a directory file contains the location of its disk block(s). The i-node of a special file contains the information that identifies the peripheral device. An i-node also contains other information, including the following:

- File access permission (read, write, and execute)
- Owner and group IDs
- File link count
- Time of the last file modification
- Time of the last file access
- Location of blocks for each regular file and directory file
- Device identification number for special files

The i-nodes are numbered sequentially.

i-nodes and Directories

i-node 2 contains the location of the block(s) that contains the root directory (/). A UNIX directory contains the list of filenames and their associated i-node numbers. When you create a directory, it automatically creates two entries, one for the `..` (dot dot) or parent directory and one for the `.` (dot) or child directory.

Filenames are stored in directories and not in the i-nodes.

7.7.2 Putting It Together

When you log in, UNIX reads the root directory (i-node 2) to find your HOME directory and saves your HOME directory i-node number. When you change your directory using **cd**, UNIX replaces this i-node number with the new directory's i-node number.

When you access a file using utilities or commands (such as **vi** or **cat**) or when a program opens a file, UNIX reads and searches the directory for the specified filename. There is an i-node associated with each filename that points to a specific i-node in the i-node list. UNIX uses your working directory i-node number to begin its search, or if you give a full pathname, it starts from the root directory, which always has the i-node 2.

Suppose your current directory is `david`, and you have a subdirectory called `memos`, and a file called `report` in `memos`, and you want to access `report`. UNIX starts searching from your current directory, `david` (with a known i-node number), and finds the filename `memos` and its i-node number. Next, it reads the `memos` i-node record from the i-node list. The `memos` i-node indicates the block that contains the `memos` directory.

Looking into the block that contains the filenames under `memos`, UNIX finds the `report` filename and its i-node number. UNIX repeats the preceding process, and reads the i-node record from the i-node list. Information in this record includes the location of the blocks on the disk that make the `report` file. (See Figure 7.13.)

Figure 7.13
The Directory Structure and i-node List

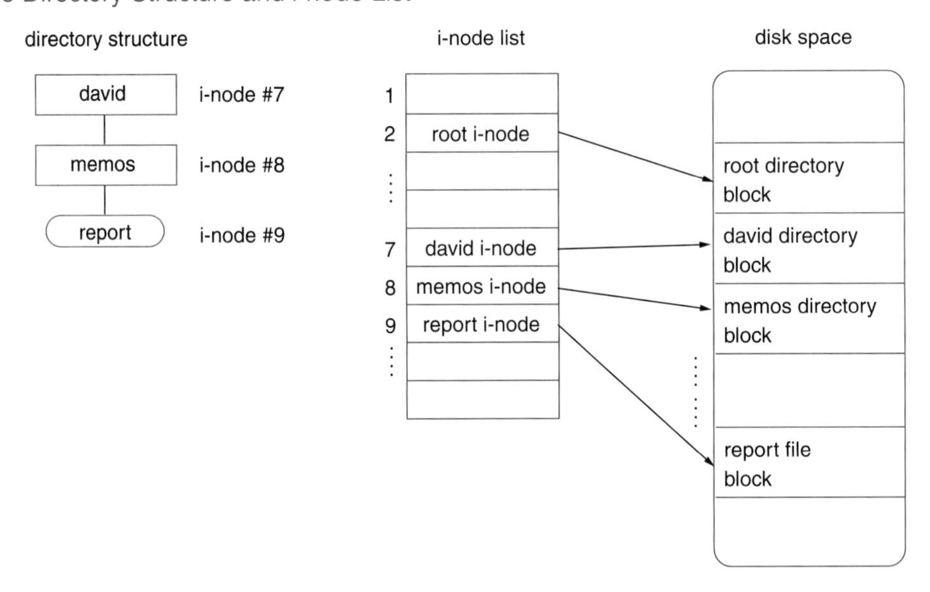

How do you determine what a file's i-node number is? You use the **ls** command with the **-i** option. For example, suppose your working directory is `david` and you have a subdirectory `memos` and a file called `report` in it.

List the filenames and their associated i-node numbers in your current directory.

```
$ ls -i
4311        memos
7446        report
$_
```

Make a copy of `report`, calling it `report.old`, and then show the i-node numbers.

```
$ cp report report.old
$ ls -i
4311        memos
7446        report
7431        report.old
$_
```

The new i-node number for `report.old` indicates that a new file has been created and a new i-node number is associated with it.

Move the `report.old` file to the `memos` directory, and then show the i-node numbers.

```
$ mv report.old memos
$ ls -i
4311        memos
7446        report
$ ls -i memos
7431        report.old
$_
```

`report.old` is moved to the `memos` directory; its i-node number remains the same, but now it is associated with the `memos` directory.

Change the name of `report.old` in `memos` to `report.sav`.

```
$ mv memos/report.old memos/report.sav
$ ls -i
4311        memos
7446        report
$ ls -i memos
7431        report.sav
$_
```

The i-node number for `report.sav` remains the same as before; only the name associated with the i-node number is changed.

To link `report` to a new filename rpt (create another filename for report) and check the i-node changes after the two files are linked, do the following:

```
$ ln report memos/rpt
$ ls -i
4311        memos
7446        report
$ ls -i memos
7431        report.sav
7446        rpt
$_
```

The i-node number for `rpt`, the new filename, is the same as `report`. Both the `report` i-node and the `rpt` i-node point to the same blocks that make up the report file.

COMMAND SUMMARY

The following commands and options have been discussed in this chapter.

cp

This command copies file(s) within the current directory or from one directory to another.

Option		Operation
UNIX	Linux Alternative	
-b	**--backup**	Makes a backup of the specified file if file already exists.
-i	**--interactive**	Asks for confirmation if the target file already exists.
-r	**--recursive**	Copies directories to a new directory.
	--verbose	Explain what is being done.
	--help	Displays the help page and exits.

cut

This command is used to "cut out" specific columns or fields from a file.

Option		Operation
UNIX	Linux Alternative	
-f	**--fields**	Specifies the field position.
-c	**--characters**	Specifies the character position.
-d	**--delimiter**	Specifies the field separator (delimiter) character.
	--help	Displays the help page and exits.
	--version	Displays the version information and exits.

find

This command locates files that match a given criterion in a hierarchy of directories. With the action options, you can instruct UNIX about what to do with the files once they are found.

Search Option	Description
-name *filename*	Finds files with the given *filename*.
size +*n*	Finds files with the size *n*.
-type *file type*	Finds files with the specified access mode.
-atime +*n*	Finds files that were accessed *n* days ago.
-mtime +*n*	Finds files that were modified *n* days ago.
-newer *filename*	Finds files that were modified more recently than *filename*.
Action Option	**Description**
-print	Prints the pathname for each file found.
-exec *command* \;	Lets you give *commands* to be applied to the files.
-ok *command* \;	Asks for confirmation before applying the command.

head

This command displays the first part of a specified file. This is a quick way to check the contents of a file. The number of lines to be displayed is an option, and more than one file can be specifed on the command line.

Option		Operation
UNIX	Linux Alternative	
-l	**--lines**	Counts by lines. This is the default option.
-c	**--chars=num**	Counts by characters.
	--help	Displays the help page and exits.
	--version	Displays the version information and exits.

ln

Creates links between an existing file and another filename or directory. It lets you have more than one name for a file.

more
This command displays files one screen at a time. This is useful for reading large files.

Option		Operation
UNIX	Linux Alternative	
-*lines*	**-num** *lines*	Displays the indicated number of *lines* per screen.
+*line-number*		Starts up at *line-number*.
+/*pattern*		Starts two lines above the line containing the *pattern*.
-c	**-p**	Clears the screen before displaying each page instead of scrolling. This is sometimes faster.
-d		Displays the prompt **[Hit space to continue, Del to abort]**.
	-- help	Displays the help page and exits.

mv
This command renames files or moves files from one location to another.

Option		Operation
UNIX	Linux Alternative	
-b	**--backup**	Makes backup of the specified file if file already exists.
-i	**--interactive**	Asks for confirmation if the target file already exists.
-f	**--force**	Removes target file if file already exists and does not ask for confirmation.
-v	**--verbose**	Explains what is being done.
	--help	Displays the help page and exits.
	--version	Displays version information and exits.

paste

This command is used to join files together line by line, or to create new files by pasting together fields from two or more files.

Option		Operation
UNIX	**Linux Alternative**	
-d	**--delimiters**	Specifies the field separator (delimiter) character.
	--help	Displays the help page and exits.
	--version	Displays the version information and exits.

pg

This comand displays files one screen at a time. You can enter the options or other commands when **pg** shows the prompt sign.

Option	Operation
-n	Does not require [Return] to complete the single-letter commands.
-s	Displays messages and prompts in reverse video.
-*num*	Sets the number of lines per screen to the integer *num*. The default value is 23 lines.
-p*str*	Changes the prompt **:** (colon) to the string specified as *str*.
+*line-num*	Starts displaying the file from the line specified in *line–num*.
+/*pattern*	Starts viewing at the line containing the first occurrence of the specified *pattern*.

The pg Command Key Operators

These keys are used when **pg** displays the prompt sign.

Key	Operation
+*n*	Advances *n* screens, where *n* is an integer number.
-*n*	Backs up *n* screens, where *n* is an integer number.
+*n***l**	Advances *n* lines, where *n* is an integer number.
-*n***l**	Backs up *n* screens, where *n* is an integer number.
n	Goes to screen *n*, where *n* is an integer number.

pr

This command provides formatted files before printing or viewing it on the screen.

Option		Operation
UNIX	Linux Alternative	
+*page*	**--pages=***page*	Starts displaying from the specified *page*. The default is page 1.
-*columns*	**--columns=***columns*	Displays output in the specified number of *columns*. The default is one column.
-a	**--across**	Displays output in columns across (rather than down) the page, one line per column.
-d	**--double-space**	Displays output double spaced.
-h*string*	**--header=***string*	Replaces the filename in the header with the specified *string*.
-l*number*	**--length=***number*	Sets the page length to the specified *number* of lines. The default is 66 lines.
-m	**--merge**	Displays all the specified files in multiple columns.
-p		Pauses at the end of each page and sounds the bell.
-*character*	**--separator=***character*	Separates columns with a single specified *character*. If character is not specified, then [Tab] is used.
-t	**--omit-header**	Suppresses the five-line header and five-line trailer.
-w*number*	**--width=***number*	Sets line width to the specified number of characters. The default is 72.
	--help	Displays help page and exits.
	--version	Displays version information and exits.

wc

This command counts number of characters, words, or lines in the specified file.

Option		Operation
UNIX	Linux Alternative	
-l	**--lines**	Reports the number of lines.
-w	**--words**	Reports the number of words.
-c	**--chars**	Reports the number of characters.
	--help	Displays the help page and exits.
	--version	Displays version information and exits.

tail

This command displays the last part (tail end) of a specified file. This is a quick way to check the contents of a file. The options give the flexibility to specify a desired part of the file.

Option		Operation
UNIX	Linux Alternative	
-l	**--lines**	Counts by lines. This is the default option.
-c	**--chars=num**	Counts by characters.
	--help	Displays the help page and exits.
	--version	Displays the version information and exits.

REVIEW EXERCISES

1. What are the symbols used for the redirection operators?
2. Explain input and output redirection.
3. What are the commands used to read a file?
4. What is the difference between moving **(mv)** a file and copying **(cp)** a file?
5. What is the command to rename a file?
6. Is it possible to have more than one name for a file?
7. What are the four regions (blocks) of a UNIX disk? Explain each part.
8. What is an i-node number, and how is it used to locate a file?
9. What is the i-node list, and what is the major information stored in each node?
10. Which one of the following commands changes or creates an i-node number?
    a. mv file1 file2
    b. cp file1 file2
    c. ln file1 file2
11. What is the command to locate a file and remove it once it is found?
12. What is the command to look at the last 10 lines of a file?
13. What is the command to select specified fields from a file?
14. What is the command to put files together line by line?
15. What is the command to read a file one screen at a time?
16. What is the command line to list all the filenames that start with *start*?
17. What is the command line to list all the filenames that end with *end*?
18. What is the command line to delete all the files that have *mid* in their filename?
19. What is the command line to copy all the files that have filenames that start with the letter *A* or *a* from your current directory a directory to called `Keep`?

20. What is the command line to copy all the files that have filenames that start with the letters *A* or *a* and end with the letters *Z* or *z* from your current directory to a directory called `Keep`?

Match the commands shown in the left column with the explanations shown in the right.

1.	wc xxx yyy	a.	Copy `xxx` to `yyy`.
2.	cp xxx yyy	b.	Rename `xxx` to `yyy`.
3.	ln xxx yyy	c.	Copy all filenames that begin with *file* and have exactly two characters (any characters) after it.
4.	mv xxx yyy	d.	Delete all files in the current directory.
5.	rm *	e.	Create another filename for `xxx` ; call it `yyy`.
6.	ls *[1-6]	f.	Display the contents of `myfile`.
7.	cp file?? source	g.	Copy `myfile` to `yyy`.
8.	pr -2 myfile	h.	Add all files that have exactly one character before the word file into one file called `yyy`.
9.	ls -i	i.	Format `myfile` in two columns.
10.	pg myfile	j.	List all files having filenames ending with digits 1 to 6.
11.	cat myfile	k.	List the current directory filenames and their i-node numbers.
12.	cat myfile > yyy	l.	Create a file called `yyy` that contains the count of characters in file `xxx` .
13.	cat ?file >> yyy	m.	View `myfile` one screen at a time.
14.	find . -name "file*" -print	n.	Save the second field from the `xyz` file in a filecalled `xxx`.
15.	find . -name xyz -size 20 -print	o.	Read `zzz` one screenful at a time.
16.	cut -f2 xyz > xxx	p.	Find all files that are called `xyz` and are exactly 20 blocks large.
17.	more zzz	q.	Find all files whose name starts with *file.*

Terminal Session

In this terminal session, you practice the commands discussed in this chapter by creating directories and then manipulating files in the directories.

1. Create a directory called `memos` in your `HOME` directory.
2. Using the vi editor, create a file called `myfile` in your `HOME` directory.
3. Using the **cat** command, append `myfile` a few times to create a large file (say, 10 pages). Call this file `large`.
4. Using the **pg** command and its options, view `large` on the screen.
5. Using the **pr** command and its options, format `large` and print it.
6. Use **cp** command to copy all files in your `HOME` directory to the `memos` directory.

7. Use the **ln** command to create another name for `large`.
8. Using the **mv** command, change the name of `large` to `larg.old`.
9. Using the **mv** command, move `large.old` to `memos`.
10. Use the **ls** command and the **-i** and **-l** options to observe the changes in the i-node numbers and the number of links when you do the following commands:
    - **a.** Change to the `memos` directory.
    - **b.** Create another name for `myfirst`; call it `MF`.
    - **c.** Copy `myfirst` to `myfirst.old`.
    - **d.** List all files whose filenames start with *my*.
    - **e.** List all files that have the extension `old`.
    - **f.** Modify `myfile`. Look at the `MF` file; `myfile` modifications are also in the `MF` file.
    - **g.** Change to your `HOME` directory.
    - **h.** Delete all files in the `memos` directory that have the word *file* as part of their file names.
    - **i.** Delete the `memos` directory with all of the files remaining in it.
    - **j.** List your `HOME` directory.
    - **k.** Remove all the files you created in this session.
11. Show the last five lines of a file.
12. Display the first five lines of a file.
13. Save the last 30 characters of a file in another file.
14. Save the list of all the files that are seven days old, starting from your `HOME` directory.
15. Find a file called `passwd`.
16. Find a file called `profile`.
17. Find all files that are exactly seven days old, starting from your `HOME` directory.
18. Find all files that are less than seven days old, starting from your `HOME` directory.
19. Find all the files that are more than 10 days old and copy them to another directory.
20. Create two files similar to those in Figures 7.11 and 7.12.
    - **a.** Use the **cut** command to cut out specific fields and columns.
    - **b.** Use the **paste** command to put the two files together.
21. Use the **more** command to read a large file.
22. Create two files called `numbers` and `characters` and type in lines of numbers and characters similar to the following:

```
  numbers              characters
111111111111         AAAAAAAAAA
222222222222         BBBBBBBBBBB
333333333333         CCCCCCCCCCC
     .                     .
     .                     .
     .                     .
101010101010         KKKKKKKKKKK
11 11 11 11 11       LLLLLLLLLLLL
```

Make many lines of numbers so you can practice the **head** and **tail** commands and options.

a. Show the first two lines of the `numbers` file.

b. Show the first two lines of `numbers` and `characters` files.

c. Show the last 10 lines of the `numbers` file.

d. Show the last two lines of the `numbers` and `characters` files.

23. Use the `numbers` and `characters` files to execute the following command lines:

a. Use the **paste** command to put together `numbers` and `characters`.

b. Use the **paste** command to show `numbers` and `characters` together and use the @ character as the field separator.

c. Use the **paste** command to show `numbers` and `characters` together using the @ character as the field separator. Save it in a file called `numbersANDcharacters`.

d. Use the **cut** command to show first five character of each line in numbers.

e. Show how many lines are in the `numbersANDcharacters` file.

f. Show how many lines, words, and characters are in the `numbers`, `characters`, and `numbersANDcharacters` files.

CHAPTER 8

Exploring the Shell

This chapter describes the shell and its role in the UNIX system and explains its features and capabilities. It discusses shell variables and explains their use and the way they are defined. The chapter also introduces more shell metacharacters and ways to make the shell ignore their special meanings and explains UNIX startup files, internal processes, and process management. This chapter continues the introduction of new commands (utilities) so you can build your vocabulary of UNIX commands.

In This Chapter

8.7 UNIX PROCESS MANAGEMENT

COMMAND SUMMARY

REVIEW EXERCISES

Terminal Session

8.1 THE UNIX SHELL

The UNIX operating system consists of two parts: the *kernel* and the *utilities*. The *kernel* is the heart of the UNIX system and is memory resident (which means that it stays in the memory from the time you boot the system until the system is shut down). All the routines that communicate directly with the hardware are concentrated in the kernel, which is relatively small in comparison with the rest of the operating system.

In addition to the kernel, other essential modules are also memory resident. These modules perform important functions such as input/output control, file management, memory management, and processor time management. Additionally, UNIX maintains several memory-resident tables for housekeeping purposes, to keep track of the system's status.

The rest of the UNIX system resides on the disk and is loaded into the memory only when necessary. Most of the UNIX commands you know are programs (called utilities) that reside on the disk. For those programs, when you type a command (request the program to be executed), the specified program is loaded into the memory.

You communicate with the operating system through a shell, and hardware-dependent operations are managed by the kernel. Figure 8.1 shows the components of the UNIX operating system.

Figure 8.1
The Components of the UNIX Operating System

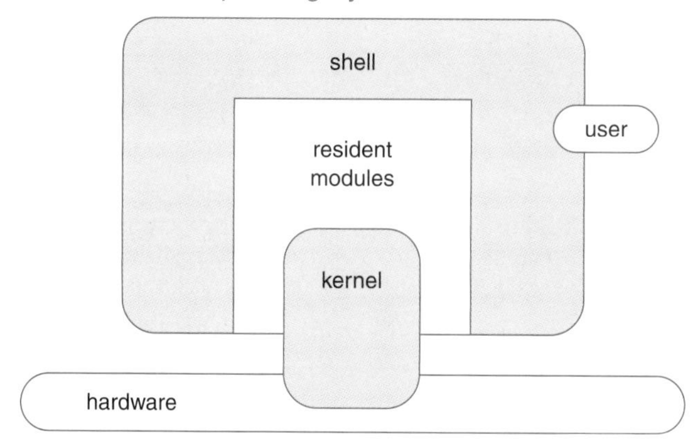

The shell is itself a program (a utility program). It loads into the memory whenever you log in to the system. When the shell is ready to receive commands, it displays a prompt. The shell itself does not carry out most of the commands that you type; it examines each command and starts the appropriate UNIX program (utility) to carry out the requested action. The shell determines what program to start (the name of the program is the same as the command you type). For example, when you type **ls** and press [Return] to list current directory files, the shell finds and starts a program called `ls`. The shell treats your application programs the same way: you type the program's name as a command, and the shell executes the program for you. Figure 8.2 shows the user interaction with the shell program.

See page x for an explanation of the icons used to highlight information in this chapter.

Figure 8.2
User Interaction with the Shell Program

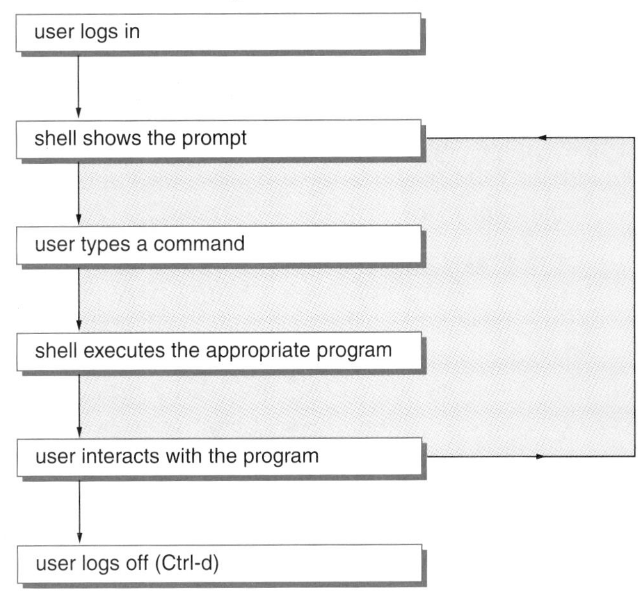

8.1.1 Starting the Shell

In Chapters 2 and 3, the shell program was explained and you learned that the shell is the main method through which a user interacts with the UNIX, and also that there is more than one shell provided with each UNIX system. But how does the shell get started?

The shell is started after the user is successfully logged in to the system and remains active until the user logs out. Each user on the system has a default shell. The default shell for each user is specified in the system password file. This file is called `/etc/passwd`. The `passwd` file is the system password file and contains, among other things, user ID for each user, an encrypted copy of each user's password, and the name of the program to run immediately after a user logs in to the system. This program does not have to be one of the shell programs; however, it usually is.

When you log in, the system determines what shell to run by looking up your entry in the `/etc/passwd` file. The last field of each entry is the name of the program to run as the default shell. Table 8.1 shows the name of the shell programs and their corresponding shell names.

Table 8.1
The Shells and Shell Program Names

Shell Program Name	Prompt Sign	Shell Name
`/bin/sh`	$	Bourne shell
`/bin/ksh`	$	Korn shell
`/bin/bash`	$	Bourne Again shell
`/bin/csh`	%	C shell
`/bin/tcsh`	%	TC shell

Built-In Shell Commands

The shell command interpreters (*sh*, *ksh*, *bash*, etc.) have special built-in functions (programs), which are interpreted by the shell as commands. These commands are part of the shell itself and are recognized and executed internally. You already know some of the built-in commands such as **cd**, **pwd**, and others. Many of these built-in commands are implemented by more than one of the shells, and some are unique to a particular shell. You can obtain the full list of the built-in commands by typing the following command line:

`$ man shell_builtins [Return]` . . Display the shell built-in commands.

Table 8.2 lists the shell built-in commands that are covered in this chapter and shows their availability under different shells.

Table 8.2
The Built-In Shell Commands

Command	Built into		
	Bourne Shell	Korn Shell	Bourne Again Shell
alias		ksh	bash
echo	sh	ksh	bash
history		ksh	bash
kill	sh	ksh	bash
set	sh	ksh	bash
unalias		ksh	bash
unset	sh	ksh	bash

8.1.2 Understanding the Shell's Major Functions

The standard UNIX system comes with more than 200 utility programs. One of these programs is `sh`, the shell itself.

The shell is the most frequently used utility program on the UNIX system. It is a sophisticated program that manages the dialogue between the user and the UNIX system. You interact with it repeatedly during work sessions. The shell is a regular executable C/C++ program that is usually stored in the `/bin` directory. When you log in, an interactive shell is invoked automatically. However, you can invoke another copy of the shell by typing **sh** (or **ksh** or **bash**, depending on what shells are available in your system) at the **$** prompt.

The shell includes the following major features. You are already familiar with some of these features, and the rest of them are explored in this chapter.

Command Execution Command (program) execution is a major function of the shell. Just about anything you type at the prompt is interpreted by the shell. When you press

[Return] at the end of the command line, the shell starts analyzing your command; if there are filename substitution characters or input/output redirection signs, it takes care of them, and then executes the appropriate program.

Filename Substitution If filename substitution (also called *filename generation*) is specified on the command line, the shell first performs the substitution and then executes the program. (The filename substitution characters—metacharacters * and ?—were discussed in Chapter 7.)

I/O Redirection The input/output redirection is handled by the shell. Again, the shell program itself is not involved, and the redirection is set up before the command execution. If input or output redirection is specified on the command line, the shell opens the file and connects it to the standard input or standard output of the program respectively. This topic was discussed in Chapter 7.

Pipes Pipes, also called *pipelines,* let you connect simple programs together to perform a more complex task. The vertical line on the keyboard (¦), is the pipe operator.

Environment Control The shell lets you customize your environment to suit your needs. By setting the appropriate variables, you may change your HOME directory, prompt sign, or other aspects of your working environment.

Background Processing The background processing capability of the shell enables you to run programs in the background while doing other jobs in the foreground. This is helpful for time-consuming, noninteractive programs.

Shell Scripts Commonly used sequences of the shell commands can be stored in files called *shell scripts.* The name of the file later can be used to execute the stored program, enabling you to execute the stored commands with a single command. The shell also includes language constructs that allow you to build shell scripts that perform more complex jobs. Shell scripts are discussed in Chapter 11.

8.1.3 Displaying Information: The echo Command

You can use the **echo** command to display messages. It displays its arguments on your terminal, the standard output device. Without the argument, it produces an empty line and by default appends a new line to the end of the output. For example, if you type **echo hello there** and press [Return] at the prompt, you will see the following:

```
hello there
$_
```

The argument string can be any number of characters long. However, if your string contains any metacharacters, the string must be enclosed in quotation marks. (This topic is discussed later in this chapter.)

echo Options

Table 8.3 lists the **echo** command options and explains how they are used.

Table 8.3
The **echo** Command Options

Option	Operation
-n	Disables output of the trailing new line.
-e	Enables interpretations of the backslash escaped characters.

-e Option This option enables the interpretation of the escaped character such as **\n**. The **echo** command usually is set with this option as the default option. The **echo** command is implemented differently in each shell. Particularly for the *bash* shell, you must use the **-e** option in order for *bash* to recognize the escape characters.

Table 8.4 shows the characters that you can use as part of a string to control the format of the message. These characters are preceded by a backslash (\) and are interpreted by the shell to produce the desired output. They are also called *escape characters.*

Table 8.4
The Escape Characters

Escape Character	Meaning
\a	Audible alert (bell)
\b	Backspace
\c	Inhibit the terminating newline
\f	Form feed
\n	Carriage return and a line feed (newline)
\r	Carriage return without the line feed
\t	Horizontal tab
\v	Vertical tab

The backslash itself is a shell metacharacter. Therefore, if it is used in your string, it must be enclosed in quotation marks.

The following command sequences show how to use the **echo** command, and the result of incorporating the escape characters in the argument string:

```
$ echo Hi, this is a test. [Return]    . . . . Show a simple message on the screen.
Hi, this is a test.
$ echo Hi, "\n" this is a test. [Return]  . . Show the same message in two lines.
Hi,
this is a test.
$_ . . . . . . . . . . . . . . . . . . . . . . Prompt.
```

1. **\n must be enclosed in quotation marks to be interpreted as the newline command.**
2. **If the echo command does not recognize the escape characters, use the -e option.**

```
$ echo -e Hi, "\n" this is a test [Return] . . . . . Using the -e option.
$ echo Hi, "\n" this is a test. > test [Return]  . . This time save the output in a file.
$ cat test [Return]  . . . . . . . . . . . . . . . . Confirm the contents of test.
Hi,
this is a test.
$_ . . . . . . . . . . . . . . . . . . . . . . . . . Prompt.
$ echo Hi, "\n" this is a test. "\c" [Return]. . . . This time do not produce the
                                                      blank line at the end of the message.
Hi,
this is a test.$
```

The prompt sign (**$**) appears right after the word *test.* That is the effect of **\c** in the argument string.

In the following command line, the extra spaces between the words are intentional:

```
$ echo This   is   a   test. [Return] . . See what happens to the
                                          blanks.
This is a test.
```

The shell interprets this command line with four arguments, and each argument is separated with a space in the output.

```
$ echo "This is   a   test." [Return]   . . See the wonderful effects
                                            of quotation marks; now
                                            the blank spaces are
                                            preserved.
This is   a   test.

$_   . . . . . . . . . . . . . . . . . . . And the prompt.
```

8.1.4 Removing Metacharacters' Special Meanings

The shell metacharacters have special meanings to the shell. But sometimes you want to override those meanings. The shell provides you with a set of characters that remove the meanings of the metacharacters. This process of removing the special meanings of the metacharacters is called *quoting* or *escaping.* The set of quoting characters is as follows:

- backslash \
- double quotation mark "
- single quotation mark '

Table 8.5 lists the characters that can be used for quoting.

The **echo** *command is used in most of the following examples to demonstrate how this process works. However, the use of quoting is applicable to other commands when you have to use any of the special characters as part of the command's argument.*

Table 8.5
Set of Quoting Characters

Quoting Character	Meaning
" " (double quotation marks)	Everything between " and " is taken literally, except for $, ` (grave accent quotation mark) and " (double quotation marks).
' ' (single quotation marks)	Everything between ' and ' is taken literally, except for ' (single back quotation mark).
\ (backslash)	Any character flowing the \ is taken literary.

Backslash The backslash (\) is used to cause the character that follows it to be interpreted as an ordinary alphanumeric character. For example, **?** is a file substitution character (wild card) and has a special meaning to the shell. But **\?** is interpreted as the real question mark.

Delete a file called `temp?` from your current directory:

```
$ rm temp? [Return]  . . . . . . . . . . .  Remove temp?.
```

The shell interprets this command as deleting all files whose filenames consist of `temp` and one character after it. Therefore, it deletes any file that matches this pattern, such as `temp`, `temp1`, `temp2`, `tempa`, `tempo`, and so on.

But if all you wanted to delete was a single file called `temp?`, then you would have typed the following:

```
$ rm temp\? [Return]  . . . . . . . . . . .  Try again, using \? to
                                             represent the ?
```

This time the shell scans the command line, finds the ****, ignores the special meaning of the question mark, and passes the filename `temp?` to the **rm** program.

Display the metacharacters:

```
$ echo \< \> \" \' \$ \? \& \¦ \\ [Return]  .  Let's display them all.
< > "' $ ? & ¦ \
$_  . . . . . . . . . . . . . . . . . . . . .  Ready for the next
                                               command.
```

To remove the special meaning of the backslash, precede it with a backslash.

Double Quotation Marks You can use the double quotation marks (") to override the meaning of most of the special characters. Any special character between a pair of double quotation marks loses its special meaning, except the dollar sign (before a variable name), the single quotation mark, and the double quotation marks. (You use the backslash to remove their special meanings.)

Double quotation marks also preserve the white-space characters (i.e., the blank space, tab, and newline characters). The use of double quotation marks for this purpose was demonstrated in the **echo** command examples.

The following command sequences show the application of the double quotation marks:

```
$ echo > [Return]  . . . . . . . . Display the > sign.
syntax error: 'newline or;' unexpected
$_ . . . . . . . . . . . . . . . . And the prompt appears.
```

The shell interprets your command as redirecting the output of the **echo** command to a file. It looks for the filename and because none is specified, it responds with a cryptic error message.

```
$ echo ">" [Return]  . . . . . . Enclose the argument with
                                 double quotation marks. The
                                 > is displayed.
>
$ ls -C [Return] . . . . . . . . Check your current directory files.
memos     myfirst     REPORTS
$ echo * [Return]  . . . . . . . Use a metacharacter as the
                                 argument.
memos     myfirst     REPORTS
```

The shell substitutes * with the names of all the files in your current directory.

```
$ echo "*" [Return]  . . . . . . Now use the double quotation
                                 marks.
*
$_ . . . . . . . . . . . . . . . And the prompt appears.
```

No substitution occurs for the character between the double quotation marks. Therefore, the special meaning of the * is removed.

Display the message "The UNIX System".

```
$ echo "\"The UNIX System\"" [Return]
"The UNIX system"
$_ . . . . . . . . . . . . . . . Prompt.
```

A backslash is necessary before a double quotation mark to override the special meaning of the double quotation marks.

Single Quotation Marks Single quotation marks (') work very much like the double quotation marks. Any special character between a pair of single quotation marks loses its special meaning, except the single quotation mark. (You use \ to remove its special meaning.)

Single quotation marks also preserve the white-space characters. The string between the single quotation marks becomes a single argument, and the space character no longer has its special meaning as argument separator.

The open and close quotation marks for this purpose are the same character, the forward quotation mark. Do not use the back quotation mark (the grave accent character `). This distinction is very important. The shell interprets the string inside the grave accent marks as an executable command.

The grave accent mark (`), which is discussed in more detail later in the chapter, is the lowercase character on the tilde (~) key at the upper left of the keyboard.

Display special characters using a pair of single quotation marks.

```
$ echo ' < > " $ ? & |'[Return] . . . Use the echo command and
                                      single quotation marks.
< > " $ ? & |
$_ . . . . . . . . . . . . . . . . . The prompt is back.
```

The spaces between the characters are preserved.

8.2 SHELL VARIABLES

The shell program handles the user interface and acts as a command interpreter. In order for the shell to service all your requests (executing commands, manipulating files, etc.), it needs to have certain information and to keep track of that information (i.e., your HOME directory, terminal type, and prompt sign).

This information is stored in what are called the *shell variables. Variables* are named items you set to specific values to control or customize your environment. The shell supports two types of variables: environment variables and local variables.

Environment Variables Environment variables are also known as *standard variables*; they have names that are known to the system. They are used to keep track of the essential things the system needs and are usually defined by the system administrator. For example, the standard variable *TERM* is assigned to your terminal type:

```
TERM=ansi
```

Local Variables Local variables are user defined; they are entirely under your control. You can define, change, or delete them as you wish.

8.2.1 Displaying and Removing Variables: The set and unset Commands

You can use the **set** command to find out what shell variables are set for your shell to use.

Type **set** at the prompt and press [Return], and the shell displays the list of the variables. Your list will be similar to but not exactly like that shown in Figure 8.3.

The names of the standard variables on the left of the equal (=) sign are shown in uppercase letters in Figure 8.3. This is not a requirement; you can use lowercase, uppercase, or any mixture for variable names. You can use characters, digits, and the underscore character in variable values, but the first letter must be a character, not a digit.

On the right side of the equal sign is the value assigned to a variable.

You must specify the exact variable name (including capitalization) when referring to a variable.

You use the **unset** command to remove an unwanted variable. If you have a variable *XYZ=10*, and you want to remove it, type **unset XYZ** and press [Return].

Figure 8.3
The Output of the **set** Command

```
$ set
HOME=/usr/students/david
IFS=
LOGNAME=david
LOGTTY=/dev/tty06
MAIL=/usr/mail/students/david
MAILCHECK=600
PATH=:/bin:/usr/bin
PS1="$"
PS2=">"
TERM=ansi
TZ=EST=EDT
$_
```

8.2.2 Assigning Values to Variables

You can create your own variables, and you can also modify the values assigned to standard variables. You assign a value to a variable by writing the variable name, followed by an equal (=) sign (the assignment operator), followed by the value you want to assign to the variable, like this:

```
age=32
```

or

```
SYSTEM=UNIX
```

The shell treats every value that you assign to a variable as a string of characters. In the preceding example, the value of the variable *age* is the string 32, and not the number 32. If your string contains embedded white-space characters (space, tab, etc.) you must enclose the entire string in a pair of double quotation marks, like this:

```
message="Save your files, and log off" [Return]
```

1. **A shell variable name must begin with a (lowercase or uppercase) letter and not a digit.**
2. **There are no spaces on either side of the equal sign.**

8.2.3 Displaying the Values of Shell Variables

To access the value stored in a shell variable, you must precede the name of the variable with a **$**. Using the previous example, *age* is the name of the variable, and **$age** is 32, the value stored in the age variable.

You use the **echo** command to display the value assigned to the shell variable.

set *displays a list of variables;* **echo** *shows the specified variable.*

Use the **echo** command to display text and the values of the shell variables:

```
$ age=32 [Return] . . . . . . . . . . . . . . Assign the value 32 to age.
$ echo Hi, nice day [Return]  . . . . . . . . Display the argument string.
Hi, nice day
$ echo age [Return]  . . . . . . . . . . . . Display the argument, the word age.
age
$ echo $age [Return]  . . . . . . . . . . . . Now the argument is $age, the value stored in age.
32
$ echo You are $age years old. [Return]  . Add some text to obtain more meaningful output.
You are 32 years old.
$_ . . . . . . . . . . . . . . . . . . . . . . Ready for the next command.
```

The shell variables are frequently used as command arguments on a command line, as in the following:

```
$ all=-lFa [Return] . . . . . . . . . . . . . Create a variable called all and assign the value
                                              (string) -lFa (hyphen, lowercase letter l, upper-
                                              case letter F, lowercase letter a) to it.
$ ls $all myfirst [Return]  . . . . . . . . . Use the variable as part of a command line.
command's output
$_ . . . . . . . . . . . . . . . . . . . . . . The prompt is displayed.
```

Variable names are preceded by **$**. Thus the shell substitutes the value **-laF** for the variable *all*. After substitution, the command becomes **ls -lFa myfirst**.

Observe the outputs of the following commands. They show the subtle differences in the way the variables are interpreted when they are between quotation marks.

```
$ age=32 [Return] . . . . . . . . . . . . . . 32 is assigned to the variable age.
$ echo $age "$age" '$age' [Return]  . . . . Display age.
32 32 $age
$_ . . . . . . . . . . . . . . . . . . . . . . Prompt.
```

8.2.4 Understanding the Standard Shell Variables

The values assigned to the standard shell variables are usually set by the system administrator. Thus, when you log in, the shell refers to these variables to keep track of things in your environment. You can change the value of these variables. However, the changes are temporary and apply only to the current session. The next time you log in, you have to set them again. If you want the changes to be permanent, place them in a file called `.profile`. The `.profile` file is explained later in this chapter.

HOME

When you log in, the shell assigns the full pathname of your HOME directory to the variable *HOME*. The HOME variable is used by several UNIX commands to locate the HOME directory. For example, the **cd** command with no argument checks this variable to determine the pathname to the HOME directory and then sets the system to your HOME directory.

To experiment with the HOME variable, try the following command sequences:

```
$ echo $HOME [Return]  . . . . . . . . . . . . . . Show your HOME directory pathname.
/usr/david
$ pwd [Return] . . . . . . . . . . . . . . . . . . Show the current directory pathname.
/usr/david/source . . . . . . . . . . . . . . . . . source subdirectory in david.
$ cd [Return]  . . . . . . . . . . . . . . . . . . No argument is specified. The default is
                                                    your HOME directory.
$ pwd [Return]  . . . . . . . . . . . . . . . . . Check your current directory. You are in
                                                    david, your HOME directory.
/usr/david
$ HOME=/usr/david/memos/important [Return] . . . Change your HOME directory pathname.
                                                    Now your HOME directory is important.
$ cd [Return]  . . . . . . . . . . . . . . . . . . Change to your HOME directory.
$ pwd [Return] . . . . . . . . . . . . . . . . . . Display your current directory; your
                                                    current directory is important.
/usr/david/memos/important
$_ . . . . . . . . . . . . . . . . . . . . . . . . And the prompt appears.
```

IFS

The Internal Field Separator (*IFS*) variable is set to a list of characters that are interpreted by the shell as separators of command line elements. For example, to get a long list of the files in your directory, you type **ls -l** and press [Return]. The space character in your command separates the command word (**ls**) from its option (**-l**).

Other separator characters assigned to the *IFS* variable are the tab character [Tab] and the newline character [Return].

The IFS characters are invisible (nonprintable) characters, so you do not see them on the right side of the equal sign. But they are there!

```
$ echo $IFS [Return] . . . . Display characters assigned to IFS.

$_
```

A blank line and prompt is back. The characters assigned to IFS are nonprintable.

Change the **IFS** field separator to the **:** (colon). The original **IFS** setting is saved before it is changed and later is restored to the original characters.

```
$ cd  $HOME [Return]  . . . . . Notice here the field separator is a space
                                 between cd and $HOME.
$ old_IFS=$IFS  [Return]  . . . Save IFS characters.
$ IFS=":" [Return]  . . . . . . Change filed separator to :.
$ cd:$HOME [Return] . . . . . . Notice that the : is used as a field
                                 separator instead of the space.
$ IFS=$old_IFS [Return] . . . . Restore the field separators back to the
                                 original ones.
$_  . . . . . . . . . . . . . . And the prompt is back.
```

It is not advisable to change the field separators. You usually do this to accommodate special circumstances. We are going to use this capability in sample programs in Chapter 12.

MAIL

The *MAIL* variable is set to the filename of the file that receives your mail. Mail sent to you is stored in this file, and the shell periodically checks the contents of this file to notify you if there is mail for you. For example, to set your mailbox to `/usr/david/mbox`, you would type **MAIL=/usr/david/mbox** and press [Return].

MAILCHECK

The *MAILCHECK* variable specifies how often the shell is to check for arrival of mail in the file set in the *MAIL* variable. The default for *MAILCHECK* is 600 (seconds).

PATH

The *PATH* variable is set to the directory names that the shell searches for the location of the commands (programs) in the directory structure. For example, **PATH=:/bin:/ usr/bin.**

The directories in the path string are separated by colons. If the very first character in the path string is a colon, the shell interprets that as **.:** (dot, colon), meaning that your current directory is first on the list and is searched first.

UNIX usually stores the executable files in a directory called `bin`. You can create your own `bin` directory and store your executable files in it. If you add your `bin` directory (or any other name you call it) to the *PATH* variable, the shell looks there for any commands that it cannot find in the standard directories.

Suppose all of your executable files are located in a subdirectory called `mybin` that is located in your HOME directory. To add it to the *PATH* variable, you type: **PATH=:/ bin:/usr/bin:$HOME/mybin** and press [Return].

PS1

The Prompt String 1 (*PS1*) variable is set to the string used as your prompt sign. The Bourne shell primary prompt sign is set to the dollar sign (**$**).

If you are tired of seeing the **$** prompt, you can easily change it by assigning a new value to the shell variable PS1.

```
$ PS1=Here: [Return] . . . . . . . . Change your prompt to Here:
Here:_ . . . . . . . . . . . . . . . . There you are.
Here: PS1="Here: " [Return]  . . . . Add an extra blank space to the end.
Here: _  . . . . . . . . . . . . . . It looks nicer!
```

If your prompt string has embedded spaces, then it must be enclosed in quotation marks.

```
Here: PS1="Next Command: " [Return] . Change the prompt sign again.
Next Command:_ . . . . . . . . . . . And it is changed.
Next Command: PS1="$ " [Return]  . . Change back to the old $ prompt.
$_ . . . . . . . . . . . . . . . . . And the $ prompt returns.
```

PS2

The Prompt String 2 (*PS2*) variable is set to the prompt sign that is displayed whenever you press [Return] before completion of the command line and the shell expects the rest of the command. You can change the *PS2* variable the same way you change the *PS1* variable. The Bourne shell secondary prompt defaults to the greater than sign (>).

The following command sequences show examples of the second prompt:

`$ echo "Good news, UNIX [Return]` . .	The command line is not complete; thus the PS2 prompt sign (>) is displayed.
`> is on CDs." [Return]`	Now the command line is complete.
`Good news, UNIX is on CDs.`	
`$ ls \ [Return]`	The command line is not complete. This is signaled by the backslash.
`>`	The shell displays the second prompt sign, and waits for the rest of the command.
`> -l [Return]`	Now the command line is complete. The shell puts it together as **ls -l** and executes it.
`$_`	And the prompt is back.

CDPATH

The *CDPATH* variable is set to a list of absolute pathnames, similar to the *PATH* variable. The *CDPATH* affects the operation of the **cd** (change directory) command. If this variable is not defined, then **cd** searches your working directory to find the filename that matches its argument. If the subdirectory does not exist in your working directory, then UNIX displays an error message. If this variable is defined, **cd** searches for the specified directory according to the pathnames assigned to the *CDPATH* variable. If the directory is found, it becomes your working directory.

For example, if you type **CDPATH=:$HOME:$HOME/memos** and press [Return], the next time you use the **cd** command, it will start searching from your current directory, then your HOME directory, and eventually the `memos` directory to find a match to the filename specified as the **cd** command argument.

SHELL

The *SHELL* variable is set to the full pathname of your login shell:

```
SHELL=/bin/sh
```

TERM

The *TERM* variable sets your terminal type:

```
TERM=ansi
```

TZ

The *TZ* variable sets the time zone that you are in:

```
TZ=EST
```

It is usually set by the system administrator.

8.3 MORE METACHARACTERS

As you remember from Chapter 7, metacharacters, or special characters, are interpreted and processed in a special manner by the shell. So far, we have discussed the file substitution and redirection metacharacters. This section explores some more of the metacharacters.

8.3.1 Executing the Commands: Using the Grave Accent Mark

The grave accent mark (`) before and after a command tell the shell to execute the enclosed command and to insert the command's output at the same point on the command line. It is also called *command substitution.* The format is as follows:

```
'command'
```

where *command* is the name of the command to be executed.

The grave accent mark character is found on the key at the far left of the keyboard, just below the [Esc] key.

The following command sequences show examples of command substitution:

```
$ echo The date and time is: 'date' [Return] . . Command date is executed.
The date and time is: Mon Nov 28 14:14:14 EDT 2005
$_ . . . . . . . . . . . . . . . . . . . . . . The prompt is back.
```

The shell scans the command line, finds the grave accent mark, and executes the command date. It replaces the `'date'` on the command line with the output from the date command and executes the echo command.

```
$ echo "List of filenames in your current directory:\n" 'ls -C' > LIST [Return]
    $ cat LIST [Return] . . . . . . . . . . . . Check what you have stored in LIST.
    List of filenames in your current directory:
    memos     myfirst     REPORT
    $_ . . . . . . . . . . . . . . . . . . . . . Ready for the next command.
```

8.3.2 Sequencing the Commands: Using the Semicolon

You can enter a series of commands on a command line, separated by semicolons. The shell executes them in sequence from left to right.

To experiment with the semicolon metacharacter, try the following:

```
$ date ; pwd ; ls -C [Return] . . . . . . . . . Three commands in sequence.
Mon Nov 28 14:14:14 EST 2005
/usr/david
memos myfirst REPORT
$ ls -C > list ; date > today; pwd [Return] . . Three commands in sequence, with the
                                                  output of two commands redirected to files.
/usr/david
$ cat list [Return] . . . . . . . . . . . . . . Check content of list.
memos     myfirst     REPORT
```

```
$ cat today [Return] . . . . . . . . . . . . . . Check contents of today.
Mon Nov 28 14:14:14 EST 2001
$_ . . . . . . . . . . . . . . . . . . . . . . . Your favorite prompt sign appears.
```

8.3.3 Grouping the Commands: Using Parentheses

You can group commands together by placing them between a pair of parentheses. The group of commands can be redirected as if they were a single command.

To experiment with the parentheses as metacharacters, try the following:

```
$ (ls -C ; date ; pwd) > outfile [Return] . . . Three commands in sequence, grouped to-
                                                gether, with the output redirected to a file.
$ cat outfile [Return] . . . . . . . . . . . .  Check contents of outfile.
memos     myfirst     REPORT
Mon Nov 28 14:14:14 EST 2005
/usr/david
$_ . . . . . . . . . . . . . . . . . . . . . .  And the prompt appears.
```

8.3.4 Background Processing: Using the Ampersand

UNIX is a multitasking system; it allows you to execute several programs concurrently. Usually, you type a command and within a few seconds the output of the command is displayed on the terminal. What if you run a command that takes minutes to execute? In that case, you have to wait for the command to finish executing before you can proceed with the next job. However, you do not need to wait during all those unproductive minutes. The shell metacharacter ampersand (&) provides you the means to run programs in the background, as long as the background programs do not require input from the keyboard. If you enter a command followed by **&**, then that command is sent to the background for execution, and your terminal is free for the next command.

The following examples show applications of the ampersand metacharacter:

```
$ sort data > sorted & [Return] . . . . . . . . Sort data and store the results in sorted.
1348 . . . . . . . . . . . . . . . . . . . . .  Process ID is displayed.
$ date [Return] . . . . . . . . . . . . . . . . The prompt is immediately displayed,
                                                ready for the next command.
```

The output of the **sort** command is redirected to another file in order to prevent **sort** from sending its output to the terminal while you are doing other tasks.

1. *The background command process ID (PID) number identifies the background process and can be used to terminate it or obtain its status.*
2. *You can specify more than one background command on a single command line.*

```
$ date & pwd & ls -C & [Return] . . . . . . . Create three background processes; three
                                              PID numbers are displayed.
2215
2217
2216
```

```
$ echo "the foreground process" [Return]  . Run the echo command in foreground.
Mon Nov 28 14:14:14 EST 2005 . . . . . . . Output from the background process date
                                           is displayed.
the foreground process . . . . . . . . . . Output from the foreground process echo
                                           is displayed.
/usr/david . . . . . . . . . . . . . . . . Output from the background process pwd
                                           is displayed.
$_ . . . . . . . . . . . . . . . . . . . . The prompt appears and then output from
                                           the background process ls -C is displayed.
memos myfirst REPORT
$_ . . . . . . . . . . . . . . . . . . . . Ready for the next command.
```

By default, the output of the background commands is displayed on your terminal. Thus, the output of the foreground program is interleaved with the output of the background program and produces quite a confusing display. You can prevent the confusion by redirecting the output of background commands to files.

8.3.5 Chaining the Commands: Using the Pipe Operator

The shell lets you use the standard output of one process as standard input to another process. You use the pipe metacharacter, ¦, between the commands. The general format is as follows:

```
command A ¦ command B
```

where command A output is introduced as input to command B. You can chain a sequence of commands together, creating what is called a *pipeline*. Let's look at some examples that give you an appreciation of this very useful and flexible shell capability.

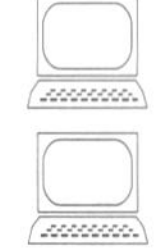

Type **ls -l¦lp** and press [Return] to send the output of the **ls -l** command to the printer.

To count the number of the files in your current directory, do the following:

```
$ ls -C [Return] . . . . . Let's see the files in your current directory.
memos myfirst REPORT
$ ls -C > count [Return]  . Now save the list of your files in count.
$ wc -w count [Return]  . . Count the number of words. You have three
                            files in your current directory.
3
ls -C ¦ wc -w [Return]  . . Use the pipe operator to obtain the number
                            of files in your current directory.
3
```

The output of the command **ls -C** (list of the files in your current directory) is passed as input to the **wc -w** command.

To save the number of users logged in the system in a file, do the following:

```
$ echo "Number of logged-in users:" 'who ¦ wc -l' > outfile [Return]
$ cat outfile [Return] . . . . . . . . . . Check what is stored in outfile.
Number of logged-in users: 20
```

In the previous command, the shell scans the command line, finds the grave accent marks, executes the **who ¦ wc -l** commands, and passes the output of **who** to **wc** as input data. If there are 20 users logged on the system, then the output is 20. The shell replaces the **`who ¦ wc -l`** with 20. Then the shell executes the **echo** command that reads *Number of the logged in users: 20* and stores the output in `outfile`.

8.4 MORE UNIX UTILITIES

These utilities give you more flexibility and control in day-to-day usage of the system. Also, some of the utilities are used in script file (program) examples in Chapters 11 and 12.

Under Linux, the **--help** and **--version** options are available for most of the commands in this chapter. Please make a habit of using the **--help** option and read the help page to familiarize yourself with the other available options. Regardless of your UNIX system, you can always use the **man** command to obtain the full usage explanation for any of these commands.

8.4.1 Timing a Delay: The sleep Command

The **sleep** command causes the process executing it to go to sleep for a specified number of seconds. You can use **sleep** to delay the execution of a command for a period of time. For example, if you type **sleep 120 ; echo "I am awake!"** and press [Return], the **sleep** command is executed, causes a two-minute delay, and then (after two minutes) the **echo** command is executed, and the string argument *I am awake!* is displayed on the screen.

8.4.2 Displaying the PID: The ps Command

You can use the **ps** (process status) command to obtain the status of the active processes in the system. When used without any options, it displays information about your active processes. This information is arranged in four columns (see Figure 8.4) with the following column headings:

- PID: the process ID number
- TTY: your terminal number that controls the process
- TIME: time duration (in seconds) that your process is running
- COMMAND: the name of the command

Figure 8.4
The **ps** Command Output Format

```
$ ps
PID      TTY      TIME      COMMAND
24059    tty11    0:05      sh
24259    tty11    0:02      ps
$_
```

ps Options

Only two of the **ps** options are discussed in this book—the **-a** option and the **-f** option—and they are summarized in Table 8.6.

Table 8.6
The **ps** Command Options

Option	Operation
-a	Displays the status of all the active processes, not just the user's.
-f	Displays a full list of information, including the full command line.

-a Option The **-a** option displays status information for all active processes. Without this option, only the user's active processes are displayed.

-f Option The **-f** option displays a full list of information including the complete command line under the command column.

Figure 8.5 shows the output of the **ps** command using both **-a** and **-f** options.

Figure 8.5
Output of the **ps** Command with **-a** and **-f** Options

```
$ PS -a -f
PID      TTY      TIME    COMMAND
24059    tty11    0:05    sh
24259    tty11    0:02    ps
24059    tty12    0:05    ksh
24259    tty12    0:02    ps
$_
```

To find the process number of a process running in the background, use the following command sequence:

```
$ ( sleep 120 ; echo "Had a nice long sleep") & [Return]
24259 . . . . . . . . . . . . . . . The background process ID number.
$ ps [Return] . . . . . . . . . . Show your processes' status.
PID   TTY    TIME COMMAND
24059 tty11  0:05 sh  . . . . . . The login shell.
24070 tty11  0:00 sleep 120 . . . The sleep command.
24150 tty11  0:00 echo  . . . . . The echo command.
24259 tty11  0:02 ps  . . . . . . The ps command.
$ Had a nice long sleep . . . . . Output from the background process.
$_  . . . . . . . . . . . . . . . And the prompt appears.
```

The **sleep** command delays the execution of the **echo** command for two minutes. The **&** at the end of the command line places the commands in the background.

Separate the commands with semicolons, and group them together by placing them between parentheses.

8.4.3 Keep On Running: The nohup Command

When you log out, your background processes are terminated. The **nohup** command causes your background processes to be immune to terminating signals. This is useful when you want your programs to continue processing after you have logged out.

If you type **nohup (sleep 120 ; echo "job done") &** and press [Return] and then log out, your command process will continue in the background. Where is the **echo** command output displayed? Since you are logged out, the process is not associated with any terminal, and the output is automatically saved in a file called `nohup.out`.

When you log in, you can check the contents of this file to determine the output of your background processes. Alternatively, you can always redirect the output of your background programs to specified files.

To experiment with the **nohup** command, try the following command sequence:

```
$ nohup (sleep 120 ; echo "job done") & [Return]
 . . . . . . . . . . . . . . Create a background job.
12235 . . . . . . . . . . .  Background job PID.
$ [Ctrl-d]  . . . . . . . .  Log out and wait a few minutes.
login: david [Return]  . . . Log in again.
password:  . . . . . . . . . Enter your password; it is not echoed to
                             the screen.
$ cat nohup.out [Return] . . Check the contents of nohup.out file.
job done
$_  . . . . . . . . . . . .  Ready for the next command.
```

8.4.4 Terminating a Process: The kill Command

Not all programs behave normally all the time. A program might be in an infinite loop or be waiting for resources that are not available. Sometimes an unruly program locks your keyboard, and then you are in real trouble! UNIX provides you with the **kill** command to terminate the unwanted process (a *process* is a running program). The **kill** command sends a signal to the specified process. The signal is an integer number indicating the kill type (UNIX is a morbid language), and the process is identified by the process ID number (PID). In order to use the **kill** command, you must know the PID of the process that you intend to terminate.

Signals Signals range from 1 to 15 and are mostly implementation dependent. However, 15 is usually the default signal value and causes the receiving process to terminate.

Some processes protect themselves from the **kill** signals. You use the signal value 9 (sure kill) to terminate these.

You can use the **kill** command with the **-l** option to obtain a list of the signals in your system:

```
$ kill -l [Return]  . . . . . . . Display list of the signals.
```

The following command sequences illustrate the use of the **kill** command and its signals.

To issue a simple **kill** command, try the following:

```
$ (sleep 120 ; echo Hi) & [Return]  . Create a background process.
22515  . . . . . . . . . . . . . . .  Process ID number.
```

```
$ ps [Return]  . . . . . . . . . . . Check the process's status. It is
                                     there:
PDI    TTY     TIME COMMAND
24059 tty11  0:05 sh . . . . . . . . The login shell.
22515 tty11  0:00 sleep 120  . . . . The sleep command.
24259 tty11  0:02 ps . . . . . . . . The ps command.
$ kill 22515 [Return]  . . . . . . . Terminate the background process.
job terminated
Hi . . . . . . . . . . . . . . . . . Output of the echo command.
$ ps [Return]  . . . . . . . . . . . Check again. The background
                                     process is terminated.
PID    TTY     TIME COMMAND
24059 tty11  0:05 sh . . . . . . . . The login shell.
24259 tty11  0:02 ps . . . . . . . . The ps command.
$_ . . . . . . . . . . . . . . . . . And the prompt.
```

No signal number is specified. The default is signal number 15, which causes the receiving process to terminate.

To make sure an unruly process has been terminated, try the following:

```
$ (sleep 120 ; echo Hi) & [Return]  . . . Create a background process.
22515  . . . . . . . . . . . . . . . . .  Process ID number.
$ kill 22515 [Return] . . . . . . . . . . A simple kill.
$ ps [Return] . . . . . . . . . . . . . . Check the process's status. It is still there:
PID    TTY     TIME COMMAND
24059 tty11  0:05 sh . . . . . . . . . .  The login shell.
22515 tty11  0:00 sleep 120 . . . . . . . The sleep command.
24259 tty11  0:02 ps . . . . . . . . . .  The ps command.
$ kill -9 22515 [Return]  . . . . . . . . A sure kill; signal value 9 is specified.
$ ps [Return]   . . . . . . . . . . . . . Check again. Sure enough, the background
                                          process is terminated.
PID    TTY     TIME COMMAND
24059 tty11  0:05 sh . . . . . . . . . .  The login shell.
24259 tty11  0:02 ps . . . . . . . . . .  The ps command.
```

You can terminate only your own processes. The system administrator is authorized to terminate anybody's processes.

To terminate all of your processes, do the following:

```
$ (sleep 120 ; echo "sleep tight" ; sleep 120) & [Return]
11234 . . . . . . . . . . . . . . . Sleep PID.
$ kill -9 0 [Return]  . . . . . . . You are logged out.
```

The PID 0 (zero) causes all processes associated with your shell to be terminated. That includes your login shell itself. Accordingly, when you use the 0 signal, you are logged out.

8.4.5 Splitting the Output: The tee Command

Sometimes you will want to look at the output of a program on the screen and also store the output in a file for later reference or obtain a hard copy of the output on the printer.

You can always do that this way: First run the command and view the output on the screen; then, using the redirection operator, save the output in a file or send it to the printer.

Alternatively, you can use the **tee** command to get the same result in less time and with less typing. The **tee** command is usually used with the pipe operator. For example, when you type **sort phone.list ¦ tee phone.sort** and press [Return], the pipe operator passes the output of the **sort** command (the sorted `phone.list`) to **tee**. Then **tee** displays it on the terminal and also saves it in `phone.sort`, the specified file.

This is an indispensable command when you want to capture the user/program dialog while running an interactive program.

Try viewing the contents of your current directory and saving the output in a file as follows:

```
$ ls -C ¦ tee dir.list [Return]  . Display the current directory file-
                                   names and also save the output in
                                   dir.list.
memos      myfile      REPORT
$ cat dir.list [Return] . . . . . Check the contents of dir.list.
memos      myfile      REPORT
$_ . . . . . . . . . . . . . . . . The prompt appears.
```

The output of **ls -C** is piped to **tee**. The **tee** command shows its input on the screen (displays the input on the default output device) and also saves it in a file called `dir.list`.

tee Options

Table 8.7 summarizes the two options of the **tee** command.

Table 8.7
The **tee** Command Options

Option	Operation
-a	Appends output to a file without overwriting an existing file.
-i	Ignores interrupts; does not respond to the interrupt signals.

To view the list of the users currently on the system and save the list in an existing file called `dir.list`, type **who ¦ tee -a dir.list** and press [Return]. If `dir.list` exists, then the output of the **who** command is added to the end of the file. If `dir.list` does not exist, it is created.

Linux Alternative Options for the tee Command

Use the Linux alternative options for the **tee** command, as suggested in the following command lines:

```
$ tee --version [Return] . . . . . Display version information.
$ tee --help [Return]  . . . . . . Display the help page.
```

Read the help page and familiarize yourself with other options available for the **tee** *command.*

8.4.6 File Searching: The grep Command

You can use the **grep** command to search for a specified pattern in a file or list of files. The pattern used by the **grep** command is called *regular expression*, hence the strange name of the command (*Global Regular Expression Print*).

grep is a file searching and selection command. You specify the filename and the pattern to be looked for in the file, and when **grep** finds a match, the line containing the specified pattern is displayed on the terminal. If no file is specified, the system searches through the input from the standard input device.

Look for the word *UNIX* in `myfile`.

```
$ cat myfile [Return] . . . Check contents of myfile.
I wish there were a better way to learn
UNIX. Something like having a daily UNIX pill.
$ grep UNIX myfile [Return] . . Find the lines that contain the word UNIX.
UNIX. Something like having a daily UNIX pill.
```

You can specify more than one file or use file substitution (wild cards) in filenames.

Look for the string **"#include <private.h>"** in all the C source files:

- Type **grep "#include <private.h>" *.c** and press [Return] to look for the pattern in all files with extension c in the current directory.

The pattern is a string with embedded spaces and metacharacters, so it is enclosed in quotation marks.

If you specify more than one file to be searched, **grep** displays the name of the file preceding each line of output.

grep Options

If you do not specify any option, **grep** displays lines in the specified file(s) that contain a match for the specified pattern. The options give you more control over the output and the way the pattern search is done. Table 8.8 summarizes the **grep** options.

Assuming you have the following three files in your current directory, the command sequences show examples of **grep** using options. You can create these files using the vi editor or the **cat** command. To get organized, create these files under the `Chapter8` directory.

FILE1	**FILE2**	**FILE3**
UNIX	unix	Unix system
11122	11122	11122
BBAA	CCAA	AADD
unix system		

Search for the word *UNIX*:

```
$ grep UNIX FILE1 [Return] . Search for the word UNIX in FILE1.
UNIX
$_ . . . . . . . . . . . . . And the prompt appears.
```

grep matches the exact pattern (whether in uppercase or lowercase letters), so it finds the word *UNIX* and not *unix* or *Unix*.

Table 8.8
The **grep** Command Options

Option		Operation
UNIX	Linux Alternative	
-c	**--count**	Displays only the count of the matching lines in each file that contains the match.
-i	**--ignore-case**	Ignores the distinction between lower- and uppercase letters in the search pattern.
-l	**--files-with-matches**	Displays the names of the files with one or more matching lines, not the lines themselves.
-n	**--line-number**	Displays a line number before each output line.
-v	**--revert-match**	Displays only those lines that do not match the pattern.
	--help	Displays help page and exits.
	--version	Displays version information and exits.

Specify more than one file as argument and use the **-i** option:

```
$ grep -i UNIX FILE? [Return] . . Use the -i option.
FILE1: UNIX
FILE1: unix system
FILE2: unix
FILE3: Unix system
$_ . . . . . . . . . . . . . . . . Your prompt appears.
```

The **-i** option tries to match the specified letter pattern, regardless of case. Thus, the specified pattern *UNIX* matches *unix, Unix,* and so on.

The name of the file is displayed when you specify more than one file as the argument.

Show the lines that do not contain the word *UNIX*:

```
$ grep -vi UNIX FILE1 [Return] . Use options -i and -v.
11122
BBAA
$_ . . . . . . . . . . . . . . . . The prompt appears.
```

Display how many lines in each file do not contain *11:*

```
$ grep -vc 11 FILE? [Return] . . Show a count of the lines in
                                 FILE1, FILE2, and FILE3 that
                                 do not contain 11.
FILE1:3
FILE2:2
FILE3:2
$_ . . . . . . . . . . . . . . . . The prompt appears.
```

Find out whether user `david` is logged in:

```
$ who ¦ grep -i david [Return] . Use grep with the pipe operator.
$_ . . . . . . . . . . . . . . . . The prompt appears.
```

The pipe makes the output of the **who** command the standard input to **grep**. Thus, **grep** scans the output of **who** for lines containing the pattern *david.* In this example, **grep** did not produce any output. Thus `david` is not in the system.

Linux Alternative Options for the grep Command

Use the Linux alternative options for the **grep** command, as suggested in the following command lines:

$ **grep --ignore-case UNIX file? [Return]** . . . . Same as **grep -i UNIX file?**.
$ **grep --revert-match --ignore-case UNIX file? [Return]** . Same as **grep -vi UNIX file?**.
$ **grep --revert-match --count file? [Return]** . . . . . . Same as **grep -vc 11 file?**.
$ **grep --version [Return]** . . . . Display version information.
$ **grep --help [Return]** . . . . Display the help page.

8.4.7 Sorting Text Files: The sort Command

You can use the **sort** command to sort the contents of a file into alphabetical or numerical order. By default, the output is displayed on your terminal, but you can specify a filename as the argument or redirect the output to a file.

The **sort** command sorts the specified file on a line-by-line basis. If the first characters on two lines are the same, it compares the second characters to determine the order of the sort. If the second characters are the same, it compares the third characters, and this process goes on until two characters differ or the line ends. If two lines are identical, then it does not matter which one is placed first.

This command sorts files alphabetically, but the order of the sort might be different from one computer to another, depending on the computer's code set. The most commonly used code set in UNIX systems is ASCII.

Many options can be used to control the sort order, but let's start with a simple example to explore the **sort** basic functions.

Suppose you have a file called `junk` in your working directory. Figure 8.6 shows the contents of `junk`. Figure 8.7 shows the output of the **sort** command, after sorting the contents of `junk`.

Figure 8.6
The `junk` File

```
This is line one
this is line two
 this is a line starting with a space character
4: this is a line starting with a number
11: this is another line starting with a number
End of junk
```

Figure 8.7
The Sorted `junk` File

```
$ sort junk
 this is a line starting with a space character
11: this is another line starting with a number
4: this is a line starting with a number
End of junk
This is line one
this is line two
$_
```

1. *ASCII values for nonalphanumeric characters (space, dash, backslash, etc.) are less than those for alphanumeric characters. Thus, in Figure 8.7, the line starting with a blank space is placed at the top of the file.*
2. *Uppercase letters are sorted before lowercase letters. Thus, in our example,* `This` *appears before* `this`.
3. *Numbers are sorted by the first digit. Thus 11 appears before 4.*

sort Options

The **sort** example showed that the result of the **sort** command, the sorted output, is probably not what you would consider sorted. The **sort** command options give you freedom to sort files in a variety of orders. Table 8.9 summarizes some of the more useful options.

Table 8.9
The **sort** Command Options

Option	Operation
-b	Ignores leading blanks.
-d	Uses dictionary order for sorting. Ignores punctuation and control characters.
-f	Ignores the distinction between lowercase and uppercase letters.
-n	Numbers are sorted by their arithmetic values.
-o	Stores the output in the specified file.
-r	Reverses the order of the sort from ascending to descending order.

-b Option The **-b** option causes **sort** to ignore the leading blanks (tabs and space characters). These characters are usually delimiters (field separators) in your file, and when you use this option, **sort** does not consider them in sort comparison.

-d Option The **-d** option, used for dictionary sorting, uses only letters, digits, and blanks (spaces and tabs) in the **sort** comparison. It ignores the punctuation and control characters.

-f Option The **-f** option considers all lowercase characters as uppercase characters; it ignores the distinction between them in **sort** comparison.

-n Option The **-n** option causes numbers to be sorted by their arithmetic values rather than by their first digit. This includes ascribing minus signs and decimal points to their arithmetic values.

-o Option The **-o** option places the output in a specified file instead of the standard output.

-r Option The **-r** option reverses the order of the sort, such as sorting from *z* to *a* instead of *a* to *z*.

Using the `junk` file again, let's see the effects of the options on the sorted output:

```
$ sort -fn junk [Return] . . . . . . . . Sort junk using the
                                          -f and -n options.
this is a line starting with a space character
End of junk
This is line one
this is line two
4: this is a line starting with a number
11: this is another line starting with a number
$_ . . . . . . . . . . . . . . . . . . . . The prompt appears.
$ sort -f -r -o sorted junk [Return] . . Sort junk using the
                                          -f, -r, and -o options
                                          and save it in sorted.
$ cat sorted [Return] . . . . . . . . . . Display sorted.
this is line two
This is line one
End of junk
4: this is a line starting with a number
11: this is another line starting with a number
this is a line starting with a space character
$_ . . . . . . . . . . . . . . . . . . . . The prompt appears.
```

A filename (`sorted`) *is specified with the* **-o** *option. Thus the output is saved in* `sorted`, *and* **cat** *is used to display the contents of* `sorted`.

8.4.8 Sorting on a Specified Field

Real files seldom contain what is in the example file `junk`. Usually files you want to sort contain lists of people, items, addresses, phone numbers, mailing lists, and so on. By default, **sort** sorts on a line-by-line basis, but you probably will want to sort files by a particular field, such as last name or area code.

You can direct the **sort** command to look at a specified field for **sort** comparison, provided that the file is set up accordingly. You specify the desired field by a number that indicates how many fields **sort** must skip to get to the field by which you want the file sorted. You set up your file by breaking each line into fields. No extra effort is needed, because in most list files each line is already divided into fields.

Create a file called `phone.list`, which contains a list of people and their phone numbers, such as the example in Figure 8.8, and then we will use the file to explore the other capabilities of the **sort** command.

Figure 8.8
Original `phone.list` File and Sorted `phone.list` File

```
$ cat phone.list
David Brown             (333) 111-1111
Emma Redd               (222) 222-2222
Tom Swanson             (111) 333-3333
Jim Schmid              (444) 444-4444
Bridget Erwin           (666) 555-5555
Mary Moffett            (555) 666-6666
Amir Afzal              (777) 777-7777

$ sort phone.list
Amir Afzal              (777) 777-7777
Bridget Erwin           (666) 555-5555
David Brown             (333) 111-1111
Emma Redd               (222) 222-2222
Jim Schmid              (444) 444-4444
Mary Moffett            (555) 666-6666
Tom Swanson             (111) 333-3333
$
```

Each line in `phone.list` consists of four fields, and the fields are separated by space or tab characters. Thus, in line one, *David* is field 1, *Brown* is field 2, and so on.

You can create the `junk` file using the vi editor or the **cat** command.

In sorting `phone.list`, no particular field is specified. Thus, the list is sorted on a line-by-line basis.

You may not want to sort the file by the first names. To sort the file in order of the last names (field 2), you must instruct **sort** to skip one field (first name) before it starts the sorting process. You specify the number of fields **sort** is to skip as part of the command argument.

To sort `phone.list` by the last names (field 2), type **sort +1 phone.list** and press [Return]. Figure 8.9 shows the result of sorting the file by the last names.

The **+1** argument indicates that **sort** must skip the first field (first name) before starting the sort process.

Figure 8.9
Sorted (by Last Names) `phone.list` File

```
$ sort +1 phone.list
Amir Afzal              (777) 777-7777
David Brown             (333) 111-1111
Bridget Erwin           (666) 555-5555
Mary Moffett            (555) 666-6666
Emma Redd               (222) 222-2222
Jim Schmid              (444) 444-4444
Tom Swanson             (111) 333-3333
$_
```

If you specify +2, then sort skips the first and second fields and starts from the third field (in this case, area code).

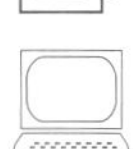

To sort phone.list by the third field, ignoring blanks, type **sort -b +2 phone.list** and press [Return]. Figure 8.10 shows the results.

Figure 8.10
Sorted (by Area Code) phone.list File

```
$ sort -b +2 phone.list
Tom Swanson                (111) 333-3333
Emma Redd                  (222) 222-2222
David Brown                (333) 111-1111
Jim Schmid                 (444) 444-4444
Mary Moffett               (555) 666-6666
Bridget Erwin              (666) 555-5555
Amir Afzal                 (777) 777-7777
$_
```

8.5 STARTUP FILES

When you log in, the login program verifies your user ID and password against the list of authorized users stored in the password file. If the login attempt is successful, the login program brings your HOME directory up on the system, sets up your user ID and group ID, and finally starts your shell. Before displaying its prompt sign, the shell checks for two special files. These two files are called *profile files*, and they contain shell scripts (programs) that the shell can execute.

8.5.1 System Profile

The *system profile* file is stored in /etc/profile. The first thing your shell does is execute this file. It typically contains commands that display the message of the day, set up systemwide environment variables, and so on. This file is usually created and maintained by the system administrator, and only the superusers can modify it.

Figure 8.11 shows an example of a system profile file. The shell executes the commands in this file, so it displays the current date and time, then the message of the day (stored in the /etc/motd file), and finally the recent news items.

Figure 8.11
An Example of a Simple Profile File

```
$ cat /etc/profile
date
cat /etc/mtd
news
mesg n
stty erase ^H
export erase
$_
```

1. *Usually, the system* `profile` *file is complex and incorporates some system administration commands and requires some programming.*
2. *You can look at your system* `profile` *file by typing* **cat /etc/profile** [Return]. *Usually, this file is a read only file. You can read it, but cannot edit it.*

8.5.2 User Profile

Each time you log in, the shell checks for a startup file called `.profile` in your HOME directory. If the file is found, then the shell commands in `.profile` are executed. Whether or not you have a `.profile` file in your HOME directory, the shell continues its process and displays its prompt.

Figure 8.12 shows an example of a `.profile` file. Usually you have a `.profile` file, courtesy of the system administrator. You can modify the existing `.profile` file or create a new one using the **cat** or **vi** utilities.

Figure 8.12
An Example of the `.profile` File

```
$ cat .profile
  echo "welcome to my super Duper UNIX"
  TERM=ansi
  PS1="David Brown:"
  export TERM PS1
            calendar
  du
$_
```

1. The **echo** command displays its argument, *welcome to my super Duper UNIX.*
2. The standard variable *TERM* (terminal type) is set to *ansi.*
3. The standard variable *PSI* (primary prompt sign) is set to *David Brown.*
4. The **export** command makes the variables *TERM* and *PSI* available (exported to all programs).
5. The **calendar** and **du** commands are explained in Chapter 13.

1. *The name of the file is* `.profile`. *The filename starts with a dot; it is a hidden file.*
2. *The* `.profile` *file must be located in your HOME directory. This is the only place that the shell checks.*
3. *The* `.profile` *file is one of the startup files you can use to customize your UNIX environment. Other startup files exist in UNIX, such as the* `.exrc` *file that customizes the vi editor (discussed in Chapter 6) and the* `.mailrc` *file that customizes your mail environment (discussed in Chapter 9).*
4. *You do not have to have a* `.profile` *file in your HOME directory. Your system works without it. However, your shell usually inherits the setup from the* `/etc/profile` *file.*

More on the export Command

The **export** command makes a list of shell variables available to subshells. When you log in, the standard variables (and variables you may have defined) are known to your login shell. However, if you run a new shell, these variables are not known to the new shell.

For example, if you want to make the variables *VAR1* and *VAR2* available to the new shell, you specify the variable names as arguments for the **export** command. To see what variables are already exported, type **export** without any arguments.

Make variables available to other shell programs:

```
$ export VAR1 VAR2 [Return]  . . Export VAR1 and VAR2.
$ export [Return] . . . . . . .  Check which variables are exported.
VAR1
VAR2
$_ . . . . . . . . . . . . . . . List of variables, then the prompt
                                 appears.
```

8.6 THE KORN AND BOURNE AGAIN SHELLS

Chapter 3 introduced the Korn shell (*ksh*) and Bourne Again shell (*bash*). Now we explain a few of the features that differ from those of the standard shell (sh). Once more, it is strongly recommended to use the **man** command and learn more about specific features of each of these shells. That is as easy as typing one of the following commands:

```
$ man sh [Return] . . . .  Display manual pages for standard shell.
$ man ksh [Return]  . . .  Display manual  pages for Korn shell.
$ man bash [Return] . . .  Display manual pages for Bourne Again shell.
```

8.6.1 The Shell Variables

The Korn shell (*ksh*) and Bourne Again shell *(bsh)* uses many of the same variables used by the standard shell (*sh*). You can define variables, redefine variables, get their values, and in general manipulate variables to customize your environment just like the *sh* shell. The following are some important variables that are used by the Korn shell and Bourne Again shell:

ENV The *ENV* variable is set to the absolute pathname of the environment file that is read by *ksh* at startup. In the following example, *ENV* is set to the pathname `($HOME/mine/my_env)` that tells *ksh* where to find the environment file:

```
ENV=$HOME/mine/my_env
```

HISTSIZE The *HISTSIZE* variable is set to the number of commands you intend to keep in your commands `history` list file. The default size is 128, but you can set it to any number of entries you wish. For example, the following command line sets the number of entries in your `history` list file to 100:

```
HISTSIZE=100
```

TMOUT The *TMOUT* variable is set to the number of seconds you want the system to wait before timing out if you do not type a command. The shell timeout logs you off if you do not provide any input within the given number of seconds. For example, the following command line sets timeout to 60 seconds:

```
TMOUT=60
```

VISUAL The *VISUAL* variable is used with command editing. If it is set to **vi**, *shell* gives you *vi-style* editing capabilities on your command lines. For example, the following command line sets the command line editing to the vi editor:

```
VISUAL=vi
```

Use the **set** *command, just like in the* sh *shell, to see the current shell variables and their values.*

8.6.2 The Shell Options

The Korn and Bourne Again shells provide a number of options that turn some of their special features *on* or *off.* To turn an option on, you use the **set** command with **-o** (option) followed by the option name. To list your options, simply type **set -o**.

noclobber The **noclobber** option prevents you from clobbering your files; that is, it prevents you from overwriting an existing file if you redirect output from a command. This can save you from losing important data by inadvertently overwriting your files.

Suppose a file called xyz exists in your current directory. The following command sequences demonstrate the use of the **noclobber** option.

```
$ set -o noclobber [Return] . . . . . Set the noclobber option.
$ who > xyz  . . . . . . . . . . . . . Redirect the output to xyz.
xyz: file exists  . . . . . . . . . . Warning message appears.
$_  . . . . . . . . . . . . . . . . . The prompt appears.
```

If you really want to overwrite an existing file, *ksh* will oblige. Type a ¦ pipe symbol after the redirection operator.

```
$ who > ¦ xyz [Return]  . . . . . . . The xyz file is overwritten.
$_  . . . . . . . . . . . . . . . . . The prompt appears.
```

To turn off the **noclobber** option, you use the **set** command with the **+o** option. For example,

```
$ set +o noclobber [Return] . . . . . The noclobber option is
                                      turned off.
$_  . . . . . . . . . . . . . . . . . The prompt appears.
```

ignoreeof The **ignoreeof** option prevents you from accidentally logging yourself off by typing [Ctrl-d]. (You already know that [Ctrl-d] typed at the beginning of the command line terminates your shell, and logs you off from the system.)

If you set this option, you should use the **exit** *command to log off.*

```
$ set -o ignoreeof [Return] . . . . . Turn on the ignoreeof option.
$ [Ctrl-d]  . . . . . . . . . . . . . [Ctrl-d] is ignored. You are
                                      not logged off.
$ set +o ignoreeof [Return] . . . . . Turn off the ignoreeof
                                      option.
$_  . . . . . . . . . . . . . . . . . The prompt appears.
```

8.6.3 Command Line Editing

The Korn shell lets you edit your command line, or any of the commands in your `history` file, using the special line version of the vi editor. This feature enhances the use of the `history` file. It enables you to correct and modify the previous commands and makes it easier to search for a specific command in your commands `history` file. In short, it is good to know this command.

Turning on the Command Line Editing Option

You can turn on the command line editing option by using the **set** command or by setting the *EDITOR* or *VISUAL* variables to the pathname of your editor command. The effect of any of the following three commands is the same:

`$ set -o vi [Return]`	Turns on the command line editing option.
`$ EDITOR=/usr/bin/vi [Return]` . . .	Turns on the command line editing option.
`$ VISUAL=/usr/bin/vi [Return]` . . .	Turns on the command line editing option.

Using the vi-Style Command Line Editor

Assuming that you have turned on the command line editing feature, and that it is set to the vi editor, this section describes how you can use this very useful feature.

The *ksh* command line editor works on your current command line and your `history` file (explained later in this chapter). When you are entering a command, you are in the vi input (*text*) mode. This is opposite of the vi editor initial mode when you edit a file. You press the [Return] key to execute your command. As in the vi editor, you can switch to command mode at any time by pressing the [Esc] key. While in the *command mode,* the vi editor commands are available to change, delete, and correct your command line.

Now, type a command line and do not press the [Return] key. Instead, press the [Esc] key. This puts you in vi command mode.

```
$ This is a test to use the command line editing [Esc]
```

Let's say you forgot to type the **echo** command at the beginning of this command line. You can use vi editor keys to get to the beginning of the line. In this case, you type **$0** and the cursor will go to the letter T. You type **i** (insert) to return to vi text mode and so on.

1. *The command line vi editor is a special built-in vi editor.*
2. *You can use the* **j** *(down) and* **k** *(up) keys to access the commands from the* `history` *file.*
3. *You can use the* **l** *(right) and* **h** *(left) keys to move the cursor left or right on the command line.*

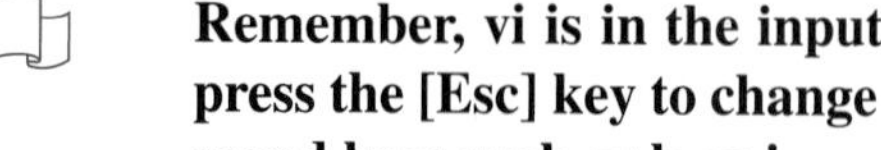

Remember, vi is in the input mode when you are entering a command. You must press the [Esc] key to change to the command mode before you can use the vi command keys such as k or j.

Table 8.10 lists some of the available commands with the built-in vi editor.

Table 8.10
The Built-In vi Editor Commands

Key	Operation
h and **l**	Moves cursor left and right one character on the command line.
k and **j**	Moves up and down one entry in the `history` list.
b and **w**	Moves cursor left and right one word on the command line.
$	Moves cursor to the end of the line.
x	Deletes the current character.
dw	Deletes the current word.
I and **i**	Inserts text.
A and **a**	Appends text.
R and **r**	Replaces text.

You can also invoke the *real* vi editor to use all the commands and features of it to edit the command line.

```
$ cp xyz xyz.bak . . .  Suppose you want to modify this command line.
[Esc] . . . . . . . .   Press the [Esc] key to get into the command mode.
v . . . . . . . . . .   Press the v key to invoke the real vi editor.
```

Now, you are using the vi editor with a file consisting of one line, your current command line. You can use vi to edit your command or add new commands. When you leave vi, the Korn shell will execute your commands.

You can use the vi editor in this manner to create a multiline sequence of commands to be executed.

8.6.4 The alias Command

You can use the **alias** command for shortening the names of frequently used commands or changing command names to names easier for you to remember. For example:

```
$ alias del=rm [Return] . . . . . .  Now del is the alias for the rm
                                     command.
$ del xyz . . . . . . . . . . . . .  Delete the file called xyz.
```

1. *Now you can type* **del** *instead of the* **rm** *command.*
2. *The* **rm** *command is not changed and you can still use it.*
3. *Aliases are defined the same way you define variables using the = sign.*

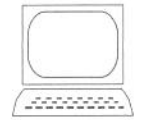

Set *ll* (*ell-ell,* for *long list*) as an alias for the command **ls -al**. Then you can type **ll** to get the listing of your current directory in long format.

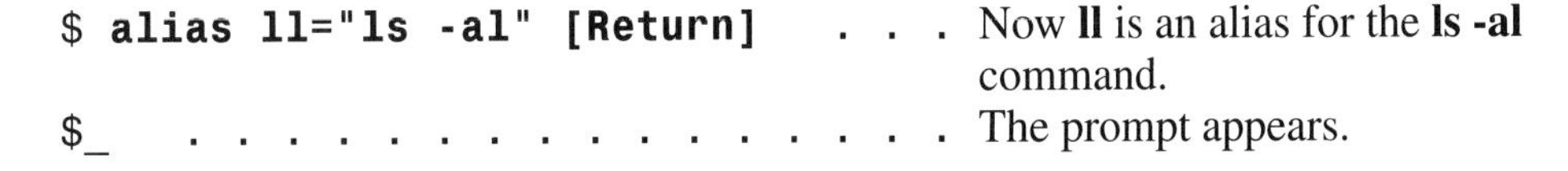

```
$ alias ll="ls -al" [Return]  . . .  Now ll is an alias for the ls -al
                                     command.
$_ . . . . . . . . . . . . . . . . . The prompt appears.
```

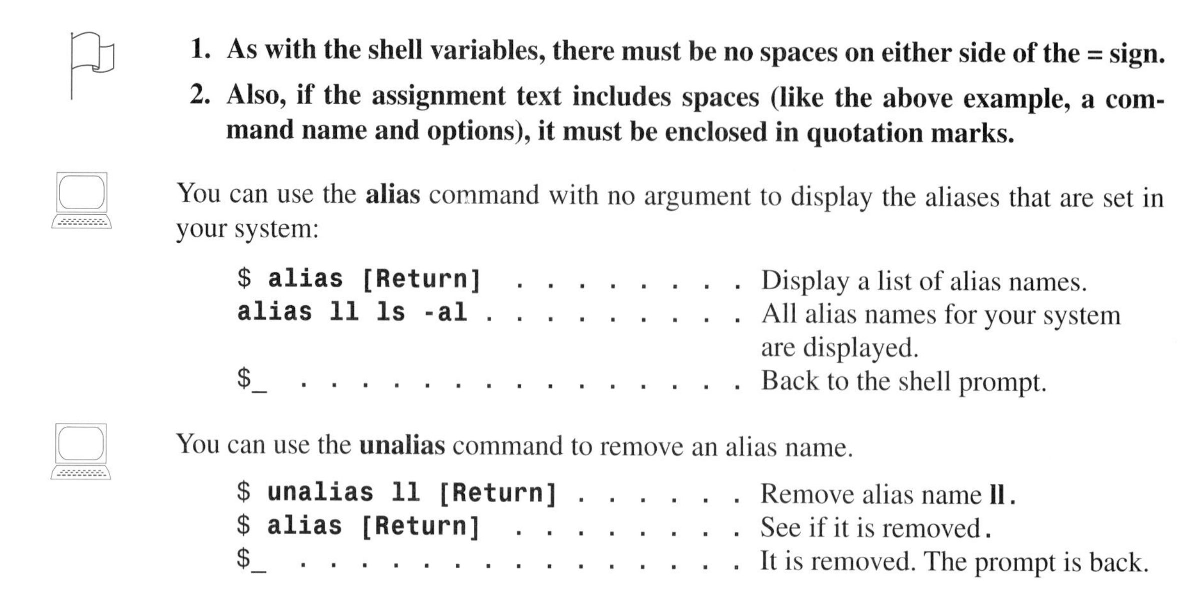

1. As with the shell variables, there must be no spaces on either side of the = sign.

2. Also, if the assignment text includes spaces (like the above example, a command name and options), it must be enclosed in quotation marks.

You can use the **alias** command with no argument to display the aliases that are set in your system:

```
$ alias [Return] . . . . . . . . . Display a list of alias names.
alias ll ls -al . . . . . . . . . . All alias names for your system
                                    are displayed.
$_ . . . . . . . . . . . . . . . . Back to the shell prompt.
```

You can use the **unalias** command to remove an alias name.

```
$ unalias ll [Return] . . . . . . Remove alias name ll.
$ alias [Return] . . . . . . . . . See if it is removed.
$_ . . . . . . . . . . . . . . . . It is removed. The prompt is back.
```

8.6.5 Commands History List: The history Command

The Korn and Bourne Again shells have a commands history feature that keeps a list of all the commands you enter during your sessions. Using this very popular feature, you are able to list your previous commands, search for a particular command that you have issued, or easily edit and redo your previous commands.

The **history** command is one of the utilities that works on the commands history list. The **history** command or sometimes just letter **h** (usually an alias for the **fc** command) is used. If the **history** command is not working in your system (*ksh* or *bash*), try the **fc** command and its options or create your own alias for the **history** command. The **fc** command is explained later in this chapter and some of the examples will show you how to create aliases for the **history**, and **fc** commands.

To display the last few commands you have entered, do the following:

```
$ history [Return] . . . . . . . . Issue the command.
101   Who am i
102   ls -l
103   pwd
104   vi myfirst
105   rm myfirst
106   history
$_ . . . . . . . . . . . . . . . . The prompt appears.
```

1. *The preceding example displays six lines of commands. The number of command lines that* ksh *or* bash *keep track of is controlled by the* HISTSIZE *variable.*
2. *The last item on the list is the last command you issued, and your earlier commands are higher on the list.*

The default history file is called .sh_history for *ksh* and .bash.history for *bash*, and is created by the system in your HOME directory. You may use another filename by setting the variable *HISTFILE* to the pathname of your desired history filename. For example:

```
HISTFILE=$HOME/history/my_hist
```

Restarting the `History` File

The history of your commands is kept from one session to another and consequently the entry numbers will become large and cumbersome to retype or refer to. If you want to restart your `history` file, remove your `.sh_history` or file from your HOME directory. The next time you log in, the system creates a new `.sh_history` or `.bash.history` file and the entries start from the first command you issue in that session.

Practicing the `History` Command

You can display the commands `history` list by starting it from a specified command in the list:

```
$ history  104 [Return]  . . . . Start from command number 104.
   104    vi myfirst
   105    cat myfirst
   106    history
   107    history 104

$ history  vi [Return] . . . . . Start from first occurrence of vi in
                                 the list.
   104     vi myfirst
   105     cat myfirst
   106     history
   107     history 104
   108     history vi
$_
```

Use the line editor to edit a command from the commands history list:

```
$ history [Return] . . . . . . . List the commands.
   104     vi myfirst
   105     cat myfirst
   106     history
   107     history 104
   108     history vi
   109     history
$_ . . . . . . . . . . . . . . . Prompt is back.
[Esc]  . . . . . . . . . . . . . Press the [Esc] key.
j  . . . . . . . . . . . . . . . Press the letter j to get to go one
                                 command up on the list.
history vi
G  . . . . . . . . . . . . . . . Show the last command in the
                                 history file.
history
105G   . . . . . . . . . . . . . Show the 105th command in the
                                 history file.
cat myfirst
[Return] . . . . . . . . . . . . Execute the command.
```

Your `history` *file probably will be different from the examples shown here. Your* `history` *file does not remain the same, and each command that you type is added to the list of commands already in the* `history` *file. In the following examples no specific* `history` *file is assumed.*

8.6.6 Redoing Commands (ksh): The r (redo) Command

You can use the **r** (*redo*) command to redo the last command you have issued. For example, suppose your last command is **rm myfirst** and you type the following:

```
$ r [Return] . . . . . Repeats the last command.
rm myfirst . . . . . . Your last command is repeated.
```

In this case, your last command is executed. The following command sequences show the **r** (*redo*) command options and features.

You can repeat other commands from the `history` file by adding the specific command name as an argument to the **r** command:

```
$ r vi [Return]   . . . Repeat vi from the history list.
vi myfirst . . . . . .  The first occurrence of the vi command in the
                        history is executed.
```

You can repeat any command from the `history` file by indicating the specific command entry number:

```
$ r 102 [Return]  . . . Repeat command number 102 from the
                        history list.
ls -l  . . . . . . . .  The specified command is executed.
```

You can also repeat commands from the `history` file by indicating the number of entries you want to go back in the list.

```
$ r -3 [Return]   . . . Go back three entries in the history list.
ls -l  . . . . . . . .  The specified command is executed.
```

8.6.7 Commands `History` List: The fc Command

The **fc** command provides capability to list, edit, and re-execute commands that were previously entered and were saved in the commands `history` list. For example, the following command line lists the previous commands from the `history` file:

```
$ fc -l [Return]  . . . Same as the history command.
```

Figure 8.13 shows the output of the **fc** command. However, your command `history` list almost certainly would be different.

Figure 8.13
Output of the **fc** Command

```
$ fc -l
101  Who a m I
102  ls -l
103  pwd
104  vi myfirst
105  cat myfirst
106  history
107  fc -l
$_
```

1. *In the commands* `history` *list, each command is referenced by a number, and the list usually starts from 1.*
2. *When the number of commands reaches the value in* HISTSIZE *(default is 128), the shell may wrap the numbers, starting the next command with 1.*
3. *If the* HISTFILE *variable is set when the shell is invoked, then this file is used to store the commands history.*
4. *If the* HISTSIZE *variable is set when the shell is invoked, then this number dictates the number of entries into the commands history. The default is 128.*

fc Options

The **fc** command is rich with options that provide you with many possibilities for editing and re-executing previous commands. Table 8.11 lists few of these options. Please use the **man** command to see a detailed list of the options.

Table 8.11
The **fc** Command Options

Option	Operation
-l	Lists the commands, with each command preceded by the command number.
-n	Suppresses command numbers when listing with **-l**.
-r	Reverses the order of the commands listed with **-l**.
-s	Re-executes the command without invoking an editor.

The following examples show the usage of the **fc** command options:

```
$ fc -l [Return] . . . . . .   Display the commands history list.
$ fc -l -n [Return]  . . . .   Display the commands history list
                               without the command numbers.
$ fc -l -r [Return]  . . . .   Display the command history list in
                               reverse order.
$ fc -s [Return] . . . . . .   Execute the previous command.
$ fc -s 107 [Return] . . . .   Execute command number 107 from the
                               history list.
$ fc -s vi [Return]  . . . .   Execute the first occurrence of vi in the
                               history list.
$ fc -s c [Return] . . . . .   Execute the first occurrence of the
                               command starting with the letter c.
```

Creating Aliases for the fc Command

The following command lines are suggestions to create aliases for the **fc** command. In case your system does not provide the **history** command, the following two commands creates aliases that behave like the **history** command:

```
$ alias r='fc -e -' [Return] . . . .   Same as the r command.
$ alias history='fc -l' [Return] . .   Same as the history command.
```

You can choose any names you want for the aliases and not necessarily the ones that duplicate the **history** command. For example:

$ **alias hr='fc -e -' [Return]** . . . Same as the **history r** command.
$ **alias h='fc -l' [Return]** . . . . . Same as the **history** command.

Now you can type a command such as:

$ **h [Return]** . . . . Same as **fc -l**.
$ **r 107 [Return]** . . . . Same as **fc -s 107**.

You can remove any of the aliases by using the **unalias** command.

8.6.8 Login and Startup

Like the standard shell (*sh*), the Korn shell (*ksh*) or Bourne Again shell (*bash*) reads the `.profile` file in your HOME directory when you log in. It executes the commands you want to run at login time, and initializes the variables that will be in effect for your login session. The `.profile` file typically includes commands such as **date**, **who**, and **calendar**, which provide information at login, terminal settings, and variable definitions that you want to export to the environment.

In addition to the `.profile` file (in your HOME directory), *ksh* or *bash* also reads your environment file. The environment file does not have a predefined name or location that *ksh* or *bash* looks for. You define its name and location with the *ENV* variable in your `.profile` file. For example, if your `.profile` file contains the line

```
ENV=$HOME/mine/my_env
```

shell will look for your environment file in the file named `my_env` in a subdirectory called `mine` in your HOME directory. Although it is not necessary, it is a good practice to call your environment file `.kshrc` or `.bashrc` (a hidden file) in your HOME directory.

```
ENV=$HOME/.kshrc
```

or

```
ENV=$HOME/.bashrc
```

1. *If your login shell is* ksh *you can specify all the shell variables and options in the* `.profile` *file in your HOME directory.*
2. *If your login shell is not* ksh *or* bash, *define all the specific shell variables and options in a file specified by the* ENV *variable. For the system to read your environment file, you must have the* ENV *variable defined in your* `.profile` *file.*

Figure 8.14 shows (using the **cat** command) an example of an environment file called `.kshrc`. This `.kshrc` file contains commands to set the vi-style command line editing, to turn on the **noclobber** and **ignoreeof** options, to set the `history` file size to 10 entries, and to set the *TMOUT* option to 600. You can create a similar file using the vi editor.

Figure 8.14
An Example of an Environment File Called `.kshrc`

```
$ cat .kshrc
   set -o vi
   set -o noclobber
   set -o ignoreeof
   HISTSIZE=10
   TMOUT=600
$_
```

8.6.9 Adding Event Numbers to the Prompt

Sometimes, it is useful to know the event number the shell gives to each command you enter. You can change your prompt to include this information. For example, if you type

```
PS1="!$"
```

the exclamation mark (!) will tell the system to read the last event number from your `history` file, add one to it, and display it. The prompt will continue displaying the event numbers as you type your commands.

Change your prompt to display the event numbers:

```
$ PS1="! $" [Return] . . . . .  Change the prompt.
6 $_  . . . . . . . . . . . . . The new prompt appears.
```

The new prompt indicates that the next command you enter will have the event number 6.

```
$ PS1="[!] $ "[Return] . . . .  Change the prompt this way.
[6] $_  . . . . . . . . . . .   The new prompt appears.
```

By adding your prompt definition to the `.kshrc` *file, you can make it appear each time you log in.*

8.6.10 Formatting the Prompt Variable (bash)

In addition to displaying static character strings in the prompts, *bash* provides a list of predefined special characters that can be used in formatting the prompt. These special characters place things such as the current time into the prompt. Table 8.12 lists some of these special character codes.

The following command lines changes the prompt display using the prompt special character codes:

```
$ PS1="[\!]$ " [Return]  . . .  Display the command number.
[72]$_
```

Table 8.12
Special Character Codes to Format the Prompt

Character	Meaning
\!	Displays the history number of the current command.
\$	Displays a $ in the prompt unless the user directory is root. When user is root, it displays a #.
\d	Displays the current date.
\s	Displays the name of the shell that is running.
\t	Displays the currant time.

```
$ PS1="[\d]$ " [Return]   . . . Display the current date between
                                brackets.
[Thu Nov 2]$_

$ PS1="\d $ " [Return] . . . . Display the current date.
Thu Nov 2 $_
$ PS1="[\!][\t]$ " [Return]   . Display the command number and
                                current time.
[79][20:33:41]$_
```

Note that the [ ] are not part of the command syntax or special codes. They are used here to make the prompt look nicer and more readable.

```
$PS1="\s$" [Return]  . . . . . Display the shell name.
bash$_
```

8.7 UNIX PROCESS MANAGEMENT

In Chapter 3, we introduced the process of booting the system. Now let's go deeper into the UNIX internal process and see how it manages the running of programs.

In this chapter, you have encountered the word *process*. The execution of a program is called a *process:* you call it a *program,* but when your program is loaded into the memory for execution, UNIX calls it a *process.*

To keep track of the processes in the system, UNIX creates and maintains a process table for each process in the system. Among other things, the process table contains the following information:

- Process number
- Process status (ready/waiting)
- Event number that the process is waiting for
- System data area address

A process is created by a system routine called **fork**. A running process calls **fork**, and in response UNIX duplicates that process, creating two identical copies. The process that calls the **fork** routine is called the *parent,* and the copy of the parent created by **fork** is called the *child.* UNIX differentiates between the parent and the child by giving them different process IDs (PIDs).

The following steps are involved in managing a process:

- The parent calls **fork**, thus starting the process.
- Calling **fork** is a system call. UNIX gets control, and the address of the calling process is recorded in the process table's system data area. This is what is called the *return address,* so the parent process knows where to start later when it gets control again.
- **fork** duplicates (copies) the process and returns control to the parent.
- The parent receives the PID of the child, a positive integer number, and the child receives the return code zero. (A negative code indicates an error.)
- The parent receiving a positive PID calls another system routine called **wait** and goes to sleep. Now the parent is waiting for the child process to finish (in UNIX terminology, waiting for the child to die).
- The child process gets control and begins to execute. It checks the return code; because the return code is zero, the child process calls another system routine called **exec**. The **exec** routine responds by overlaying the child process area with the new program.
- The new program's first instruction is executed. When the new program gets to the end of the instruction, it calls yet another system routine called **exit**, and the child process dies. The death of the child awakens the parent, and the parent process takes over.

This process is depicted in Figures 8.15 through 8.18. An example is in order to shed some light on this apparently confusing process. Imagine that the shell (the `sh` program) is running, and you type a command, say **ls**. Let's explore the steps UNIX takes to run your command.

The shell is the parent process, and when created, the **ls** program becomes the child process. The parent process (shell) calls **fork**. The **fork** routine duplicates the parent (shell) process, and if the creation of the child process is successful, assigns the child process a PID and adds it to the system process table. Next, the parent receives the child PID, the child receives code zero, and control is returned to the parent. The shell calls the **wait** routine and goes to the wait state (goes to sleep). Meanwhile, the child gets control and calls **exec** to overlay the child process area with the new program—in this case **ls**, the command you typed. Now **ls** carries out the command. It lists your current directory filenames, and when it is finished processing, it calls **exit**. Thus the child dies. The death of the child generates an event signal. The parent process (shell) is waiting for this event. It is awakened and gets control. The shell program continues, starting execution from the same address it was at before going to sleep (recall that this address was stored in the process table system data area as return address), and the prompt is displayed.

What happens if the child is a background process? In that case, the parent (shell) does not call the **wait** routine; it continues in the foreground, and you see the prompt right away.

What creates the first parent and child processes? When UNIX is booted, the **init** *process is activated. Next,* **init** *creates one system process for each terminal. Thus,* **init** *is the original ancestor to all the processes in the system. For example, if your system supports 64 concurrent terminals, then* **init** *creates 64 processes. When you log in to one of these processes, the login process executes the shell. Later, when you log out (when the shell dies),* **init** *creates a new login process.*

Figure 8.15
Events Happening When **fork** Is Called

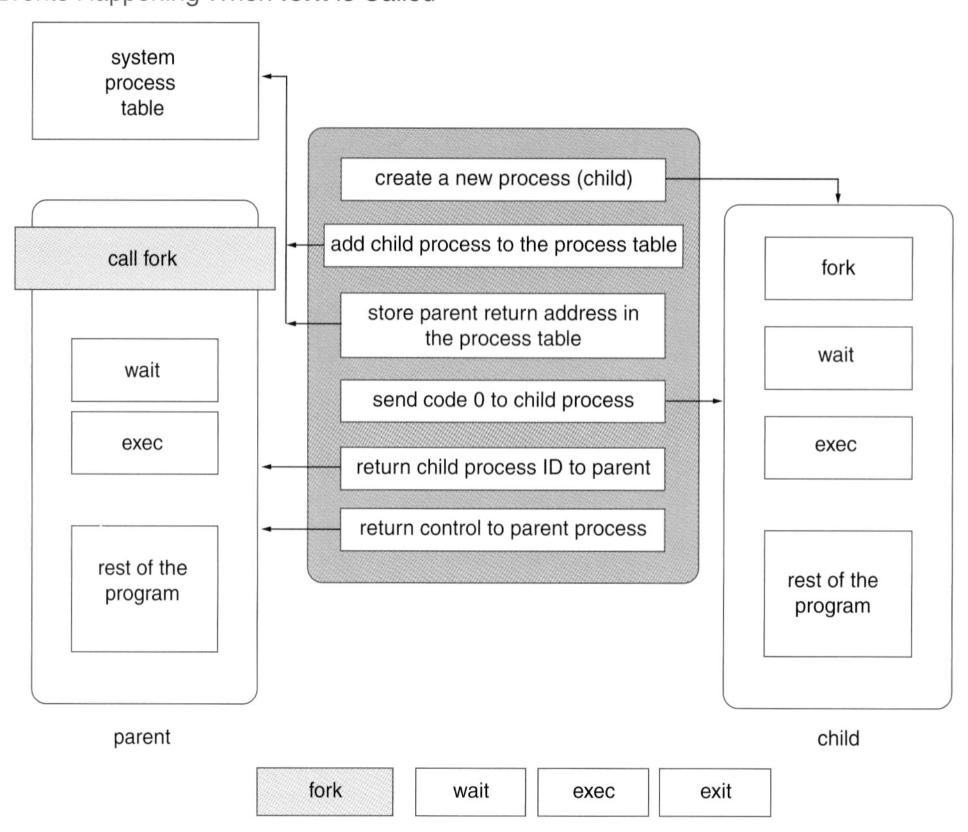

Figure 8.16
Events Happening After **wait** Is Called

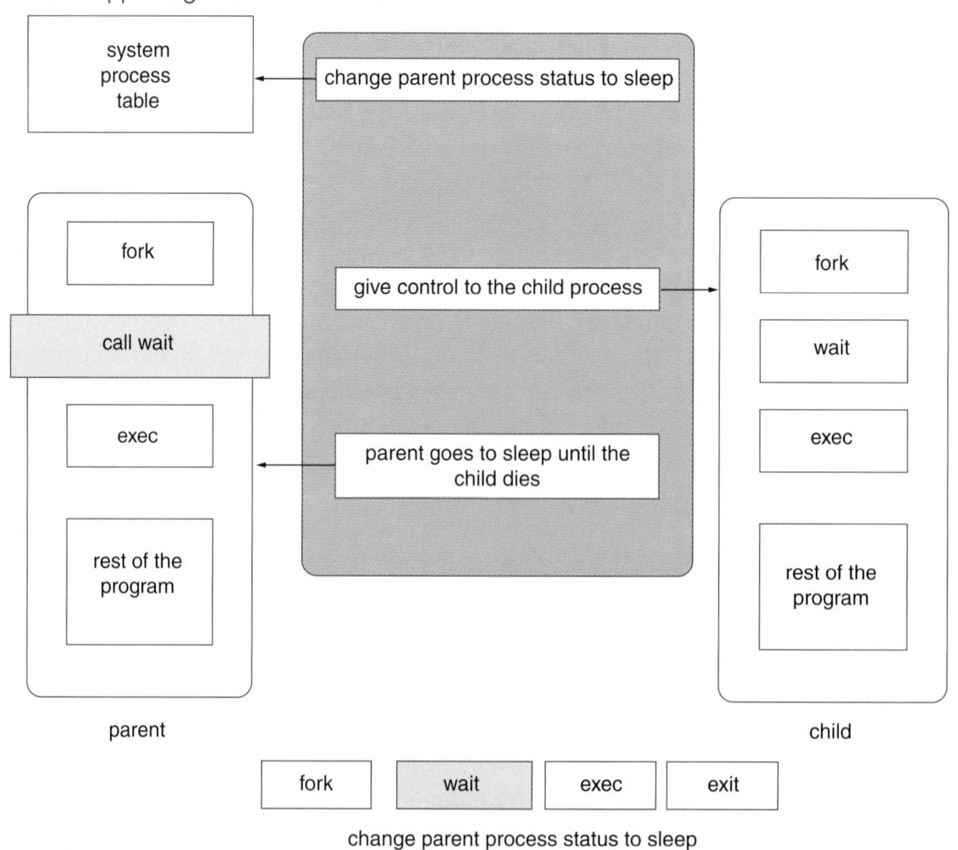

Figure 8.17
Events Happening When **exec** Is Called

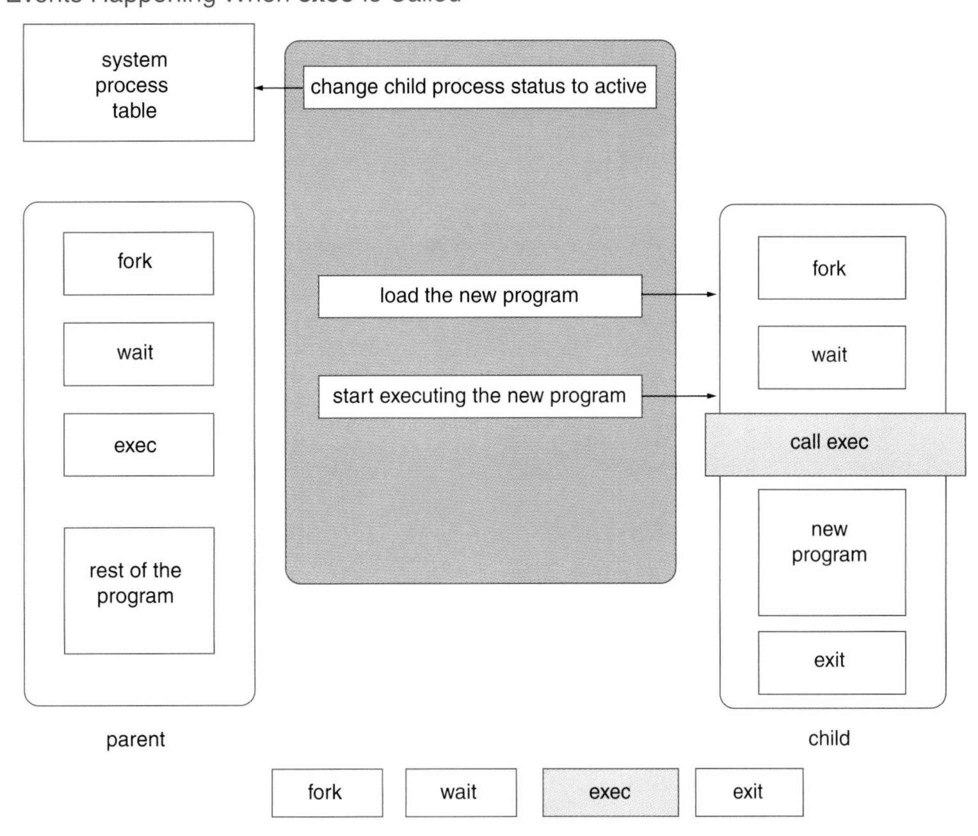

Figure 8.18
Events Happening When Child Calls **exit**

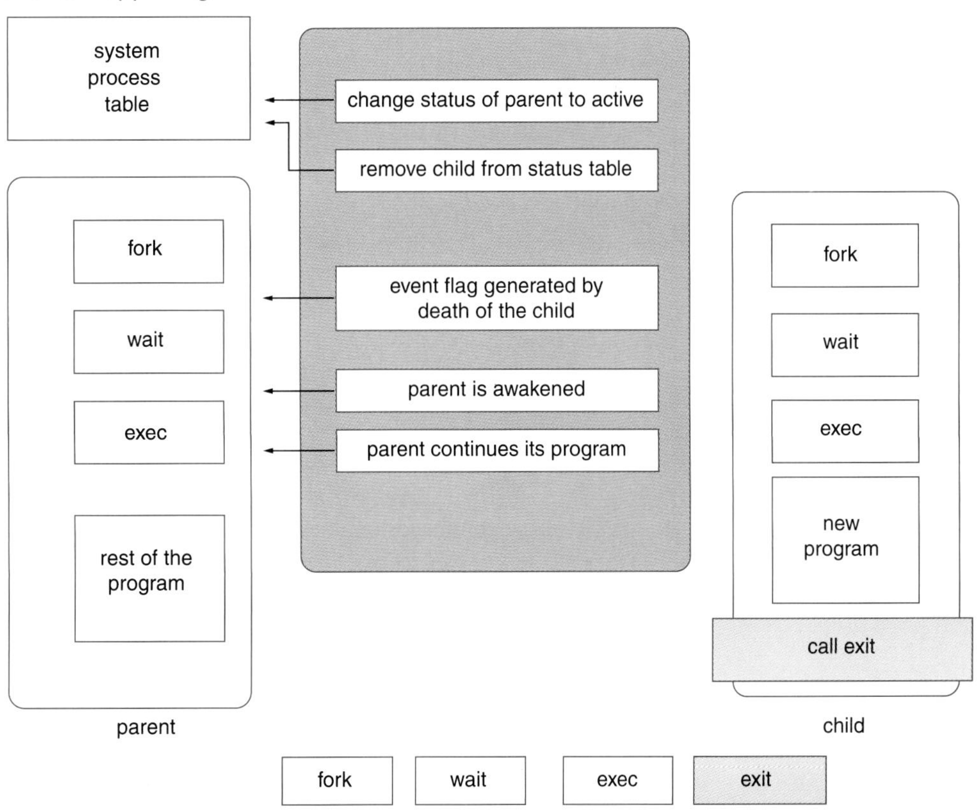

COMMAND SUMMARY

The following UNIX commands have been discussed in this chapter.

alias

This command creates aliases (names) for commands.

echo

This command displays (echoes) its arguments on the output device.

Escape Character	Meaning
\a	Audible alert (bell)
\b	Backspace
\c	Inhibit the terminating newline
\f	Form feed
\n	Carriage return and a line feed (newline)
\r	Carriage return without the line feed
\t	Horizontal tab
\v	Vertical tab

export

This command makes a specified list of variables available to other shells.

fc

This command provides capability to list, edit and re-execute commands that previously were entered and were saved in the history file.

Option	Operation
-l	Lists the commands, with each command preceded by the command number.
-n	Suppresses command numbers when listing with -l.
-r	Reverses the order of the commands listed with -l.
-s	Re-executes the command without invoking an editor.

grep (Global Regular Expression Print)

This command searches for a specified pattern in file(s). If the specified pattern is found, the line containing the pattern is displayed on your terminal.

Option		Operation
UNIX	Linux Alternative	
-c	**--count**	Displays only the count of the matching lines in each file that contains the match.
-i	**--ignore-case**	Ignores the distinction between lower and uppercase letters in the search pattern.
-l	**--files-with-matches**	Displays the names of the files with one or more matching lines, not the lines themselves.
-n	**--line-number**	Displays a line number before each output line.
-v	**--revert-match**	Displays only those lines that do not match the pattern.
	--help	Displays help page and exits.
	--version	Displays version information and exits.

history

This command is a Korn and Bourne Again shell feature that keeps a list of all the commands you enter during your sessions.

kill

This command terminates an unwanted or unruly process. You have to specify the process ID number. The process ID 0 kills all programs associated with your terminal.

nohup

This command prevents the termination of the background process when you log out.

ps (process status)

This command displays the process ID of the programs associated with your terminal.

Option	Operation
-a	Displays the status of all the active processes, not just the user's.
-f	Displays a full list of information, including the full command line.

r *(redo)*

This command is a Korn shell command that repeats the last command or commands from the history file.

set

This command displays the environmental/shell variables on the output device. The command **unset** removes the unwanted variables.

sleep

This command causes the process to go to sleep (wait) for the specified time in seconds.

sort

This command sorts text file(s) in different orders.

Option	Operation
-b	Ignores leading blanks.
-d	Uses dictionary order for sorting. Ignores punctuation and control characters.
-f	Ignores the distinction between lowercase and uppercase letters.
-n	Numbers are sorted by their arithmetic values.
-o	Stores the output in a specified file.
-r	Reverses the order of the sort from ascending to descending.

tee

This command splits the output. One copy is displayed on your terminal, the output device, and another copy is saved in a file.

Option	Operation
-a	Appends output to a file without overwriting an existing file.
-i	Ignores interrupts; does not respond to the interrupt signals.

REVIEW EXERCISES

1. What are the major functions of the shell?
2. What is the name of your system shell program, and where is it stored?
3. What are the metacharacters? How does the shell interpret them?
4. What are the quoting characters?
5. What are the shell variables?
6. What is the command to display the environment/shell variables?
7. What is the command to remove a variable?
8. Name some of the environment/standard variables.
9. What are the variables, and what role do they play?
10. How do you run a program in the background?
11. How do you terminate a background process?
12. What is the process ID number, and how do you know the process ID of a particular process?
13. What is the pipe operator, and what does it do?
14. How do you prevent termination of your background process after you log off?
15. What is the command for searching for a specified pattern in a file?
16. How do you delay the execution of a process?
17. What is the operator that groups the commands together?
18. What is the startup file?
19. What is the `.profile` file, and what is the `profile` file?
20. What are the parent and child in reference to UNIX process management?
21. What is a process?
22. What is the command to activate the command line editor?
23. What variable is set to change the size of the `history` file?
24. How do you make your `history` file start from event 1?
25. Which file does the Korn shell read at startup?
26. What is the command to repeat your last command?
27. What is the command to repeat event number 105 from your `history` file?
28. What is the command to set an alias for a command name?
29. What is the command to export a list of variables to the other shells?
30. What is the command to list aliases?
31. What is the command to display a list of the files in your directory and save a copy in another file?
32. What is the command to obtain a detailed description of the **alias** command?
33. What is the effect of setting the shell option **noclobber**?
34. What is the effect of setting the shell option **ignoreeof**?

Terminal Session

In this terminal session, you will practice the commands explained in this chapter. The following exercises are only some suggestions of how to use the commands. Use your own examples, and devise different scenarios to master the use of these commands.

1. Use the **echo** command to produce the following outputs:
   a. Hello There
   b. Hello
      There
   c. "Hello There"
   d. These are some of the metacharacters:
      ? * [ ] & ( ) ; ><
   e. Filename: file? Option: all
2. Use the **echo** command and other commands to produce the following outputs:
   a. Display the contents of your current directory. Have a header that shows a short prompt and the current date and time before listing your directory.
   b. Show the message "I woke up" with a two-minute time delay.
3. Change your primary prompt sign.
4. Create a variable called *name* and store your first and last names in it.
5. Display the contents of the variable name.
6. Check whether you have a `.profile` file in your HOME directory.
7. Create a `.profile` file or modify your existing one to produce the following output each time you log in:
```
Hello there
I am at your service David Brown
Current Date and Time: [the current date and time]
Next Command:
```
8. Create a background process, check its process ID, and then terminate it.
9. Create a background process. Use the **nohup** command to prevent the termination of the background process.
10. Create a phone list. Let's say you gather the names and phone numbers of ten of your classmates. Use the **sort** command to sort this list in different orders: by first name, by last name, by phone number, in reverse order, and so on.
11. Use **grep** and its options to find a particular name in your phone list.
12. Use the **kill** command to log off.
13. If your shell is the Korn shell (*ksh* or *bash*), set up the following variables and practice the *ksh* commands:
    a. Set the `history` file size to 50.
    b. Activate the command line editor.
    c. Use the built-in vi editor commands to access commands in your `history` file.
    d. Use the built-in vi editor commands to edit/change the command line.

e. Repeat your last command.
f. Display the list of your previous commands.
g. Repeat the first command in the `history` file.
h. Start a new `history` file.
i. Create an alias for the delete command with the confirmation option (**rm -i**).
j. Create a `.kshrc` file that contains the setup for *ksh* variables and aliases.

14. Change your prompt to show the command number.
15. Change your prompt to show the name of the shell.
16. Create an alias called **ls** for **ls -l**.
17. Create an alias called **rm** for **rm -i**.
18. Set the online editor shell option. Check it to see if it works.
19. Set the **noclobber** shell option. Check it to see if it works.
20. Set the **ignoreeof** shell option. Check it to see if it works.

CHAPTER 9

UNIX Communication

This chapter concentrates on the UNIX communication utilities. It describes the commands available for communicating with other users on the system, reading the news about the system, and broadcasting messages to all users. It explains the UNIX electronic mail (e-mail) facilities and shows the commands and options available. This chapter also describes how the shell and other variables affect your e-mail environment, and it shows you how to make a startup file that customizes use of the e-mail utilities.

In This Chapter

9.1 WAYS TO COMMUNICATE

UNIX provides an array of commands and capabilities for communicating with other users. You can have a simple interactive communication with another user by sending and receiving mail through the mail delivery system or you can broadcast messages to everyone on the system.

Be sure to follow some basic guidelines for communication with other users in the system:

- Be polite; do not use profanity.
- Think before sending. Do not send mail that you may regret later.
- Save a copy of all your outgoing mail.

9.1.1 Using Two-Way Communication: The write Command

You can use the **write** command to communicate with another user. This communication is interactive, from your terminal to another terminal, so the receiving terminal must be a logged-on user. The message you send appears on the receiving user's screen. Then that user can send you a reply by issuing the **write** command from his or her terminal. Using the **write** command, two users can effectively have a conversation through their terminals.

Let's follow an example step by step to see how **write** works. Suppose your user ID is `david`, and you want to chat with Daniel, whose user ID is `daniel`.

Type **write daniel** and press [Return].

If Daniel is not logged on, you see the message

```
daniel not logged on.
```

If Daniel is logged on, he sees a message similar to this on his screen:

```
Message from david on (tty06) [Thu Nov 9:30:30]
```

On your terminal, the cursor is placed on the next line, and the system waits for you to type your message. Your message may contain many lines, and each line you type is transmitted to Daniel when you press [Return]. You signal the end of your message by pressing the [Ctrl-d] key at the beginning of an empty line. This terminates your **write** and sends an EOT (end of transmission) message to Daniel.

Figure 9.1 shows two screens, depicting a typical conversation. The top screen is David's terminal and the bottom one is Daniel's.

To use the **write** *command, you must know the user ID of the person with whom you want to communicate. You use the* **who** *command (discussed in Chapter 3) to obtain the user IDs of the logged-on users.*

Daniel can reply by using the **write** command from his terminal, but he does not have to wait for you to finish your message. When he sees the initial message that you are writing to him, he can send you messages while you send him messages if he types **write david** and presses [Return].

See page x for an explanation of the icons used to highlight information in this chapter.

Figure 9.1
Typical Screen Conversation: The Top Screen is David's and the Bottom Screen is Daniel's

```
$write daniel [Return]
Hello Dan [Return]
Is today's meeting still on? [Return]
[Ctrl-d]
<EOT>
$
```

```
Message from david on (tty06) [Thu Nov 9:30:30] ...
Hello Dan
Is today's meeting still on?
<EOT>
$
```

With **write** simultaneously active on both terminals, you and Daniel can carry on a two-way conversation. Sometimes this two-way exchange becomes confusing, so it is useful to establish a protocol for using **write**. The common protocol for UNIX users is to end message lines with the character *o* (for over) to inform the receiving party that a message is finished and you are (possibly) waiting for a reply. When you intend to end the conversation, type **oo** (for over and out).

If you are receiving a **write** message from another user, the message appears on your terminal, regardless of what you are doing. If you are using the vi editor and are in the middle of an editing job, the write message appears on the screen where the cursor is. But don't be alarmed. This is a terminal-to-terminal communication, and what **write** produces does not damage your editing file. It simply overwrites information on your screen, and you can continue with your editing or whatever other job you were doing.

Nevertheless, receiving messages while you are concentrating on a job is inconvenient, not to mention the mess it makes on your screen. You can prevent your terminal from accepting messages coming from the **write** command.

9.1.2 Inhibiting Messages: The mesg Command

You can use the **mesg** command as a toggle to stop receiving messages from the **write** command or to reactivate receiving messages. **mesg** without an argument shows the current status of your terminal in this respect.

The following command sequence shows you how to protect yourself from annoying messages:

```
$ mesg [Return] . . . . . . Check status of your terminal.
is y . . . . . . . . . . .  It is set to YES, accepting messages.
$ mesg n [Return] . . . . . Set it to NO, and stop receiving messages.
$ mesg [Return]   . . . . . Check again.
is n . . . . . . . . . . .  Now it is set to NO.
$_ . . . . . . . . . . . .  Return to the prompt.
```

To practice the communication commands, you usually want to have another user to participate in the exercises. However, you can practice most of these commands by yourself. You can send and receive messages using your own user ID.

9.1.3 Displaying News Items: The news Command

You can use the **news** command to find out what is happening in the system. **news** gets its information from the system directory where the news files are placed, usually in /usr/news. Without any options, **news** displays all the files you have not seen from the news directory. It refers to and updates a file called .news_time in your HOME directory. This file is created in your HOME directory the first time you use the **news** command and remains an empty file. The **news** command uses its access time to determine the last time you gave the **news** command.

1. *You press the interrupt key (usually [Del]) to stop displaying one news item and continue with the next item.*
2. *You press the interrupt key twice to quit (terminate) the* **news** *command.*

To check the latest news, type **news** and press [Return]. Figure 9.2 shows some sample news items.

Figure 9.2
The **news** Command

```
$ news [Return]
david  (root) Mon Nov     28    14:14:14  2005
   Let's congratulate david; he got his B.S. degree.
   Friday night party on second floor. Be there!

books (root) Mon Nov      28    14:14:14  2005
   Our technical library is growing.
   New set of UNIX books is now available.
$_
```

Each news item has a header that shows the filename, file owner, and the time that the file was placed in the news directory.

news Options

The **news** options are summarized in Table 9.1. The options do not update your .news_time file.

The following command sequences show how the **news** command options work:

- List the current news items, using the **-n** option: type **news -n** and press [Return]. UNIX shows only the headers of the files that contain the news items. The header shows the filename, the owner of the file, and the time it was created. Figure 9.3 shows the output of the command.

Table 9.1
The **news** Command Options

Option	Operation
-a	Displays all the news items, old or new files.
-n	Lists only the names of the files (headers).
-s	Displays the number of the current news items.

Figure 9.3
The **news** Command and the **-n** Option

```
$ news -n [Return]
   david  (root) Mon Nov  28  14:14:14  2005
   books  (root) Mon Nov  28  14:14:14  2005
$_
```

□ To display a specific news item, type **news david** and press [Return], for example. Figure 9.4 shows the output of the sample news item.

Figure 9.4
The **news** Command with Specified News Item

```
$ news david [Return]
david  (root) Mon Nov     28  14:14:14  2005
   Let's congratulate david; he got his B.S. degree.
   Friday night party on second floor. Be there!
$_
```

The name of a news item is its filename in the `news` *directory.*

9.1.4 Broadcasting Messages: The wall Command

You can use the **wall** (write all) command to send messages to all currently logged on users. The **wall** command reads from the keyboard (standard input) until you enter [Ctrl-d] at the beginning of a blank line to signal the end of the message. The **wall** executable file is usually placed in the `/etc` directory and is not defined in the *PATH* standard variable—which means you have to type the full pathname to invoke it.

The **wall** command is usually used by the system administrator to warn users of imminent events. You may not have access to it.

Assuming your login name is `david` (and you have access to **wall**), send a message to all users:

□ Type **/etc/wall** and press [Return].

□ Type **Alert ...** and press [Return].

□ Type **Lab will be closed in 5 minutes. Time to log out.** and press [Return].

□ Press [Ctrl-d]. The system responds as shown in Figure 9.5.

Figure 9.5
Invoking the **wall** Command

```
$ /etc/wall [Return]
   Alert ... [Return]
   Lab will be closed in 5 minutes. Time to log out. [Return]
[Ctrl-d]
$_

Broadcast message from david
   Alert ...
   Lab will be closed in 5 minutes. Time to log out.
```

1. *The message is also sent to its sender, so you see your own broadcast message.*
2. *The* **wall** *message you send is not received by currently logged-on users who set* **mesg** *to* **n**.
3. *The system administrator can override access denial.*
4. *Your messages carry your user ID; thus you cannot send an anonymous message.*

9.1.5 Using Two-Way Communication: The talk Command

You can use the **talk** command to communicate with another logged-on user. This command is similar to the **write** command. When you enter **talk**, your display will be divided into two sections. The top section displays what you type and the bottom section displays what the other person types. For example, if you type **talk daniel** and press the [Return] key, your screen splits into two sections, and the following message appears on your screen:

```
[waiting for your party to respond]
```

and the following message appears on Daniel's screen:

```
Message from Talk_daemon@xyz at 111:22 ..
talk: Connection requested by david@xyz
talk: Respond with: talk david@xyz
```

and Daniel responds by typing

```
talk david@xyz
```

Now, Daniel's screen is divided into two sections, and the conversation begins.

To end the session, either of the parties involved can press [Ctrl-c] and the following message is displayed on both terminals.

```
[Connection closing. Exiting]
```

The following message is displayed on your terminal if Daniel has used the **mesg** command to block messages:

```
[Your party is refusing messages]
```

The following message is displayed on your terminal if Daniel is not logged on:

```
[Your party is not logged on]
```

Once more, if you are David and want to communicate with Daniel, Figure 9.6 shows the screen conversation.

Figure 9.6
Screen Conversation: The Top Screen is David's; the Bottom One is Daniel's

```
[Connection established]                    david's screen
Hi Dan!
Is today's meeting still on?
------------------------------------------------------------
Hi Dave
Yes. Be there!
Bye.
```

```
[Connection established]                    daniel's screen
Hi Dave!
Yes. Be there!
Bye.
------------------------------------------------------------
Hi Dan!
Is today's meeting still on?
```

9.2 ELECTRONIC MAIL

Electronic mail (*e-mail*) is an essential part of the contemporary office environment. E-mail gives you the capability to send and receive messages, memos, and other documents to and from other users. The main difference between sending mail using the e-mail service and using the **write** command is that with the **write** command you see only the messages sent to you if you are logged on. But with e-mail, your mail is automatically kept for you until you issue the command to read it. E-mail service is more convenient and is faster than conventional mail service, and it does not interrupt the receiving party the way phone calls do.

Under UNIX, either the **mail** or **mailx** command can be used to send or read e-mail. The **mailx** utility is based on Berkeley UNIX **mail** and has more powerful features than **mail**, enabling you to manipulate (review, store, dispose of, etc.) your mail easily and efficiently. **mailx** is the e-mail command discussed in this book. It has a large number of features and options; using some of them requires advanced UNIX experience. In this chapter, we describe enough to make you feel comfortable using it and interested in looking for more information about it.

Where do you find more information? Well, how about the **man** *command? (In case you have forgotten,* **man** *was discussed in Chapter 3.)*

You use the **mailx** command to send mail to other users or read mail that is sent to you. The **mailx** operation involves a number of files; the way it appears and functions depends on the environment variables that are set up in files, and it needs files for storing your mail.

9.2.1 Using Mailboxes

In the UNIX mail system, you have two kinds of mailboxes: a system mailbox and a private mailbox.

Your System Mailbox

Every user of the system has a *mailbox*, which is a file with the same name as your login name (user ID). This file is typically stored in `/usr/mail`. Mail sent to you is stored in this file, and when you read a message, **mailx** reads from your mailbox. If your login name is `david`, the full pathname to your mailbox could be `/usr/mail/students/david`.

You can use the **set** command (discussed in Chapter 8) to see your system mailbox pathname. The variable *MAIL* is set to the filename that receives your mail.

Your Private Mailbox: The *mbox* File

After you read your mail, **mailx** automatically appends a copy of it to a file called `mbox` in your HOME directory. If an `mbox` file does not exist, then **mailx** creates one in your HOME directory the first time you read your mail. This file contains mail that you have read but that `$HOME/mbox` has not deleted or saved elsewhere. The variable *MBOX* controls the filename. The default value is `HOME/mbox`. For example, the following command changes the default setting and sets up your private mailbox in the `email` directory.

```
MBOX=$HOME/EMAIL/mbox
```

The explicit saving of a message or using the x (exit) command to exit mailx disables the automatic saving of your messages.

The Customizable mailx Environment

You can customize your **mailx** environment by setting up appropriate variables in two startup files: the `mail.rc` file in the system directory and the `.mailrc` file in your HOME directory.

When you call **mailx**, it first checks for a startup file called `mail.rc`. The full pathname to this file is similar to the following:

```
/usr/share/lib/mailx/mail.rc
```

This file is usually created and maintained by the system administrator. Variables set in this file are applicable to all the system users.

The second file that **mailx** looks for is a file called `.mailrc` in your HOME directory. You can change the **mailx** environment that the system administrator has set up in the `mail.rc` file by setting variables in your .`mailrc` file. This file is not necessary, and **mailx** works fine without it, as long as you are happy with the system administrator's arrangement. Ways to customize your **mailx** environment are discussed in more detail in Section 9.5.

9.2.2 Sending Mail

In order to send mail to another person, you must know that person's login name. For example, if you want to send mail to a user identified by the login name `daniel`, you type **mailx daniel** and press [Return].

By default, input to **mailx** (your message) comes from the keyboard (standard input). Depending on how your system environment variables are set, **mailx** may show the **Subject:** prompt. If it does, you type the subject for your message, and **mailx** changes to the *input mode* and waits for you to enter the rest of your message. You signal the end of your message by pressing [Ctrl-d] at the beginning of a line. **mailx** shows **<EOT>** (for end of transmission), and your message is transmitted.

While in the input mode, **mailx** provides you with a large number of commands, enabling you to compose your message with ease and efficiency. All input mode commands start with a tilde (~), and they are called *tilde escape commands.* (The tilde escape commands are explained later in this chapter.)

To send a message to Daniel (login name `daniel`), do the following:

```
$ mailx daniel [Return] . . . . Send a message to daniel.
Subject: meeting [Return] . . . Enter the subject.
Hi, Dan [Return]
Let me know if tomorrow's meeting is still on. [Return]
Dave [Return]
[Ctrl-d] . . . . . . . . . . . Signal the end of the message.
EOT . . . . . . . . . . . . . . mailx shows end of transmission.
$_ . . . . . . . . . . . . . . Return to the prompt.
```

1. *The subject field is optional. Just press [Return] to skip the* **Subject:** *prompt.*
2. *Signal the end of a message by pressing [Ctrl-d] at the beginning of a blank line.*
3. *Mail is delivered to the other user's mailbox regardless of whether he or she is logged on.*
4. *Recipients are informed that they have mail as soon as they log in. The following message appears on their terminal:*

   ```
   You have mail
   ```

9.2.3 Reading Mail

In order to read your mail, you type **mailx** with no argument. If you have mail in your mailbox, then **mailx** shows two lines of information followed by a numbered list of headers of messages in your mailbox and then the **mailx** prompt, which by default is a question mark. At this point, **mailx** is in command mode, and you can issue commands to delete, save, or reply to messages. You press **q** at the **?** prompt to exit **mailx**.

The list of headers consists of one line for each mail item in your mailbox. The format is as follows:

```
> status message # sender date lines/characters subject
```

Each field in the header line conveys certain information about your mail:

- The > indicates that the message is the current message.
- The *status* is **N** if the message is new. This means that you have not read this mail.

- The *status* is **U** (unread) if the message is not new. That means you have seen the message header before, but you have not read the mail itself yet.
- The *message #* indicates the sequence number of the mail in your mailbox.
- The *sender* is the login name of the person who sent you the mail.
- The *date* shows the date and time that mail arrived in your mailbox.
- The *line/characters* shows the size of your mail by the number of lines and number of characters.

Suppose you are Daniel and you have just logged in and you want to read your mail.

□ The system informs you that you have mail:

```
You have mail
```

□ To read your mail, type **mailx** and press [Return]; UNIX responds:

```
mailx version 4.0 Type ? for help.
"/usr/students/mail/daniel": 1 message 1 new
> N  1  david .... Thu Nov 28 14:14 8:126 meeting
```

1. *The first header line shows your* **mailx** *version number and informs you that you can press* **?** *to get help.*
2. *The second header line shows* `/usr/mail/daniel`*, your system mailbox, followed by the number and status of your messages. In this case, you have one message, and* **N** *indicates that this is the first time you are reading it.*

The **?** prompt shows that **mailx** is in command mode. You specify the mail you want to read by typing its associated message number. You can also press [Return] to start reading from the current mail (indicated by the > sign on the header) and continue reading your mail in sequence by pressing [Return] after the **?** prompt.

```
? . . . . . . . . . mailx is in command mode.
? 1 [Return] . . . Display message 1, the only message in your mailbox.
Message 1:
From: david  Mon, 28 Nov 01 14:14 EDT  2005
To: daniel
Subject: meeting
Status: R
Hi, Dan
Let me know if tomorrow's meeting is still on.
Dave
? . . . . . . . . . Ready for the next command.
? q [Return] . . . Exit mailx.
Saved 1 message in /usr/students/david/mbox
$_  . . . . . . . . Return to the shell.
```

When you use **q** *to quit from* **mailx**, *it saves a copy of the mail that you have read in* `mbox`, *your private mailbox in your HOME directory. (Thus, at this point, your system mailbox is empty.)*

Suppose you want to read your mail again. Type **mailx** and press [Return]; UNIX responds:

```
No mail for daniel
```

Check what you have in mbox.

```
$ cat mbox [Return] . . . . . . Check what you have in your mbox.
From: david  Mon, 28 Nov 01 14:14 EDT  2005
To: daniel
Subject: meeting
Status: R
Hi, Dan
Let me know if tomorrow's meeting is still on.
Dave
```

As you expected, mbox contains a copy of your mail.

9.2.4 Exiting mailx: The q and x Commands

You can exit **mailx** by typing the **q** (*quit*) or **x** (*exit*) command at the **?** prompt. Although both commands cause you to exit from the **mailx**, they do so in a different manner.

The **q** command causes the automatic removal of the mail that you have read from your system mailbox. By default, a copy of the removed mail is kept in your private mailbox (any filename assigned to the MAIL variable).

The **x** command does not remove the mail you have read from your system mailbox. In fact, when you use **x**, nothing changes in your mailbox. Even deleted messages remain intact.

mailx Options Table 9.2 summarizes the **mailx** options. These options are used in the command line when you invoke **mailx** for reading or sending mail. The following command sequences show the use of the **mailx** options:

Table 9.2
The **mailx** Command Options

Option	Operation
-f *filename*	Reads mail from the specified *filename* instead of the system mailbox. If no file is specified, it reads from mbox.
-H	Displays a list of the message headers.
-s *subject*	Sets the subject field to the string *subject*.

```
$ mailx -H [Return]   . . . . . Display the message headers only.
N 1 daniel         Thu ESP 30  12:26 6/103  Room
N 2 susan          Thu Sep 30  12:30 6/107  Project
N 3 marie          Thu Sep 30  13:30 6/70   Welcome
$_ . . . . . . . . . . . . . . . Back to the shell.
```

Type **mailx -f mymail** and press [Return] to read mail from the specified file mymail instead of your system mailbox; UNIX responds:

```
/usr/students/daniel/mymail: No such file or directory
```

The **mailx** by default reads mail from your system mailbox. With the **-f** option, it reads your mail from a specified file, such as your old mail files. In this case, you specified `mymail`, and the message shows `mymail` is not in your current directory.

If you type **mailx -f** and press [Return], with no filename specified, UNIX defaults to your private mailbox, so you see a display like the following:

```
mailx version 4.0 Type ? for help.
  "/usr/students/mail/daniel": 1 message 1 new
> N  1  david Thu Nov 28 14:14 17/32 meeting
```

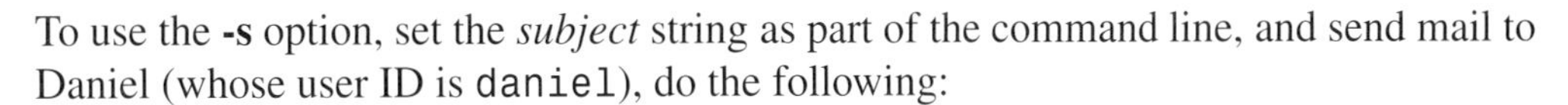

To use the **-s** option, set the *subject* string as part of the command line, and send mail to Daniel (whose user ID is `daniel`), do the following:

`$ mailx -s meeting daniel [Return]`	Send mail to `daniel`; set the subject field to the string *meeting*.
`_`	Compose your message to Daniel.
`[Ctrl-d]`	End your message.
`EOT`	**mailx** shows end of transmission.
`$_`	Return to the shell.

When Daniel reads your message, the subject field shows **Subject: meeting**.

Use quotation marks around the subject string if it contains spaces.

9.3 mailx INPUT MODE

While **mailx** is in the *input mode* (composing the mail to be sent), a variety of commands are at your disposal. These commands all start with the tilde (~); that lets you temporarily escape from the input mode and issue certain commands, which is why they are referred to as *tilde escape commands.* Table 9.3 summarizes some of these commands.

Some of these commands are quite important. Imagine you want to write a message that contains more than just a few lines. Using the primitive **mailx** editor is cumbersome and just not up to the job. Instead of using it, you can invoke the vi editor, compose your message using all the ease and power of vi, and then, when you have finished composing your message, exit vi and return to the **mailx** input mode. Then you can give other commands or send the message.

1. **Tilde escape commands are applicable only when mailx is in the input mode.**
2. **All tilde commands must be entered at the beginning of a line.**

The following command sequences show the use of the tilde escape commands while **mailx** is in the input mode. Assume your login name is `david` and you are sending mail to yourself.

`$ mailx -s "Just a Test" david [Return]` . .	Send mail to yourself.

mailx is invoked with the **-s** option. The quotation marks are necessary because of the embedded spaces in the specified *subject* string.

Table 9.3
The **mailx** Tilde Escape Commands

Command	Operation
~?	Displays a list of all tilde escape commands.
~! *command*	Lets you invoke the specified shell *command* while composing your message.
~e	Invokes an editor. The editor to be used is defined in the mail variable called *EDITOR*. vi is the default.
~p	Displays the message currently being composed.
~q	Quits input mode. Saves your partially composed message in the file called `dead.letter`.
~r *filename*	Reads the file named *filename* and adds its contents to your message.
~< *filename*	Reads the file named *filename* (using the redirection operator) and adds its contents to your message.
~<! *command*	Executes the specified command and places its output into the message.
~v	Invokes the default editor, the vi editor, or uses the value of the mail variable *VISUAL*, which can be set up for other editors.
~w *filename*	Writes the message currently being composed to the specified *filename*.

```
$_ . . . . . . . . . . . . . . . mailx is in the input mode.
~! date [Return] . . . . . . . . Invoke the date command.
Mon, Nov 28  16:16 EDT 2005
_  . . . . . . . . . . . . . . . mailx is ready for input.
```

At this point you are using `~!` and executing the **date** command. You can execute any command you wish. The output of the command does not become part of the message you are composing.

```
$~< ! date [Return] . . . . . . . Invoke the date command and redirect the output of the
                                  date command to be included in your message.
"date" 1/29 . . . . . . . . . . . Feedback message is given.
_ . . . . . . . . . . . . . . . . Ready.
```

The feedback message indicates that one line consisting of 29 characters (the output of the **date** command) is added to your text.

```
This is a test message to explore mailx. [Return]
$~v [Return] . . . . . . . . . . Now use the vi editor to compose the
                                 rest of your message.
```

At this point, you have invoked the vi editor and your partially composed message is the input file to the vi editor.

```
Mon, Nov 28  16:16 EDT 2005
This is a test message to explore mailx capabilities.
~
~
~
"/tmp/Re26485"  2 lines, 69 characters
```

Now all the power and flexibility of the vi editor are at your disposal. You can delete, modify, or save your text. You can execute commands or import another file and continue composing your message.

```
Mon, Nov 28  16:16 EDT 2005
This is a test message to explore mailx capabilities.
This message is composed using the vi editor.
:wq [Return] . . . . . . . . . .  Exit vi.
3 lines, 115 characters   . . .  Get vi feedback.
(continue) . . . . . . . . . . .  Feedback message.
_  . . . . . . . . . . . . . . .  You are back to mailx input mode.
~w first.mail [Return]  . . . .  Save your mail in a file called
                                  first.mail.
"first.mail" 3/115  . . . . . .  Get feedback message.
```

The feedback message indicates the size of `first.mail`: It contains 3 lines and 115 characters.

The **~w** is used to save the currently composed message in the specified file `first.mail`. This is a good habit to practice so that you have a copy of your transmitted messages. If you set the **mailx** record variable, then your outgoing mail is automatically saved.

```
~q [Return]   . . . . . . . . .  Quit mailx input mode.
$_  . . . . . . . . . . . . . .  Return to the shell.
```

Using ~ **q** to quit **mailx** input mode also saves your partially composed message in a file called `dead.letter` in your HOME directory (or any filename assigned to the DEAD variable).

Let's start again and complete the sending of your mail. At this point the scenario is this: you are `david` and you want to send mail to yourself. You have a copy of your message in a file called `first.mail` (you used ~ **w** to save it) and another copy in `dead.letter` (you used ~ **q** to exit input mode).

Send `first.mail` to `david` using the input redirection sign (<) on the command line.

```
$ mailx david < first.mail [Return] . . . .  Send mail.
$_  . . . . . . . . . . . . . . . . . . . .  Done.
```

The < sign directs the shell to pass the specified filename (`first.mail`) as input to the **mailx** command.

You can specify more than one input file.

By default, the Bourne shell checks every 10 minutes for your new mail, so you have to wait a while before you can read the mail you just sent to yourself. When there is new mail in your mailbox, UNIX displays the message *you have mail* before the next prompt.

To send `first.mail` to `david` using the tilde escape command, do the following:

```
$ mailx david [Return]  . . . .  Send mail to yourself.
Subject:  . . . . . . . . . . .  Ready for you to type your message.
~< first.mail [Return]  . . . .  Read in the contents of the file
                                  first.mail.
"first.mail" 3/115  . . . . . .  Get feedback message.
```

```
[Ctrl-d] . . . . . . . . . . . . Transmit it.
EOT . . . . . . . . . . . . . . mailx shows EOT.
$_ . . . . . . . . . . . . . . . Return to the shell.
```

The ~ < reads in the specified file, in this case `first.mail`*, and adds its contents to the message you are currently composing. You can also use the ~* **r** *command to achieve the same results.*

Get your partially composed message that **mailx** saved in the `dead.letter` file, complete the message, and send it to `david`:

```
$ mailx david [Return] . . . . . . . . . . . . . Send message to david.
Subject: . . . . . . . . . . . . . . . . . . . . Ready to go.
~r dead.letter [Return] . . . . . . . . . . . . . Read in (import) dead.letter file.
~v [Return] . . . . . . . . . . . . . . . . . . . Invoke the vi editor.
Mon, 28 Nov 01 16:16 EDT 2005
  This is a test message to explore mailx capabilities.
  This message is composed using the vi editor.
  ~
  ~
  "/tmp/Re265" 3 lines and 115 characters . . . . . . The vi editor feedback message.
```

Let's assume you want to add a few lines to your message:

```
Mon, 28 Nov 01 16:16 EDT 2005 . . . . . . . . . Complete the message.
This is a test message to explore mailx capabilities.
This message is composed using the vi editor.
This is the first time I am using e-mail. Maybe I should save
this text, get a copy of it, and frame it!
~
~
:wq [Return] . . . . . . . . . . . . . . . . . . Save and quit.
(continue) . . . . . . . . . . . . . . . . . . . Get feedback message.
```

At this point your message is not displayed, but you can see what you have composed before sending it by typing **~p** and pressing [Return]; UNIX displays the whole message one page at a time:

```
~p [Return] . . . . . . . . . Display the composed message.
Mon, 28 Nov 01 16:16 EDT 2005
This is a test message to explore mailx capabilities.
This message is composed using the vi editor.
This is the first time I am using e-mail. Maybe I should
save this text, get a copy of it, and frame it!
[Ctrl-d] . . . . . . . . . . . Indicate the end of your message.
EOT . . . . . . . . . . . . . . This is the end of transmission.
$_ . . . . . . . . . . . . . . . Return to the shell.
```

9.3.1 Mailing Existing Files

You do not have to compose your messages using the **mailx** editor at all. Maybe you have a memo already written that you want to send to another user. In that case, you use the shell input redirection operator to redirect the **mailx** input from the default input device (the keyboard) to an existing file.

Send a file called memo to Daniel, whose user ID is daniel.

```
$ mailx daniel < memo [Return]  . . Mail memo to daniel.
$_  . . . . . . . . . . . . . . . . Return to the shell.
```

Assuming that your message is in a file called memo and you want to send it to the user ID daniel, this command will do the job.

9.3.2 Sending Mail to a Group of Users

What if you want to send your memo file to Daniel and a few others? It would be very inconvenient to type the **mailx** command to send the same message to 10 different users. In that case, you specify a list of the user IDs of the people you intend to receive the mail, and **mailx** sends the mail to all of them.

To send memo to user IDs daniel, susan, and emma, type:

```
mailx daniel susan emma < memo [Return]
```

The user IDs are separated by spaces.

If you send mail often to a specific group of people, you can save yourself a lot of typing by defining a name for your list and using the defined name instead of typing the whole list of user IDs every time. To do this, you use the **alias** command, and the format looks like this:

```
alias [name] [userID01] [userID02] [userID03] [userID04]
[userID05] [userID06]
```

where *name* is the name you want to type when you want to send mail to all the people on this list.

For example, if you type:

```
alias friends daniel david gabe emma [Return]
```

UNIX assigns the name friends to this list of user IDs. Now, instead of typing all those user IDs, you can just type **friends**. If you place **alias** commands like this in your .mailrc file, they become part of your **mailx** environment and you do not need to assign them every time you want to use **mailx**.

Assuming your friends' user IDs are assigned to the name friends, mail your memo to your friends by typing **mailx friends < memo** and pressing [Return].

9.4 mailx COMMAND MODE

When you are reading mail, **mailx** is in command mode, and the question mark prompt means it is waiting for your commands. While **mailx** is in command mode, a large number of commands are at your service, enabling you to copy, save, or delete your

Table 9.4
The **mailx** Commands Available in Command Mode

Command	Operation
!	Lets you execute the shell commands (the shell escape).
cd *directory*	Changes to the specified directory, or to the HOME directory if none is specified.
d	Deletes the specified messages.
f	Displays the headlines of the messages.
q	Exits **mailx** and removes the messages from the system mailbox.
h	Displays active message headers.
m *users*	Sends mail to specified *users*.
R *messages*	Replies to the sender of the *messages*.
r *messages*	Replies to the sender of the *messages* and all the other recipients of the same *messages*.
s *filename*	Saves (appends) the indicated messages to the *filename*.
t *messages*	Displays (types) the specified *messages*.
u *messages*	Undeletes the specified *messages*.
x	Exits **mailx**; does not remove messages from system mailbox.

mail. You can reply to the sender of a message or send mail to a specific user without leaving the command mode. Table 9.4 summarizes some of these commands.

Scenario The command sequences in the following sections show how **mailx** reads your mail and what commands and options are applicable in the command mode. They assume that your user ID is `david`, you have three messages in your system mailbox, and you want to read, display, and delete your mail. The **mailx** command provides you with the capabilities to manipulate your mail in different ways. Your only problem would be the selection of appropriate commands for the job at hand.

9.4.1 Ways to Read/Display Your Mail

The **mailx** command lets you read and display your mail in several different ways: each piece of mail one at a time, as a specified range of messages, or as a specified single message. Generally, regardless of how you plan to read your mail, you will want to display a list of the mail, which you do simply by invoking **mailx**. Pressing [Return] at the **?** prompt shows the current message.

Read/display your mail:

```
$ mailx [Return] . . . . . . Invoke mailx and display a list of the mail.
mailx version 4.0  Type ? for help.
"/usr/student/mail/david": 3 messages 3 new
```

```
> N 1  daniel                Fri Sep 30 12:26 6/103 Room
  N 2  susan                 Fri Sep 30 12:30 6/107 Project
  N 3  marie                 Fri Sep 30 13:30 6/79  Welcome
? [Return] . . . . . . . . Display the current message.
Message 1:
From: daniel Fri Sep 30 12:26 EDT 2005
To: david
Subject: Room
Status: R
Room 707 is reserved for your meetings.
? . . . . . . . . . . . . Ready for the next command.
```

Pressing [Return] at the **?** prompt shows the current message, in this case message 1.

```
? 3 [Return] . . . . . . . Display message 3.
Message 3:
From: marie Fri Sep 30 12:30 EDT 2005
To: david
Subject: Welcome
Status: R
Welcome back!
? . . . . . . . . . . . . Ready for the next command.
? t 1-3 [Return] . . . . . Type messages 1 through 3.
Message 1:
From: daniel Fri Sep 30 12:26 EDT 2005
To: david
Subject: Room
Status: R
Room 707 is reserved for your meetings.
Message 2:
From: susan Fri Sep 30 12:26 EDT 2005
To: david
Subject: Project
Status: R
Your project is in trouble! See me ASAP.
Message 3:
From marie Fri Sep 30 12:30 EDT 2005
To: david
Subject: Welcome
Status: R
Welcome back!
? . . . . . . . . . . . . You are ready for the next command.
```

The **t** (type) command displays messages one after another. The range of messages is indicated by the low and high limit numbers separated by a hyphen. In this case, **t 1-3** means to display message numbers 1, 2, and 3.

```
? t 1 3 [Return] . . . . . Display messages 1 and 3.
Message 1:
From: daniel Fri Sep 30 12:26 EDT 2005
To: david
Subject: Room
Status: R
Room 707 is reserved for your meetings.
```

```
Message 3:
From: marie Fri Sep 30 12:30 EDT 2005
To: david
Subject: Welcome
Status: R
Welcome back!
?_ . . . . . . . . . . . . . You are ready for the next command.
```

*The **t** command also shows any specified mail indicated by the mail sequence numbers. If more than one mail sequence number is indicated, the numbers must be separated by a space. In this case, **t 1 3** means display messages 1 and 3.*

```
? n [Return] . . . . . . . . Show the next message.
At EOF . . . . . . . . . . Get feedback message.
?_ . . . . . . . . . . . . . Prompt returns for the next command.
```

The **n** (next) command shows the next message in your mailbox, just like pressing [Return]. In this case, there is no next message; thus, the *End of the File* message is displayed.

```
? n10 [Return] . . . . . . Show the tenth message.
10: . . . . . . . . . . . invalid message number.
```

You can indicate a message number if you want a specific message to be displayed. In this case, **n10** means to display message 10. But there is no message 10, so the error message is displayed.

```
? f [Return] . . . . . . . . Show the headline of the current message.
. 3 marie Fri Sep 30 12:30 EDT 2005
?_ . . . . . . . . . . . . . The prompt is back.
```

The **f** command shows the headline of your current message, in this case message 3.

```
? x [Return] . . . . . . . . Exit mailx.
$_ . . . . . . . . . . . . . Return to the shell prompt.
```

If you use the **x** command to exit from **mailx**, all your messages remain intact in your mailbox.

9.4.2 Ways to Delete Your Mail

The **mailx** command lets you delete your mail one message at a time, delete all of the messages at one time, delete a specified range of messages, and recover messages you deleted by mistake. Let's look at some examples. The scenario remains the same. You are David, and there are three messages in your system mailbox.

Delete your mail:

```
$ mailx [Return] . . . . . Read your mail.
mailx version 4.0  Type ? for help.
"/usr/student/mail/david": 3 messages 3 new
> N 1  daniel                Fri Sep 30 12:26 6/105  Room
  N 2  susan                 Fri Sep 30 12:30 6/107  Project
  N 3  marie                 Fri Sep 30 13:30 6/79   Welcome
? d [Return] . . . . . . . . Delete the current message.
```

```
? d3 [Return]   . . . . . . Delete message 3.
? h  . . . . . . . . . . . . Show only the headlines.
> N 2  . . . . . . . . . . . susan Fri  Sep 30  12:30  6/107 Project
```

The **h** command displays the headlines of the messages in your mailbox. You have deleted messages 1 and 3, and the remaining message (message 2) is displayed.

```
? u1 [Return] . . . . . . . Undelete message 1.
? u3 [Return] . . . . . . . Undelete message 3.
```

The **u** *command undeletes the specified message, in this case messages 1 and 3.*

```
? h [Return]  . . . . . . . Check to see whether all messages are in
                            your mailbox.
  1 daniel          Fri Sep 30 12:26 6/103 Room
  2 susan           Fri Sep 30 12:30 6/107 Project
  3 marie           Fri Sep 30 13:30 6/79  Welcome
? d 1-3 [Return]  . . . . . Delete messages 1 through 3.
? h [Return]  . . . . . . . Check if they are deleted.
No applicable messages
? u* [Return]   . . . . . . Undelete all the deleted messages.
```

Using the **u** command with the * wild card undeletes all messages in your mailbox.

```
? d/vacation [Return] . . . Delete all messages with the word vacation
                            in their subject field.
No applicable messages
```

Using */string* with the delete command, you can delete mail that has the specified string as part of the subject field. In this case, no message in your mailbox matches the string *vacation*, and so no message is removed.

```
? d daniel [Return] . . . . Delete all messages from daniel.
```

You can specify the sender's user ID with the delete command to delete all the mail sent by the specified person.

```
? x [Return]  . . . . . . . Exit mailx.
$_  . . . . . . . . . . . . Back to the shell prompt.
```

There are many ways to delete unwanted mail. However, when you exit from **mailx** using the **x** command, all messages remain in your mailbox, *even the messages you have deleted.* If you use the **q** command to exit **mailx**, then your mailbox is permanently updated according to the commands you have issued.

After you quit mailx using q, you can no longer use the undelete command to restore deleted messages, so make sure that you have deleted only the messages you intended to delete before you quit mailx.

9.4.3 Ways to Save Your Mail

The **mailx** command lets you save your messages in a specified file while you are reading them. You can save all your messages, a single message, or a range of messages. Let's look at some examples. The scenario remains as before: `david` is your login name, and there are three messages in your system mailbox.

```
$ mailx [Return] . . . . . . Read your mail.
mailx version 4.0  Type ? for help.
"/usr/student/mail/david": 3 messages  3 new
> N 1  daniel               Fri Sep 30 12:26 6/103 Room
  N 2  susan                Fri Sep 30 12:30 6/107 Project
  N 3  marie                Fri Sep 30 13:30 6/79  Welcome
? s mfile [Return] . . . . . Append the current message to mfile.
"mfile" [New file] 12/86 . . Get feedback message.
```

The **s** command saves your indicated message in the specified file. At this point `mfile` contains your current message, message 1, indicated by the > sign.

```
? s 2 3 mfile [Return] . . . Append messages 2 and 3 to mfile.
"mfile" [Appended] 18/286  . Get feedback message.
? s 1-3 mfile [Return] . . . Append messages 1 through 3 to mfile.
"mfile" [Appended] 18/286  . Get feedback message.
? x [Return] . . . . . . . . Exit; mailbox remains the same.
$_ . . . . . . . . . . . . . Return to the shell prompt.
```

At this point, all your messages are appended to `mfile`. You can use the **mailx** command option **-f** to read your mail from `mfile`.

1. *Do not confuse* **mailx** *command option* **-f** *(to which you add a filename to read a specified file) with* **mailx** *command option* **f** *(which displays only the headline of only the current message).*
2. *Do not confuse* **mailx** *command option* **-s** *(which sets the subject field to the specified characters) with* **mailx** *command option* **s** *(which saves your messages in the specified file).*

9.4.4 Ways to Send a Reply

You can send a reply to the sender of a message while reading your mail. This is convenient because you can send a reply right after you read a message.

To send a reply while in the **mailx** read mode, do the following:

```
$ mailx [Return] . . . . . . Read your mail.
mailx version 4.0  Type? for help.
"/usr/student/mail/david": 3 messages 3 new
> N 1  daniel               Fri Sep 30 12:26 6/103 Room
  N 2  susan                Fri Sep 30 12:30 6/107 Project
  N 3  marie                Fri Sep 30 13:30 6/79 Welcome
? R [Return] . . . . . . . . Reply to the current message.
Subject: RE: Room  . . . . . The subject prompt refers to Room.
_ . . . . . . . . . . . . .  The cursor appears at the beginning of
                             the line, ready for you to type your reply.
Thank you  . . . . . . . . . Compose your reply.
[Ctrl-d] . . . . . . . . . . End your message.
EOT  . . . . . . . . . . . . Message transmitted.
?  . . . . . . . . . . . . . Ready for the next command.
? R3 [Return]  . . . . . . . Respond to message 3.
? r3 [Return]  . . . . . . . Respond to message 3 and all the others
                             who received a copy of it.
```

Reply command **R** sends your reply only to the author (originator) of the message or a specified list of users. Reply command **r** sends your reply to the author of the message and to all other users who have received the same message.

You can also use the **m** command to send mail to other users. The **m** command places **mailx** into input mode so you can compose your mail. With the **m** command, you can specify one user or a list of users to receive your message.

While in the **mailx** command mode, send mail to a specific user by doing the following:

```
? m daniel [Return] . . . . .      Send mail to daniel.
? m daniel susan [Return] . .      Send mail to daniel and susan.
? x [Return] . . . . . . . . .     Exit mailx.
$_ . . . . . . . . . . . . . .     The shell prompt appears.
```

9.5 CUSTOMIZING THE mailx ENVIRONMENT

You can customize the **mailx** environment by setting the **mailx** variables in the `.mailrc` file. To define these variables, you use the **mailx set** command. The **mailx** command also recognizes some of the shell's standard variables.

9.5.1 Shell Variables Used by mailx

Some of the standard shell variables are used by **mailx**, and their values affect the **mailx** behavior: *HOME* (which defines your HOME directory) and *MAILCHECK* (which defines the frequency with which **mailx** checks your mail). For example, if you want arrival of mail in your mailbox to be checked once a minute, then you define *MAILCHECK* as follows:

```
MAILCHECK=60
```

MAILRC is another shell variable used by **mailx**. This variable defines the startup file that **mailx** checks each time you invoke it. If this variable is undefined, then the default value of `$HOME/.mailrc` is used. For example, you can define *MAILRC* in your `.profile` file (the startup file for your login shell) as follows:

```
MAILRC=$HOME/E-mail/.mailrc
```

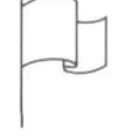

The *HOME*, *MAILCHECK*, and *MAILRC* shell variables are used by mailx, but you cannot change them while in mailx.

A large number of **mailx** variables can be manipulated to tailor the **mailx** environment to your specifications. You can set these variables from within **mailx** or set them in `.mailrc`. You use the **set** command to set up **mailx** variables and the **unset** command to reverse their settings. The format of the **set** command and the way you create and set up your .`mailrc` file are similar to those for `.exrc` (the vi editor startup file).

These variables can be set in the startup file `.mailrc` *or internally from* **mailx**.

append When you terminate reading your mail, if **append** is set, then **mailx** appends messages to the end of the `mbox` file instead of the beginning.

You have two choices for the **append** setting:

- Type **set append** and press [Return] to add messages to end of `mbox`.
- Type **unset append** and press [Return] to add messages to the beginning of `mbox`. This is the default.

asksub When **asksub** is set, **mailx** prompts you for the **Subject:** field. It is set by default.

crt and PAGER The *crt* variable is set to the number of lines on your screen. Messages that have lines more than the set number are piped through the command defined by the *PAGER* variable.

To set up **mailx** so that, if a message is longer than 15 lines, it is piped to the **pg** command and displayed one page at the time, do the following:

- Type **set PAGER='pg'** and press [Return]. (This is the default value of the *PAGER* variable, unless your system administrator has changed it.)
- Type **set crt=15** and press [Return] to set the number of lines to 15.

You can use the **pg** command options to scan messages up and down (see Chapter 7).

DEAD Your partially composed messages that were interrupted (or messages that for some reason or another are undeliverable) are stored in the specified filename. The default is as follows:

```
set DEAD=$HOME/dead.letter
```

To change the default file specified by the *DEAD* variable, you could type

```
set DEAD=$HOME/E-mail/dead.mail [Return]
```

EDITOR The *EDITOR* variable sets the editor invoked when you use the edit (or **~e**) commands. The default is as follows:

```
set EDITOR=ed
```

To change the default *EDITOR* variable, you could type

```
set EDITOR=ex [Return]
```

It is changed to an editor called *ex*.

escape The *escape* variable lets you change the **mailx** escape character. The default value is the tilde (~).

To change the default *escape* variable character to @, you can type

```
set escape=@ [Return]
```

folder The *folder* variable makes a specified directory the standard directory for **mailx**. All mail files are saved in the directory specified by the folder variable. There is no default value for the folder variable.

To specify the `EMAIL` file in the HOME directory for mail files, type

```
set folder=$HOME/EMAIL [Return]
```

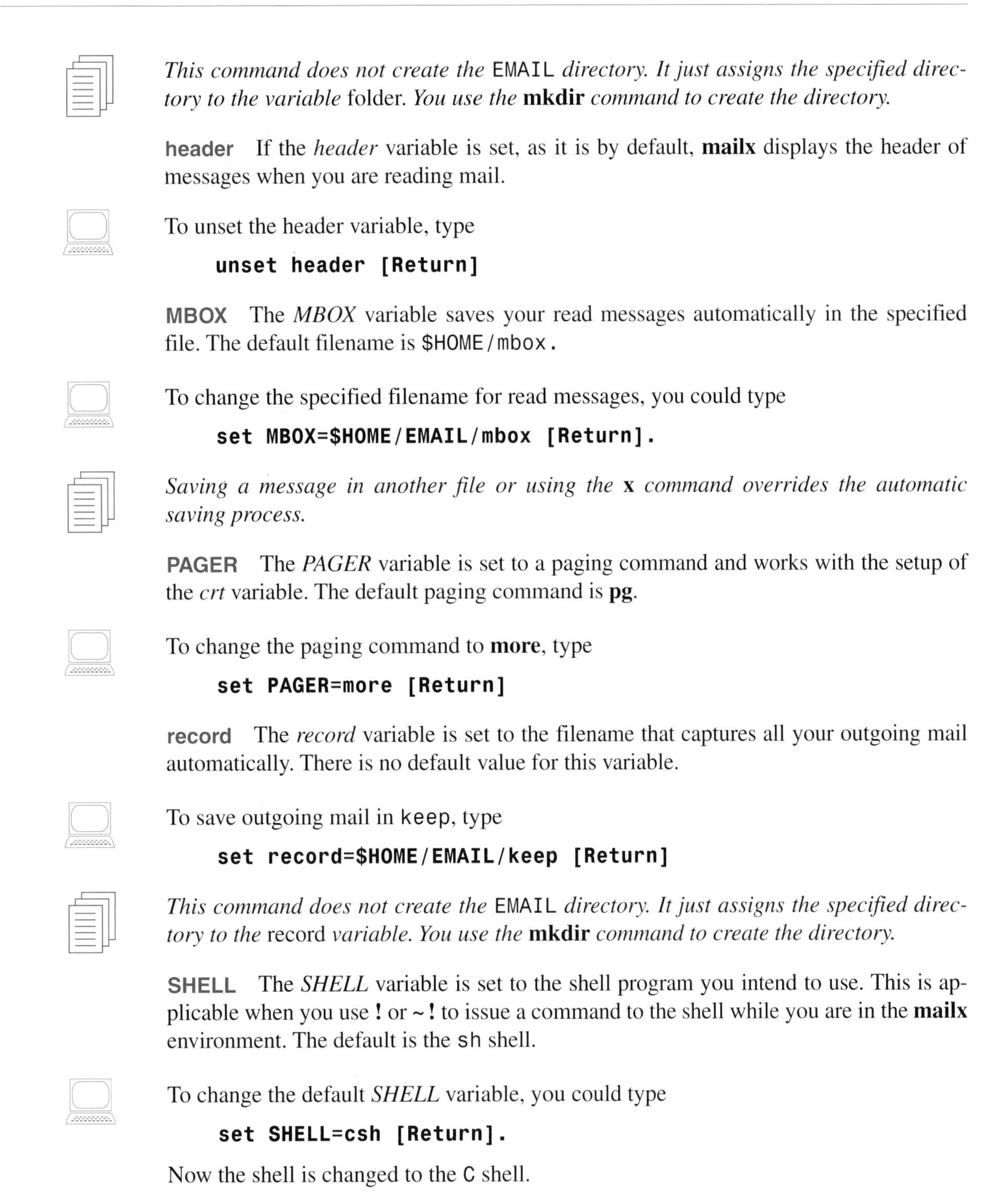

This command does not create the `EMAIL` *directory. It just assigns the specified directory to the variable* folder. *You use the* **mkdir** *command to create the directory.*

header If the *header* variable is set, as it is by default, **mailx** displays the header of messages when you are reading mail.

To unset the header variable, type

unset header [Return]

MBOX The *MBOX* variable saves your read messages automatically in the specified file. The default filename is `$HOME/mbox`.

To change the specified filename for read messages, you could type

set MBOX=$HOME/EMAIL/mbox [Return].

Saving a message in another file or using the **x** *command overrides the automatic saving process.*

PAGER The *PAGER* variable is set to a paging command and works with the setup of the *crt* variable. The default paging command is **pg**.

To change the paging command to **more**, type

set PAGER=more [Return]

record The *record* variable is set to the filename that captures all your outgoing mail automatically. There is no default value for this variable.

To save outgoing mail in `keep`, type

set record=$HOME/EMAIL/keep [Return]

This command does not create the `EMAIL` *directory. It just assigns the specified directory to the* record *variable. You use the* **mkdir** *command to create the directory.*

SHELL The *SHELL* variable is set to the shell program you intend to use. This is applicable when you use **!** or **~!** to issue a command to the shell while you are in the **mailx** environment. The default is the `sh` shell.

To change the default *SHELL* variable, you could type

set SHELL=csh [Return].

Now the shell is changed to the `C` shell.

VISUAL The *VISUAL* variable is set to the screen editor you intend to use when **mailx** is in the input mode and you use the **~v** command. The default is the vi editor.

9.5.2 Setting Up the `.mailrc` File

The `.mailrc` startup file in your HOME directory contains the commands and variables you set to tailor the **mailx** according to your preference. You can use the vi editor or the **cat** command to create a `.mailrc` file.

Figure 9.7 shows a .mailrc sample file. In the sample file, *friends* and *chess* are the names assigned to two groups of user IDs. The sample file is also set up so that if a message is longer than 20 lines, it is piped to the **pg** command (the default PAGER value). Finally, your private mailbox, as set up in the sample file, is in the EMAIL directory, and your outgoing mail is saved in a file called record in the EMAIL directory.

Figure 9.7
The .mailrc Sample File

```
$ cat .mailrc
alias friends daniel david marie gabe emma
alias chess emma susan gabe
set crt=20
set MBOX=$HOME/EMAIL/mbox
set record=$HOME/EMAIL/record
$_
```

9.6 COMMUNICATIONS OUTSIDE THE LOCAL SYSTEM

This chapter has discussed the UNIX communication utilities that enable you to send mail to users who have login accounts on your local host. You can also send mail to users on other UNIX computers. If you are on a UNIX network, as big companies and universities usually are, you can use the same commands. However, you must give more information. For example, the destination of your message needs to have the name of the computer (node name on the network) in addition to the user ID of the person on the specified computer. For example, if the computer nodes between you and David on the network were named X, Y, and Z, then you would type the following to send mail to David:

```
mailx X\!Y\!Z\!david
```

The details of communicating with other UNIX systems are outside the scope of this book. The commands in this chapter are all the basic skills you need; the rest is just a matter of looking up the commands in a reference book.

COMMAND SUMMARY

This chapter focused on the UNIX communication utilities and discussed the following commands and options.

mailx

This utility provides the electronic mail system for the users. You can send messages to other users on the system, regardless of whether they are logged on or not.

Option	Operation
-f *filename*	Reads mail from the specified *filename* instead of the system *mailbox*. If no file is specified, it reads from `mbox`.
-H	Displays a list of the message headers.
-s *subject*	Sets the subject field to the string *subject*.

mailx command mode

When you invoke **mailx** to read your mail, it places itself in command mode. The prompt for this mode is the question mark (**?**).

Command	Operation
!	Lets you execute the shell commands (the shell escape).
cd *directory*	Changes to the specified directory, or to the HOME directory if none is specified.
d	Deletes the specified messages.
f	Displays the headlines of the current message.
q	Exits **mailx** and removes the messages from the system mailbox.
h	Displays active message headers.
m *users*	Sends mail to specified *users*.
R *messages*	Replies to the sender of the *messages*.
r *messages*	Replies to the sender of the *messages* and all the other recipients of the same *messages*.
s *filename*	Saves (appends) the indicated messages to the file named *filename*.
t *messages*	Displays (types) the specified *messages*.
u *messages*	Undeletes the specified *messages*.
x	Exits **mailx** does not remove messages from the system mailbox.

mailx tilde escape commands

When you invoke **mailx** to send mail to others, it places itself in input mode, ready for you to compose your message. The commands in this mode start with a tilde (~) and are called tilde escape commands.

Command	Operation
~?	Displays a list of all the tilde escape commands.
~! *command*	Lets you invoke the specified shell *command* while composing your message.
~e	Invokes an editor for editing your message. The editor to be used is defined in the mail variable called *EDITOR*.
~p	Displays the message currently being composed.
~q	Quits input mode. Saves your partially composed message is saved in the file called `dead.letter`.
~r *filename*	Reads the file named *filename* and adds its contents to your message.
~< *filename*	Reads the file named *filename* (using the redirection operator) and adds its contents to your message.
~<! *command*	Executes the specified command and places its output into your message.
~v	Invokes the default visual editor, the vi editor, or uses the value of the mail variable *VISUAL*, which can be set up for other editors.
~w *filename*	Writes the message currently being composed to the specified *filename*.

mesg

This command is set to **n** to prohibit unwanted **write** messages. It is set to **y** to receive messages.

news

This command is used to look at the latest news in the system. It is used by the system administrator to inform others of the events happening.

Option	Operation
-a	Displays all the news items, whether they are in old or new files.
-n	Lists only the names of the news files (headers).
-s	Displays the number of the current news items.

talk

This command is used for terminal-to-teminal communication. The receiving party must be logged on.

wall
This command is used mostly by the system administrator to warn users of some imminent events.

write
This command is used for terminal-to-terminal communication. The receiving party must be logged on.

REVIEW EXERCISES

1. What is the command for terminal-to-terminal communication? What is the key used to signal the end of the communication?
2. What command is usually used by the system administrator to inform users about everyday events?
3. What command broadcasts a message to everyone on the system?
4. How do you make your terminal immune to unwanted messages?
5. What is the command used to read mail?
6. How do you know you have mail?
7. What is the purpose of the `.mailrc` file?
8. What is the command in **mailx** command mode to
   - **a.** display the list of the message headers?
   - **b.** read mail from a specific filename?
   - **c.** change the subject field?
   - **d.** display a list of tilde escape commands?
   - **e.** read a specific file and add it into your message?
   - **f.** invoke the default editor?
9. What is the purpose of the following shell variables?
   - **a.** *DEAD*
   - **b.** *record*
   - **c.** *PAGER*
   - **d.** *MBOX*
   - **e.** *LISTER*
   - **f.** *header*
10. What is the command to communicate with another logged-on user?

Terminal Session

In this terminal session, you practice sending messages to other users. Send the messages to yourself to practice the commands. Then, when you feel comfortable using the commands, select another user as your partner and practice sending mail.

The following exercises are recommended. In order to master the many commands of the UNIX communication utilities, you must spend time at your terminal and try all combinations of the commands.

1. Create a directory `EMAIL` in your HOME directory.
2. Create a `.mailrc` file in your HOME directory. If one already exists, modify it.
3. Set up **mailx** as follows:
   - **a.** Use the **alias** command to assign names to a group of users.
   - **b.** Set up a file in the `EMAIL` directory to save your outgoing mail automatically.
   - **c.** Set your `mbox` in the `EMAIL` directory.
4. Send the following mail to yourself:

   ```
   So little time and so much to do.
   Could it be reversed?
   So much time and so little to do.
   ```

5. While **mailx** is in the input mode, use the vi editor to compose your mail.
6. Read the current date and time and append it to the end of your message.
7. Save the message in a file before transmitting it.
8. Compose a few more messages in the same manner, using vi and other commands, and send them to yourself. This should give you enough messages to practice reading your mail.
9. Read your mail.
10. Use all the commands in the **mailx** command mode, including delete, undelete, save, and so on.
11. Read your mail and exit **mailx** using the **x** command. Then read your mail and exit **mailx** using the **q** command. Observe the UNIX messages.
12. Use **mailx** and look at your `mbox`.
13. Now find another partner and have a chat using the **write** command.
14. Set the **mesg** to **n**, then to **y**, and observe the effect with your partner.
15. Sending mail to others may be one of the less boring parts of learning UNIX. Practice and have fun.

CHAPTER 10

Program Development

This chapter describes the essentials of program development. It explains the steps in the process of creating a program and provides a general description of the available computer programming languages. It gives an example of a simple C program and walks you through the process of writing source code to make an executable program. The chapter also explains the use of the shell redirection operator to redirect the output and error messages of programs.

In This Chapter

10.1 PROGRAM DEVELOPMENT

Chapter 1 discussed software and computer programming in general. You learned the important role of software in making computers do all these wonderful things (open to discussion!). You also learned that there are two categories of software: application software and system software.

A *program* consists of a set of instructions that guide the computer in performing its basic arithmetic and logical operations. Each instruction tells the machine to perform one of its basic functions, and usually consists of the operation code and one or more operands. The *operation code* specifies the function to be performed, and the *operands* specify the location or data elements to be manipulated. Figure 10.1 shows a typical instruction.

Figure 10.1
The Format of a Simple Instruction

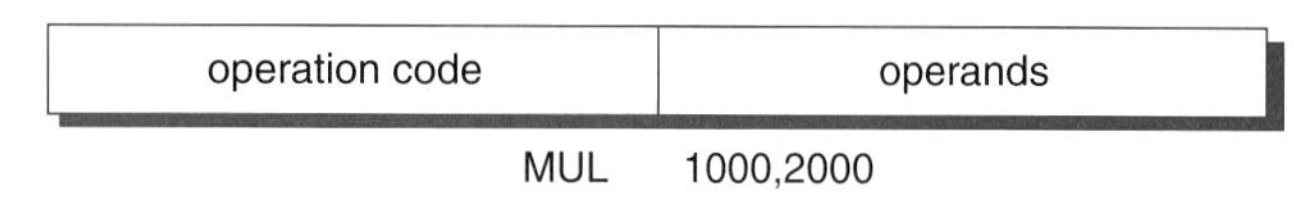

A computer is controlled by programs that are stored in the computer's memory. Memory storage is capable of storing only zeros and ones, so programs in memory must be in *binary form.* That means programs must be written in zeros and ones or converted to zeros and ones. Do programmers really write programs in zeros and ones? Fortunately, they do not have to anymore. Early programmers did not have a choice; they had to code programs in zeros and ones, for which they deserve a lot of respect. Programmers write programs to create both categories of software. To write a program you need a programming language, and there are quite a few you can choose from.

10.2 PROGRAMMING LANGUAGES

Programming is the process of writing instructions (a program) for a computer to solve a problem. These instructions must be written in a computer programming language. A program can be anything from a simple list of instructions that adds a series of numbers together, to a large and complex structure with many sections that calculates the payroll of a big corporation.

Like computer hardware, programming languages have evolved in generations. Each new generation of languages improved on the previous ones and incorporated more capabilities for programmers.

Figure 10.2 shows the hierarchy and generations of the programming languages. Let's briefly explore the different levels of this hierarchy.

See page x for an explanation of the icons used to highlight information in this chapter.

Figure 10.2
The Hierarchy of the Programming Languages

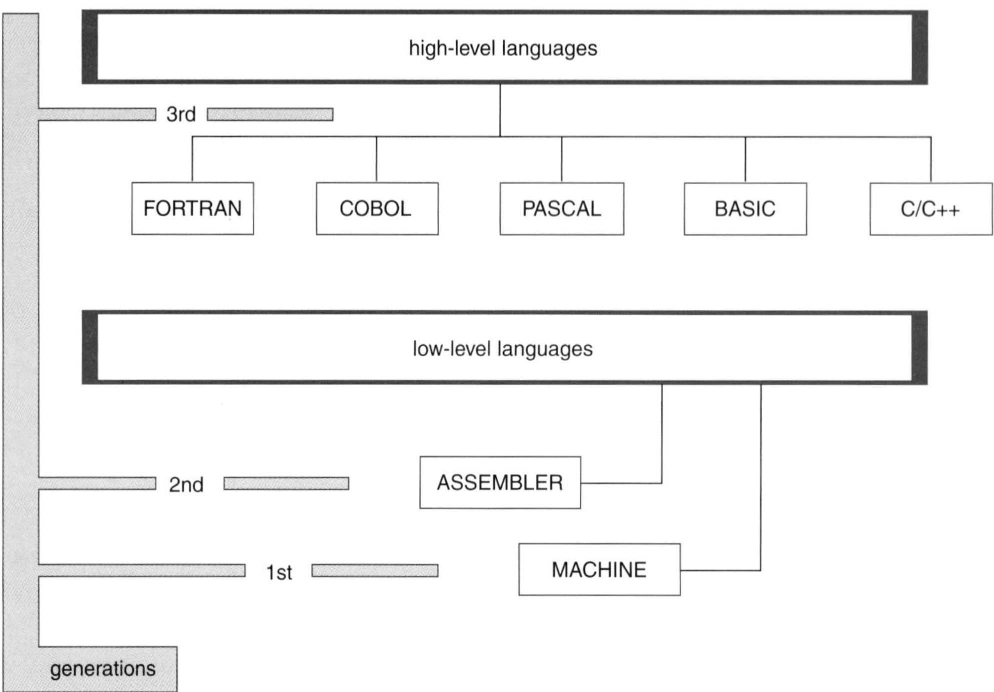

10.2.1 Low-Level Languages

Machine Language In machine language, instructions are coded as a series of zeros and ones. It is cumbersome and difficult to write programs in machine language. Machine language programs are written at the most basic level of computer operation; it is the only language computers can understand and execute. Programs written in other programming languages must be translated into the machine language of the computer on which the program is to be executed. (Programs called *compilers* do this translation and are discussed in the next section.)

Assembler Language Like machine language, assembler language is unique to a particular computer, but the instructions are represented differently. Instead of a series of zeros and ones, assembler language uses some recognized symbols, called *mnemonics* (memory aids), to represent instructions. For example, the mnemonic MUL is used to represent a multiply instruction. Because computers understand only zeros and ones, you must change a program in assembler language to machine language format for execution. This translation is done by invoking a program called an *assembler.* The assembler program translates the mnemonics in your program back to zeros and ones.

10.2.2 High-Level Languages

No matter what computer and which high-level language you use to write your program, the program must be translated into machine language before it can be executed. This conversion of a program from a high-level programming language to low-level machine language is the job done by the software programs called *compilers* and *interpreters.* High-level programming languages are simply a programmer's convenience;

they cannot be executed in their original form (source code). This section describes a few of the major programming languages.

COBOL The COBOL (COmmon Business Oriented Language) programming language was introduced in 1959. Developed in response to business community requirements, it was used mostly on mainframe computers providing data-processing services to large companies. More efficient and faster general-purpose computer programming languages exist in the market, but COBOL is going to remain with us because more than half of all business application programs are written in COBOL.

FORTRAN The FORTRAN (FORmula TRANslator) programming language was developed in 1955. It is most suitable for scientific or engineering programming, and remains the most popular scientific language. Of course, FORTRAN has been modified to meet the demands of new computers. The current version of FORTRAN is FORTRAN77, which is a general-purpose language capable of dealing with numeric and symbolic problems. As with COBOL, a large body of existing code was written in FORTRAN, and it is still used in engineering and science.

Pascal The Pascal programming language was developed in 1968 and is named after the seventeenth-century French mathematician Blaise Pascal. The idea of good style and habits in programming was around and the buzzwords "structured programming" were taking shape when Pascal was created, which incorporated *structured programming* ideas. Pascal was designed as a language to help students learn structured programming and develop good habits in programming. However, the use of Pascal is not limited to educational institutions; it is used in industry to create readable and maintainable codes.

BASIC The BASIC (Beginners All-purpose Symbolic Instruction Code) programming language was developed in 1964 to help students with no computer background learn about computers and programming. It is accepted as the most effective programming language for instructional purposes. Since then, BASIC has evolved into a general-purpose language, and its use is not limited to the education field. BASIC is widely implemented and is used in business applications.

C The C programming language was developed in 1972, and it is based on principles practiced in Pascal programming. It is mainly for targeted system programming, creating operating systems, compilers, and so on. Most of the UNIX operating system is written in C. C is a fast, efficient, general-purpose language. It is also a portable language; it is relatively machine independent. A C program written for one type of computer can be run on another with little or no modification. C is by far the language of choice for many programmers developing programs for business, scientific, and other applications.

C++ In the 1980s, C++ was developed to add to the C language the tools needed to make it an object-oriented language. *Object-oriented programming* (OOP) is one of the many programming paradigms. The OOP techniques make programmers' lives easier and reduce the program development time, especially when programs reach a certain size. C++ provides the language mechanism for implementing the OOP concept.

Java The Java programming language was developed by Sun Microsystems, Inc. in 1990 and was released to the public in 1995. Originally, it was designed for use in control systems of consumer electronics devices such as television sets, microwave oven, and so forth. Later, it was embedded into Internet browsers and programs for Web pages. Java is syntactically similar to C++, but many of the confusing features of C++ are discarded.

10.3 PROGRAMMING MECHANICS

To write a program, you have to choose a computer programming language. The choice of programming language depends on the nature of the application. Many general-purpose and specialized programming languages exist that fit any requirement. The exact steps you must follow to produce a program depend on the computer environment. The following discussion traces the steps in a typical program's development.

10.3.1 Steps to Creating an Executable Program

Regardless of the computer operating system and the programming language you use, the following steps are necessary to create an executable program:

1. Create the source file (source code).
2. Create the object file (object code/object module).
3. Create the executable file (executable code/load module).

Source Code You usually use an editor (such as the vi editor) to write a program and save what you write in a file. This file is the *source code.* The source code is written in the programming language of your choice. By itself, your computer does not understand it. The goal is to convert the source code file to an executable file.

A source file is a text file, which is also called an ASCII file. You can display it on the screen, modify it using one of the available editors, or send it to the printer to obtain a hard copy of your program's source code.

Object Code Source code is incomprehensible to the computer. Remember, computers understand only machine language (the pattern of zeros and ones). Thus, your source code must be translated to machine-understandable language. This is the job of a compiler or an interpreter. They produce the object code. The *object code* is the machine language translation of your source code. However, it is not an executable file yet; it lacks some necessary parts. These necessary parts are programs that provide the interface between the program and the operating system. They are usually grouped together in files called *library files.*

You cannot send an object file to the printer. It is a file of zeros and ones. If you display it, you may lock your keyboard or hear beeps as some of the zeros and ones are translated into the ASCII codes that represent codes for locking the keyboard, the beep at your terminal, stop scrolling, and so on.

Executable Code Your object code might refer to other programs that are not part of your *object module.* Before your program can be executed, these references to other programs must be resolved. This is the job of the *linker* or *link editor.* It creates the *executable code,* the *load module.* The *load module* is a complete, ready-to-be-executed program with all its parts put together.

The load module, like the object file, is not a file to be sent to the printer or displayed on your terminal.

In some systems, you invoke the compiler, and after the compilation process is completed, you invoke the linker to create the executable file. Other systems have the compiler start the linker automatically. On those systems, you just give the compile command, and, if your program does not have any compilation error, the linker is invoked to link your program.

Exactly how this works varies among systems, depending on the system environment/ configuration setup as done by the system administrator.

This process, starting from a source file and ending with creating an executable file, is depicted in Figure 10.3.

Figure 10.3
Program Development Steps

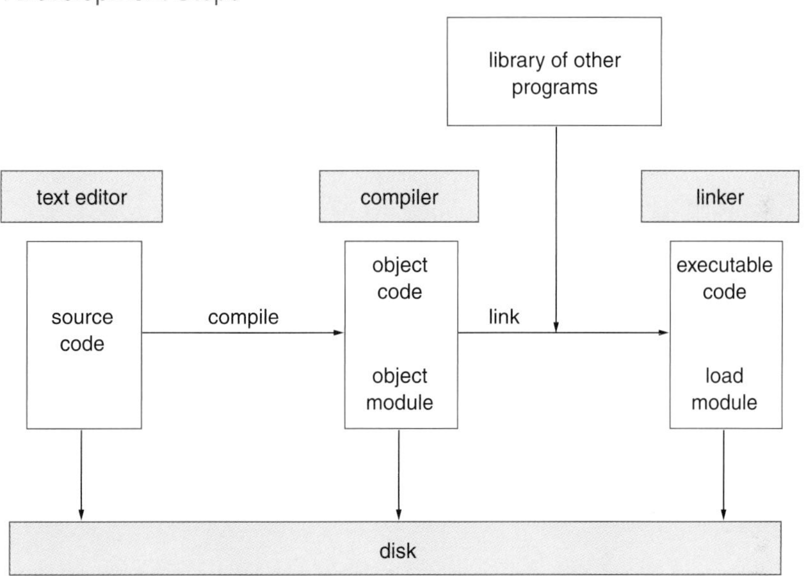

10.3.2 Compilers/Interpreters

The main function of a compiler or an interpreter is to translate your source code (program instructions) to machine code so the computer can understand your instructions. The compiled languages and interpreted languages represent two different categories of computer languages. Each has its advantages and disadvantages.

Compiler The compiler is a system software program that translates high-level program instructions, such as a Pascal program, into machine language that the computer can interpret and execute. It compiles the entire program at one time and does not give you any feedback until it compiles the entire program.

A separate compiler is required for each programming language you intend to use on your computer system. To execute C and Pascal programs, you must have a C compiler and a Pascal compiler. Compilers produce a better and more efficient object code than interpreters, so a compiled program runs faster and needs less space.

Interpreter Like the compiler, the interpreter translates a high-level language program into machine language. However, instead of translating the entire source program, it translates a single line at a time. An interpreter gives you immediate feedback. If the code contains an error, the interpreter picks it up as soon as you press [Return] to finish a line of code. Therefore, you can correct errors during program development. The interpreter does not produce a separate object code file, and it must perform the translation process each time a program is executed. Interpreters usually are used in an educational environment, and the executable code they produce is less efficient than the code produced by a compiler.

10.4 A SIMPLE C PROGRAM

Let's write a simple C program and go through the steps to see how this process works. The goal is to understand and practice the compilation process, not to learn the C language. However, you will have to understand some very basic characteristics of the C language to be able to write a simple C program and compile it successfully.

Using the vi editor (or any other editor), create a source file called `first.c`. Figure 10.4 shows the contents of the file.

Figure 10.4
A Simple C Program

```
$ cat first.c
/* my first C program */
# include <stdio.h>
int main()
{
  printf ("Hi there!\n");
  printf ("This is my first C program \n");
return 0;
}
$_
```

1. **Type the source code in lowercase letters. Like UNIX, the C language is a lowercase language, and all the keywords must be typed in lowercase.**
2. **In most systems, the extension `.c` at the end of the source code filename is mandatory.**

The next step is to compile your source code. The command is as follows:

```
$ cc first.c [Return]
```

The **cc** command compiles your source code and, if you do not have any errors, automatically invokes the linker. The end result is that you have an executable file called `a.out`. By default, `a.out` is the name of the executable file. Now if you want to execute the file to see the program output, you type

```
$ a.out [Return] . . . . . . Execute your program.
Hi there!
This is my first C program.
$_ . . . . . . . . . . . . . And the prompt appears.
```

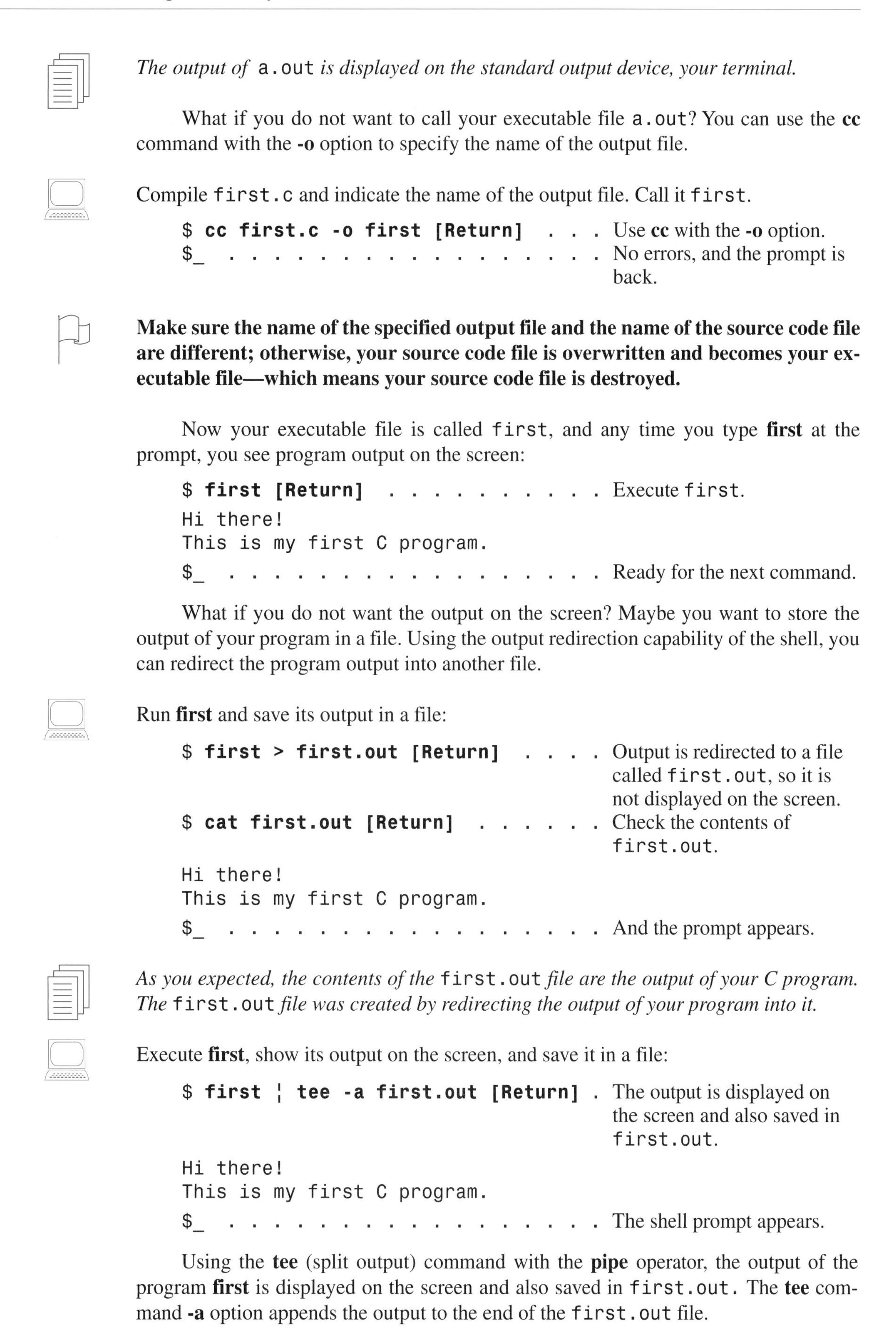

The output of `a.out` *is displayed on the standard output device, your terminal.*

What if you do not want to call your executable file `a.out`? You can use the **cc** command with the **-o** option to specify the name of the output file.

Compile `first.c` and indicate the name of the output file. Call it `first`.

```
$ cc first.c -o first [Return] . . . Use cc with the -o option.
$_ . . . . . . . . . . . . . . . . . No errors, and the prompt is
                                     back.
```

Make sure the name of the specified output file and the name of the source code file are different; otherwise, your source code file is overwritten and becomes your executable file—which means your source code file is destroyed.

Now your executable file is called `first`, and any time you type **first** at the prompt, you see program output on the screen:

```
$ first [Return] . . . . . . . . . . Execute first.
Hi there!
This is my first C program.
$_ . . . . . . . . . . . . . . . . . Ready for the next command.
```

What if you do not want the output on the screen? Maybe you want to store the output of your program in a file. Using the output redirection capability of the shell, you can redirect the program output into another file.

Run **first** and save its output in a file:

```
$ first > first.out [Return] . . . . Output is redirected to a file
                                     called first.out, so it is
                                     not displayed on the screen.
$ cat first.out [Return] . . . . . . Check the contents of
                                     first.out.
Hi there!
This is my first C program.
$_ . . . . . . . . . . . . . . . . . And the prompt appears.
```

As you expected, the contents of the `first.out` *file are the output of your C program. The* `first.out` *file was created by redirecting the output of your program into it.*

Execute **first**, show its output on the screen, and save it in a file:

```
$ first ¦ tee -a first.out [Return] . The output is displayed on
                                      the screen and also saved in
                                      first.out.
Hi there!
This is my first C program.
$_ . . . . . . . . . . . . . . . . . The shell prompt appears.
```

Using the **tee** (split output) command with the **pipe** operator, the output of the program **first** is displayed on the screen and also saved in `first.out.` The **tee** command **-a** option appends the output to the end of the `first.out` file.

10.4.1 Correcting Mistakes

Assuming you did not make any mistake in writing your first C program, things are fairly simple. You compile and execute your program. But this is not always the case when you write large programs. The chances are good that you will make syntax and logical mistakes that you will have to correct before being able to run a program successfully. The C compiler recognizes syntax errors and displays them on the screen with a line number reference.

Suppose you modify `first.c` to look like the file in Figure 10.5, so that it has no semicolon (;) at the end of the first **printf**( ) statement. This time when you compile `first.c`, the compiler is going to complain.

Figure 10.5
A Simple C Program with a Syntax Error

```
$ cat first.c
/* my first C program */
# include <stdio.h>
int main()
{
  printf ("Hi there!\n")  /* semi colon omitted intentionally.*/
  printf ("This is my first C program \n");
return 0;
}
$_
```

Now compile `first.c`.

```
$ cc first.c -o first [Return] . . . . Compile and call the output
                                       file first.
"first.c", line 6: syntax error.
"first.c", line 6: illegal character: 134(octal)
"first.c", line 6: cannot recover from earlier error: goodbye!
$_ . . . . . . . . . . . . . . . . . . Prompt.
```

The error messages indicate that you have some sort of syntax error on line 6. The compiler is not clever enough to recognize the exact error and its location; it just directs you to the vicinity of the error in the source code.

At this point, the compiler has not produced an object code file. To correct the errors, you must go back to the source code, `first.c`, and add the semicolon. Then, to get the correct executable file, you have to recompile the file.

It is easy to look at one or two error lines on the screen, but when you have a large amount of source code, the chances are good that you will have more than a few lines with errors. Remembering the errors and their line numbers when you want to edit your source file to correct the errors is not a trivial task. Therefore, it is desirable to save the compiler error messages in a file for easy reference. The redirection capability of the shell becomes handy again!

The default error device is usually the same as the output device, your terminal, so the errors are displayed on your terminal. Suppose you want to redirect the errors to a specified file.

Compile first.c and save compilation errors, if any, in another file:

```
$ cc first.c -o first 2> error [Return]. . Recompile.
$_ . . . . . . . . . . . . . . . . . . . . The shell prompt appears.
```

The digit 2 before the > sign is necessary and indicates the redirection of the standard error device.

Where does the digit 2 come from? The command just shown needs some explanation. Let's go back to the UNIX redirection concept and explore the origins of the digit 2.

10.4.2 Redirecting the Standard Error

The shell interprets the > as standard output redirection. The notation **1>** is the same as > and tells the shell to redirect the standard output. The number *1* in **1>** is the file descriptor number; by default, file descriptor 1 is assigned to the *standard output device.* For example, the following two commands do the same job: They redirect the output of the **ls** command to a file.

```
$ ls -C > list [Return]
```

or

```
$ ls -C 1> list [Return]
```

File descriptor 2 is assigned to the *standard error device.* The shell interprets the notation **2>** as the redirection of the error output. For example, suppose you do not have a file called **Y** in your current directory, and you issue the following command:

```
$ cat Y > Z [Return] . . . . . . . . . . Copy Y to Z.
cat: cannot open Y . . . . . . . . . . . Error message.
```

The error message appears on the screen. Now redirect the error output to a file.

```
$ cat Y > Z 2> error [Return] . . . . .  Copy Y to Z. Keep errors
                                         in a file called error in
                                         the current directory.
$_  . . . . . . . . . . . . . . . . . .  No error messages. The
                                         prompt is back.
$ cat error [Return] . . . . . . . . . . Show the contents of the
                                         error file.
cat: cannot open Y  . . . . . . . . . .  As you expected.
$_  . . . . . . . . . . . . . . . . . .  And the prompt appears.
```

Now back to your C program and its compilation errors; you can redirect the compilation errors to another file by typing:

```
$ cc first.c -o first 2> [Return] error. . Error message redirected
                                           to the error file.
$_ . . . . . . . . . . . . . . . . . . . . No error is displayed.
$ cat error [Return] . . . . . . . . . . . Check the compilation
                                           errors.
"first.c", line 6: Syntax error.
"first.c", line 6: Cannot recover from earlier error: goodbye!
$_ . . . . . . . . . . . . . . . . . . . . Prompt.
```

10.5 UNIX PROGRAMMING TRACKING UTILITIES

Other computer language compilers are available that work in a UNIX environment. You can obtain compilers for nearly every language to run under UNIX.

The goal of this chapter is to introduce you to programming development under UNIX—not to present you with a comprehensive list of languages, compilers, and UNIX programming. However, it is important to know that UNIX provides you with utilities to help you organize your program development process. These utilities become especially useful and important when you are developing large-scale software. Following is a very brief explanation of these utilities and their functions.

10.5.1 The make Utility

The **make** utility is useful when your program consists of more than one file. **make** automatically keeps track of the source files that are changed and that need recompilation, and relinks your programs if required. The **make** program gets its information from a control file. The *control file* contains rules that specify source files' dependencies and other information.

10.5.2 The SCCS Utility

The **SCCS** (Source Code Control System) is a collection of programs that helps you to maintain and manage the development of your programs. If your program is under **SCCS** control, then you can create different versions of your program easily. The **SCCS** keeps track of all the changes among different versions.

REVIEW EXERCISES

1. Explain the steps necessary to write a program and produce an executable file.
2. What is a source code?
3. What is the function of a compiler?
4. What is the difference between a compiler and an interpreter?
5. What is the command to compile a C program, and what is the default name of the executable file?
6. Why can't you send your executable file to a printer?

Terminal Session

In this terminal session, you are to write a sample C program. It is not expected that you know C programming. Copy the simple C program example in this chapter or one from any C programming book. The purpose is to familiarize yourself with the process of program development.

1. Write a simple C program.
2. Compile it.

3. Run the program.
4. Compile it again and specify the executable file.
5. Run it again.
6. Save the output of your program in another file.
7. Modify your source code and make an intentional syntax error.
8. Compile again.
9. Observe the error messages. See if you can decipher them.
10. Recompile your program and save the error messages in a file.
11. Look at the file that contains the compilation errors.

CHAPTER 11

Shell Programming

This chapter concentrates on shell programming. It explains the capabilities of the shell as an interpretive high-level language. It describes shell programming constructs and particulars. It explores shell programming aspects such as variables and flow control commands. It shows the creation, debugging, and running of shell programs, and it introduces a few more shell commands.

In This Chapter

11.1 UNDERSTANDING UNIX SHELL PROGRAMMING LANGUAGE: AN INTRODUCTION

Command languages provide the means for writing programs using a sequence of commands, the same commands that you type at the prompt. Most of today's command languages provide more than just the execution of a list of commands. They have features that you find in traditional high-level languages, such as looping constructs and decision-making statements. This gives you a choice of programming in high-level languages or command languages. *Command languages* are interpreted languages, unlike *compiled languages* such as C and C++. Programs written in a command language are easier to debug and modify than programs written in compiled languages. But you pay a price for this convenience: These programs typically take much longer to execute than the compiled ones.

The UNIX shell has its own built-in programming language, and all shells (Bourne shell, Korn shell, C shell, etc.) provide you with this programming capability. The shell language is a command language with a lot of features common to many computer programming languages, including the structured language constructs: sequence, selection, and iteration. Using the shell programming language makes it easy to write, modify, and debug programs, and this language does not need compilation. You can execute a program as soon as you finish writing it.

The shell program files are called *shell procedures*, *shell scripts*, or simply *scripts*. A *shell script* is a file that contains a series of commands for the shell to execute. As you execute a shell script, each command in the script file is passed to the shell to execute, one at a time. When all the commands in the file are executed, or if an error occurs, the script ends.

You do not have to write shell programs. But as you use UNIX, you will find that sometimes you want UNIX to perform functions that it does not have a command for, or to perform several commands simultaneously. UNIX commands are numerous, difficult to remember, and involve a lot of typing. If you do not want to have to remember all that strange UNIX command syntax, or if you are not a good typist, then you may want to write script files and use them instead of complicated commands.

1. *The* **cat -n** *command is used to display most of the script (program) file examples on the screen. You can use vi to create these files. Also, you can use the* **cat** *command to create any one of these files quickly. Remember, the* **cat** *command is good for creating small files with few lines of code.*
2. *You can create these script files in the* `$HOME/Chapter11` *directory, under the appropriate subdirectories. For example, create programs in section 11.2 in the* `$HOME/Chapter11/11.2` *directory or programs in section 11.5 under the* `$HOME/Chapter11/11.5` *directory. Of course, you can create these files in any directory you wish and this is only a suggestion to make your directory structure and files organized according to the chapters and sections of your textbook.*

See page x for an explanation of the icons used to highlight information in this chapter.

11.1.1 Writing a Simple Script

You do not need to be a programmer to write simple script files. For example, suppose you want to know how many users are currently on the system. The command and output look like this:

```
who ¦ wc -l [Return] . . . . . . Type the command.
6 . . . . . . . . . . . . . . . . UNIX will display the number.
```

The output of the **who** command is passed as input to the **wc** command, and the **-l** option counts the lines that indicate the number of current users on the system.

You can write a simple script file to do the same thing. Figure 11.1 shows a simple shell script called won (Who is ON) that uses the **who** and **wc** commands to report the number of people currently logged on the system. Shell scripts are stored in UNIX text files, so you use the vi editor (your favorite text editor) or the **cat** command to create them.

Figure 11.1
A Simple Shell Script

```
$ cat -n won
1 #
2 # won
3 # Display # of users currently logged in
4 #
5 who ¦ wc -l
$_
```

A # at the beginning of a line indicates that the line is a remark line and is for documentation. The shell ignores lines starting with #.

11.1.2 Executing a Script

There are two ways to execute a shell script: You can use the **sh** command, or you can make the shell script file an executable file.

Invoking Scripts

You can use the **sh** command to execute script files. Every time you type **sh** (or the name of any other shell such as **ksh** or **bash**) you invoke another copy (instance) of the shell. The shell script won is not an executable file, so you must invoke another shell to execute it. You specify the file name, and the new shell reads the script, executes the commands listed in it, and ends when all the commands have been executed (or when it reaches an error).

Figure 11.2 shows how the won file can be executed using this method. Typing **sh** (or **ksh**, **bash**, etc.) to invoke another shell each time you want to execute a script file is not very convenient. However, this method has its advantages, especially when you write complex script applications that need debugging and tracing tools. (These tools are discussed later in this chapter.) Under normal circumstances, the second method, making the script executable, is preferred. After all, the less typing, the better.

Figure 11.2
Using the **sh** Command to Run a Script

```
$ sh won
6
$_
```

As was mentioned before, you can use the name of your current shell or any other shell in your system instead of **sh**. For example, the following commands all invoke a shell script called `won`:

$ **sh won [Return]** . . . . Using *sh*.
$ **ksh won [Return]** . . . . Using *ksh*.
$ **bash won [Return]** . . . . Using *bash*.

Making Files Executable: The chmod Command

The second method of executing a shell program is to make the script file an executable file. In this case, you do not need to invoke another shell; all you do is type the name of the script program, just as you do any other shell program (command). This is the preferred method.

To make a file executable, you must change the access permissions for that file. You use the **chmod** command to change the mode of a specified file. Table 11.1 shows the **chmod** options. Assuming that you have a file called `myfile`, the following example shows how you can change its access mode.

Table 11.1
chmod Command Options

Character	Who is Affected?
u	User/owner.
g	Group.
o	Others.
a	All; can be used instead of the combination of **ugo**.
Character	**Permission Category**
r	Read permission.
w	Write permission.
x	Execute permission.
-	No permission.
Operator	**Specific Action to be Taken on Permission Categories**
+	Grants permission.
-	Denies permission.
=	Sets all permissions for a specified user.

The following command makes `myfile` an executable file:

```
$ ls -l myfile [Return]  . . . . Check myfile mode.
- rw- rw- r-- 1   david   student   64 Oct 18 15:45 myfile
$ chmod u+x myfile [Return]  . . Change myfile mode.
$ ls -l myfile [Return]  . . . . Check myfile mode again.
- rwx rw-r-- 1   david   student   64 Oct 18 15:45 myfile
$_  . . . . . . . . . . . . . . . The shell prompt appears.
```

The **ls** command is used to verify the fact that its access mode is changed. The **u** indicates that the user has access to `myfile`; the **+x** indicates that the `myfile` access mode is to be changed to executable.

Assuming user, group, and others have read and write access on `myfile` file, the following command removes the write access for all other users:

```
$ chmod o-w myfile [Return]  . . Change myfile mode.
$_  . . . . . . . . . . . . . . . The prompt reappears.
```

You and members of your group still have read and write access privileges, but others can only read from your file. (You can use the **ls** command to verify the changes.) The letter **o** indicates others, and **-w** indicates that the others do not have write access to `myfile`.

The following command changes the access privileges for all users (owner, group, and others) to read and write:

```
$ chmod a=rw myfile [Return] . . Change myfile mode.
$_  . . . . . . . . . . . . . . . Return to prompt.
```

Now anyone can read or write to `myfile`. (Again you can use the **ls** command to verify the changes.) The letter **a** indicates all users, and **rw** indicates read and write access.

The following commands remove all access privileges for the group and other users:

```
$ ls -l myfile [Return]  . . . . Display myfile mode.
- rwx rw- r-- 1   david   student   64 Oct 18 15:45 myfile
$ chmod go= myfile [Return]  . . Change myfile mode. Notice that
                                 there is at least one space between
                                 "=" and "myfile."
$ ls -l myfile [Return]  . . . . Check the mode changes.
- rwx------1      david   student   64 Oct 18 15:45 myfile
$_  . . . . . . . . . . . . . . . Return to the prompt.
```

The owner (in this case you) is the only one who has access to `myfile`. The **ls** command verifies the changes. The **go** command indicates group and others, so **go=** indicates removal of all access privileges to `myfile`.

Returning to your `won` shell script, let's make it an executable file:

```
$ ls -l won [Return] . . . . . . Check mode set for won.
-rw- rw- r-- 1   david   student   64 Oct 18 15:45 won
$ chmod u+x won [Return] . . . . Change mode.
$ ls -l won [Return] . . . . . . Verify the change.
-rwx rw- r-- 1   david   student   64 Oct 18 16:45 won
$_  . . . . . . . . . . . . . . . Return to the prompt.
```

Now won, your shell script file, is an executable file, and you do not need to invoke another shell program to execute it. You execute won like any other command (executable file) by simply typing the filename and pressing [Return].

```
$ won [Return] . . . . . . . . . Execute won.
6  . . . . . . . . . . . . . . . The output shows there are 6 users.
$_ . . . . . . . . . . . . . . . The shell prompt appears.
```

11.2 WRITING MORE SHELL SCRIPTS

Shell programming is relatively simple, and it is a powerful tool at your disposal. You can place any command or sequence of commands in a file, make the file executable, and then execute its contents simply by typing its name at the **$** prompt.

Table 11.2 lists the shell built-in commands that are covered in this chapter and shows their availability under different shells.

Table 11.2
The Shell Built-in Commands

Command	Built into		
	Bourne Shell	Korn Shell	Bourne Again Shell
exit	sh	ksh	bash
for	sh	ksh	bash
if	sh	ksh	bash
let		ksh	bash
read	sh	ksh	bash
test	sh	ksh	bash
until	sh	ksh	bash
while	sh	ksh	bash

1. *Most UNIX systems include more than one shell. The commands and script files (programs) in this book should work with the* sh, ksh, *and* bash *shells. However, there are many variations with subtle differences in the installation. Use the* **man** *command to see how a particular command is used in your version of the shell.*
2. *You can change your current shell by typing* **sh**, **ksh**, *or* **bash** *to invoke any of the three shells, the Bourne shell, Korn shell, or Bourne Again shell. This does not change your login shell. Typing* **exit** *terminates the current shell and you are back to your login shell.*
3. *The standard variable* SHELL *is set to your login shell. You can find out what is your login shell by typing the following command line:*

```
$ echo $SHELL [Return] . . . . . Display content of the SHELL variable.
```

Let's modify the won file and add a few more commands to it. Figure 11.3 shows won2, the modified version of the won script file.

Figure 11.3
A Shell Script Sample won2 Program

```
$ cat -n won2
    1 #
    2 # won2
    3 # won version 2
    4 # Displays the current date and time, # of users currently
    5 # logged in, and your current working directory
    6 #
    7 date                # displays current date
    8 who ¦ wc -l         # displays # of users logged in
    9 pwd                 # diplays current working directory
$_
```

Assuming the won access mode is changed and it is an executable file, Figure 11.4 shows the output of the second version of the won program.

Figure 11.4
The Output of the won2 Program

```
$ won2

Wed Nov 30 14:00:52 EDT 2005
    14
/usr/students/david

$_
```

The output of the second version of the won program is cryptic and certainly can be improved. Let's use a few **echo** commands here and there to make the output more meaningful. Figure 11.5 shows the won script, third version, called won3.

Figure 11.5
Another Version of the won3 Program

```
$ cat -n won3
    1 #
    2 # won3
    3 # won version 3 - The user-friendly version
    4 # Displays the current date and time, # of users currently
    5 # logged in, and your current working directory
    6 #
    7 echo                     # skip a line
    8 echo "Date and Time:\c"
    9 date                     # displays current date
   10 echo "Number of users on the system:\c"
   11 who ¦ wc -l              # displays # of users logged in
   12 echo "Your current directory:\c"
   13 pwd                      # displays your current directory
   14 echo                     # skip a line
$_
```

The **echo** command with no argument is one way to output a blank line. The **echo** command ends its output with a new line.

Figure 11.6 shows the output of the won program, third version (won3). It is more informative and looks better.

Figure 11.6
The Output of the won3 Shell Program

```
$ won3

Date and Time:Wed Nov 30 15:00:52 EDT 2005
Number of users logged in:14
Your current directory:/usr/students/david

$_
```

11.2.1 Using Special Characters

As was discussed in Chapter 8, the **echo** command recognizes special characters, called *escape characters.* They all start with the backslash (\), and you can use them as part of the **echo** command argument string. These characters give you more control over the format of your output. For example, the following command produces four new lines:

$ **echo "\n\n\n" [Return]** . . . . Produce four blank lines.

Three blank lines are produced by the three **\n** codes, and one blank line is produced by the **echo** command default next line (return key) at the end of the output string.

Table 11.3 summarizes these escape characters. The following command sequence shows examples of using the escape characters.

Table 11.3
The **echo** Command Special Characters

Escape Character	Meaning
\b	A backspace.
\c	Inhibits the default next line ([Return] key) at the end of the output string.
\n	A carriage return and a form feed next line ([Return] key).
\r	A carriage return without a line feed.
\t	A tab character.
\0*n*	A zero followed by 1-, 2-, or 3-digit octal number representing the ASCII code of a character.

If these commands do not work, make sure you are using the **-e** *option to make the* **echo** *command to recognize escape characters. For example, the previous command is typed as*

$ echo -e "\n\n\n"[Return] . . . Using the **-e** option.

This note is applicable to the rest of the programs that use the **echo** *command with escape characters such as* **\c**, **\b**, *and so on.*

Use the \n escape character:

```
$ echo "\nHello\n" [Return] . . . Use \n escape character.
              . . . . . . . . . . . . . . A blank line is produced by the
                                              first \n.
hello . . . . . . . . . . . . . . . . The word hello.
              . . . . . . . . . . . . . . A blank line is produced by the
                                              second \n.
              . . . . . . . . . . . . . . A blank line is produced by the
                                              echo command.
$_ . . . . . . . . . . . . . . . . . . The prompt reappears.
```

Beeping Sound [Ctrl-G] produces a beeping sound on most terminals, and the ASCII code for it is the octal number 7. You can use the **\0n** format to sound a beep on your terminal.

Use the **\0***n* escape code to produce a beeping sound:

```
$ echo "\07\07WARNING" [Return] . . . Use \07 for the bell sound.
WARNING
$_ . . . . . . . . . . . . . . . . . . . . . . The prompt reappears.
```

You hear the terminal bell sound twice (beep beep), and the **WARNING** *message is displayed.*

Clearing Screen When you write scripts, especially interactive ones, you usually need to clear the screen before doing anything else. One way to clear the screen is to use the **\0***n* escape character format with the **echo** command to send the clear screen code to the terminal. The code for clearing the screen is terminal dependent; you may need to ask your system administrator or look it up in the terminal user/technical manual.

Assuming [Ctrl-z] clears your terminal screen (octal number 32), use the **echo** command to clear the screen.

```
$ echo "\032" [Return] . . . . . Clear the screen.
$_ . . . . . . . . . . . . . . . . . . Return to the prompt.
```

1. *The screen is cleared, and the prompt appears at the top left corner of the screen.*
2. *If this command doesn't clear your terminal screen, it means [Ctrl-z] is not the clear screen code for your terminal. For most shells you can simply type* **clear** *and press the [Return] key to clear the screen.*

11.2.2 Logging Off in Style

Normally, you press [Ctrl-d] or use the **exit** command to log off and end your session. Suppose you want to change that: You want to type **bye** to log off.

Write a script file called `bye` that contains one line of code, the command **exit**:

```
$ cat bye [Return] . . . . . . . . Show contents of bye.
exit . . . . . . . . . . . . . . . . One line of code.
$_ . . . . . . . . . . . . . . . . . . Prompt.
```

Change bye to an executable file and run it to log out:

```
$ chmod u+x bye [Return]    . . . Change the file mode.
$ bye [Return]    . . . . . . . . Log out.
$_  . . . . . . . . . . . . . . . You are still in UNIX.
```

Why didn't **exit** work? When you give a command to the shell, it creates a child process to execute the command (as described in Chapter 8). Your login shell is the parent process, and your script file bye is the child process. The child process (bye) gets the control and executes the **exit** command that consequently terminates the child process (bye). When the child is dead, control goes back to the parent (your login shell). Therefore the **$** prompt is displayed.

To make the bye script work, you must prevent the shell from creating a child process (by using the dot command as described in the following section), so that the shell executes your program in its own environment. Then the **exit** command terminates the current shell (your login shell), and you are logged out.

11.2.3 Executing Commands: The dot Command

The **dot** command (**.**) is a built-in shell command that lets you execute a program in the current shell and prevents the shell from creating the child process. This is useful if you want to test your script files, such as the .profile startup file. The .profile file in your home directory contains any command that you want to execute whenever you log in. You do not need to log out and log in to activate your .profile file. Using the dot command, you can execute .profile, and the commands in it are applied to the current shell, your login shell.

Returning to your bye script file, let's execute it again, this time using the **dot** command:

```
$ . bye [Return]   . . . . . . . Use the dot command.
UNIX System V Release 4.0
login:_  . . . . . . . . . . . . Receive login prompt.
```

As with other commands in UNIX, there is a space between the dot command and its argument (in this case, your bye program).

Let's try another version of the bye script that does not require the **dot** command so you can simply type **bye** and press [Return] to log off.

In Chapter 8, you learned about the **kill** command. The following command placed in the bye script file means to kill all processes, including the login shell.

```
$ cat bye [Return]   . . . . . . bye new version.
kill -9 0  . . . . . . . . . . . Place command to kill all processes.
$_  . . . . . . . . . . . . . . . Receive prompt.
```

Only one problem remains. You want to be able to use the bye script regardless of your current directory. If bye is in your HOME directory, you have to either change to your HOME directory every time you want to execute bye or type the full pathname to the bye file.

Path Modification You can modify the *PATH* shell variable and add your HOME directory to it. Thus your HOME directory is also searched to find the commands.

Change the *PATH* variable to add your HOME directory.

```
$ echo $PATH [Return]   . . . . .  Check the PATH assignment.
PATH=:/bin:/usr/bin
$ PATH=$PATH:$HOME [Return] . . .  Add your HOME directory.
$ echo $PATH [Return] . . . . . .  Check again.
PATH=:/bin:/usr/bin:/usr/students/david
$_ . . . . . . . . . . . . . . .   Receive the prompt.
```

Now everything is ready; type **bye** and press [Return] at the **$** prompt, and you are logged off.

To make your changes permanent, place the new PATH *in your* `.profile` *file. Then, any time you log in, the* PATH *variable is set to your desired path.*

Command Substitution You can place the output of a command in the argument string. The shell executes the command surrounded by a pair of grave accent marks (`) and then substitutes the output of the command in the string passed to the **echo**. For example, if you type this:

```
$ echo Your current directory: 'pwd' [Return]
```

the output looks like this:

```
Your current directory: /usr/students/david
```

In this case, **pwd** *is executed, and its output, your current directory name, is placed in a string that is passed to the* **echo** *command.*

11.2.4 Reading Inputs: The read Command

One way to assign values to variables is to use the assignment operator, the equal sign (=). You can also store values in variables by reading strings from the standard input device.

You use the **read** command to read the user input and save it in a user-defined variable. This is one of the most common uses of the user-defined variables, especially in interactive programs in which you prompt the user for information and then read the user response. When **read** is executed, the shell waits until the user enters a line of text; then it stores the entered text in one or more variables. The variables are listed after the **read** on the command line, and the end of user input is signaled when the user presses [Return].

Figure 11.7 is a sample script named `kb_read` that shows how the **read** command works.

The **read** command is usually used in combination with **echo**. You use the **echo** command to prompt the user for what you want to be entered, and the **read** command waits for the user to enter something.

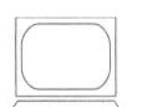

Invoke the `kb_read` shell script:

```
$ kb_read [Return]  . . . . . . .  Run it.

Enter your name:  . . . . . . . .  User is prompted.
david [Return]  . . . . . . . . .  Enter a name.
Your name is david  . . . . . . .  Name is echoed back.

$_ . . . . . . . . . . . . . . .   Receive prompt.
```

Figure 11.7
A Shell Script Sample: `kb_read`

```
$ cat -n kb_read
    1 #
    2 # kb_read
    3 # A sample program to show the read command
    4 #
    5 echo                        # skip a line
    6 echo "Enter your name:\c"   # prompt the user
    7 read name                   # read from keyboard and save it in name
    8 echo "Your name is $name"   # echo back the inputted data
    9 echo                        # skip a line
$_
```

1. *Your input string is stored in the variable name and then displayed.*
2. *It is a good idea to put the variables in quotation marks because you cannot anticipate the user input, and you do not want the shell to interpret special characters such as [*], [?], and so on.*

The **read** command reads one line from the input device. The first input word is stored in the first variable, the second word in the second variable, and so on. If your input string contains more words than the number of variables, the leftover words are stored in the last variable.

The characters assigned to the IFS *shell variable (see Chapter 8) determine the words' delimiters, in most cases the space character.*

Figure 11.8 shows a simple script, `read_test`, that reads the user response and displays it back on the screen. In this example, your input string consists of five words. The **read** command has three arguments (variables *Word1*, *Word2*, and *Rest*) to store your entire input. The first two words go to the first two variables, and the rest of the input string is stored in the third and last variable, *Rest*. If the **read** command in this example had only one parameter, say the variable *Word1*, the entire input string would have been stored in *Word1*.

Figure 11.8
A Simple Script, `read_test`, to Test the **read** Command

```
$ cat -n read_test
  1 #
  2 # read_test
  3 A sample program to test the read command.
  4 #
  5 echo                              # skip a line
  6 echo "Type in a long sentence:\c" # prompt user
  7 read Word1 Word2 Rest             # read user input
  8 echo "$Word1 \n $Word2 \n $Rest"  # display the contents of the variables
  9 echo "End of my act! :-)"          # signal the end of the program
 10 echo                              # skip a line
$_
```

Let's run the `read_test` script:

```
$ read_test [Return] . . . . . . . . . . . Assume read_test is an executable file.

Type in a long sentence: . . . . . . . . . Prompt is displayed.
Let's test the read command. [Return] . . . Show your input.
Let's  . . . . . . . . . . . . . . . . . . Content of $Word1.
test . . . . . . . . . . . . . . . . . . . Content of $Word2.
the read command . . . . . . . . . . . . . Content of $Rest.
End of my act! :-)

$_ . . . . . . . . . . . . . . . . . . . . Receive prompt.
```

11.3 EXPLORING THE SHELL PROGRAMMING BASICS

Now that you know you can place a sequence of commands in a file and create a simple script file, let's explore the shell script as a full-powered programming language that you can use to write applications.

Like any complete programming language, the *shell* provides you with commands and constructs that help you write well-structured, readable, and maintainable programs. This section explains the syntax of these commands and constructs.

11.3.1 Comments

Documentation is important when writing computer programs, and writing a shell script program is no exception. Documentation is essential to explain the purpose and logic of the program and commands used in the program that are not obvious. Documentation is for you and anybody else who reads your program. If you look at a program a few weeks after you have written it, you will be surprised at how much of your code you do not remember.

The shell recognizes **#** *as the comment symbol; therefore, characters after the* **#** *are ignored.*

The following command sequences show examples of a comment line:

```
#  . . . . . . . . . . . . . . . . . . . . . This is a comment.
# program version 3  . . . . . . . . . . . . This is also a comment line.
date   # show the current date . . . . . . . This is a comment that begins in the
                                             middle of the line.
```

11.3.2 Variables

Like other programming languages, the UNIX shell lets you create variables to store values in them. To store a value in a variable, you simply write the name of the variable, followed by the equal sign and the value you want to store in the variable, like this:

```
variable=value
```

Spaces are not permitted on either side of the equal sign.

The UNIX shell does not support data types (integer, character, floating point, etc.) as some programming languages do. It interprets any value assigned to a variable as a string of characters. For example, **count=1** means store character *1* in the variable called *count.*

The variable names follow the same syntax and rules that apply to file names as described in Chapter 5. Briefly, to refresh your memory, names must begin with a letter or an underscore character (_); you can use letters and numbers for the rest of the name.

The shell is an interpretive language, and the commands or variables you place in your script file can be directly typed at the **$** prompt. Of course, this method is a one-time process; if you want to repeat the sequence of commands, you have to type them again.

The following examples are valid variable assignments, and they remain in effect until you change them or log off:

```
$ count=1 [Return] . . . . . . . . . Assign character 1 to count.
$ header="The Main Menu" [Return] . Store the string in header.
$ BUG=insect [Return]  . . . . . . . Make another variable assignment.
```

If your value string contains embedded white spaces, then you must place it between quotation marks.

Variables in your shell script stay in the memory until the shell script is finished/terminated, or you can erase the variables by using the **unset** command. To do this, you type **unset**, specify the name of the variable you want to erase, and press [Return]. For example, the following command erases the variable called *XYZ*:

```
$ unset XYZ [Return]
```

Displaying Variables

You know from Chapter 8 that the **echo** command can be used to display the contents of variables. The format is as follows:

```
echo $variable
```

Let's show the contents of the three variables assigned in the previous examples:

```
$ echo $count $header $BUG [Return]  . Show the values stored in the
                                        variables.
1 The Main Menu insect
$_  . . . . . . . . . . . . . . . . . . The shell prompt appears.
```

Command Substitution

You can store the output of a command in a variable. You enclose a command in a pair of grave accent marks (`), and the shell executes the command and replaces the command with its output, like this:

```
$ DATE=`date` [Return] . . . . . . . Save the output of the date
                                     command in the variable DATE.
$ echo $DATE [Return] . . . . . . . . Check what is stored in DATE.
Wed Nov 29 14:30:52 EDT 2001
$_ . . . . . . . . . . . . . . . . . . The shell prompt appears.
```

The output of the **date** *command is stored in the* DATE *variable, and the* **echo** *command is used to display* DATE.

11.3.3 The Command Line Parameters

The shell scripts can read up to 10 command line parameters (also called *arguments*) from the command line into the special variables (also called *positional variables*, or *parameters*). *Command line arguments* are the items you type after the command, usually separated by spaces. These arguments are passed to the program and change the behavior of the program or make it act in a specific order. The *special variables* are numbered in sequence from 0 to 9 (no number 10) and are named *$0, $1, $2*, and so on. Table 11.4 shows the list of the special (positional) variables.

Table 11.4
The Shell Positional Variables

Variable	Meaning
$0	Contains the name of the script, as typed on the command line.
$1, $2,... $9	Contains the first through ninth command line parameters, respectively.
$#	Contains the number of command line parameters.
$@	Contains all command line parameters: "$1 $2 . . . $9".
$?	Contains the exit status of the last command.
$*	Contains all command line parameters: "$1 $2 . . . $9".
$$	Contains the PID (process ID) number of the executing process.

Using Special Shell Variables

Let's look at examples to understand how these special shell variables are used and arranged. Suppose you have a shell script named `BOX`, whose mode you have changed to an executable file, using the **chmod** command. Figure 11.9 shows your `BOX` script file.

1. The special variable *$0* always holds the name of the script typed on the command line.
2. Special variables *$1, $2, $3, $4, $5, $6, $7, $8*, and *$9* hold arguments 1 through 9, respectively. Command line arguments after the ninth parameter are ignored.
3. The special variable *$** holds all of the command line arguments passed to the program in a single string. It can store more than nine parameters.
4. The special variable *$@* holds all of the command line arguments, just like the $*. However, it stores them with quotation marks around each command line argument.
5. The special variable *$#* holds the count of the command line arguments that are typed on the command line.
6. The special variable *$?* holds the exit status received from the **exit** command in your script file. If you have not coded an **exit**, then it holds the status of the last non-background command executed in your script file.
7. The special variable *$$* holds the process ID of the current process.

Figure 11.9
A Shell Script File Named BOX

```
$ cat -n BOX
   1 #
   2 # BOX
   3 # A sample program to show the shell variables
   4 #
   5 echo                   # skip a line
   6 echo "The following is output of the $0 script: "
   7 echo "Total number of command line arguments: $#"
   8 echo "The first parameter is: $1 "
   9 echo "The second parameter is: $2"
  10 echo "This is the list of all is parameters: $* "
  11 echo                   # skip a line
$_
```

The following command sequences show different invocations of the BOX program (with or without the command line arguments), and the output of each run is investigated.

Invoke BOX with no command line arguments; just type the name of the file:

```
$ BOX [Return] . . . . . . . No command line arguments are given.

The following is output of the BOX script:
Total number of command line arguments: 0
The first parameter is:
The second parameter is:
This is the list of all parameters:

$_ . . . . . . . . . . . . . Receive prompt.
```

The variable *$0* holds the name of the script that is invoked. In this case, BOX is stored in *$0*. There are no command line arguments, so the *$#* holds *$0*, and no value is stored in variables *$1*, *$2*, and *$**. The **echo** command shows the prompt strings.

Invoke BOX with two command line arguments.

```
$ BOX IS EMPTY [Return] . . . Command line has two arguments.

The following is output of the BOX script:
Total number of command line arguments: 2
The first parameter is: IS
The second parameter is: EMPTY
This is the list of all parameters: IS EMPTY

$_ . . . . . . . . . . . . . Receive prompt.
```

The script name is stored in *$0*; in this case, BOX is stored in *$0*. The first argument is stored in special variable *$1*, the second argument in *$2*, and so on. In this case, **IS** is the first command line argument (stored in *$1*), and **EMPTY** is the second one (stored in *$2*). The *$** holds the list of all the command line arguments. In this case, it holds **IS EMPTY**.

If a script with more than nine command line arguments is invoked, any argument after the ninth one is ignored. However, you can capture all of them when you use the special variable *$**.

Assigning Values

Another way to assign values to the positional variables is to use the **set** command. What you type as the **set** command arguments are assigned to the positional variables.

Examine the following examples:

```
$ set One Two Three [Return] . .   Assign three arguments.
$ echo $1 $2 $3 [Return]. . . .    Display the values assigned to $1–$3.
One  Two  Three
$ set 'date' [Return] . . . . .    This is another example.
$ echo $1 $2 $3 [Return]. . . .    Display the positional variables.
Wed Nov 29
$_. . . . . . . . . . . . . . .    Back to the prompt.
```

The **date** command is executed, and its output becomes the argument to the **set** command. If the output of the **date** command is

```
Wed Nov 30  14:00:52 EDT 2005
```

then the **set** command has six arguments (the space character is the delimiter), and the *$1* variable holds *Wed* (the first argument), *$2* holds *Nov* (the second argument), and so on.

Scenario Now let's write a script file and make use of these positional variables. Suppose you want to write a script file that stores the specified file in a directory called keep in your HOME directory, and then invoke the vi editor to edit the specified file. The commands to do this job are

```
$ cp xyz $HOME/keep [Return]  . . .  Copy the specified file xyz to keep.
$ vi xyz [Return] . . . . . . .      Invoke the vi editor.
```

Figure 11.10 shows a script file called svi that also does the job. Commands in the svi file are the same commands you type at the prompt. The positional variables are used to make svi a versatile program. It saves and edits any specified file passed to it on the command line argument.

If svi works, the xyz file is saved in the keep directory, the vi editor is invoked, and you are able to edit the xyz file. You can use the **ls** command to check if a copy of the xyz file is saved in keep.

Figure 11.10
The svi Script File

```
$ cat -n svi
   1 #
   2 # svi
   3 # save and invoke vi
   4 A sample program to show the usage of the shell variables
   5 #
   6 DIR=$HOME/keep        # assign keep pathname to DIR
   7 cp $1 $DIR            # copy specified file to keep
   8 vi $1                 # invoke vi with the specified filename
   9 exit 0                # end of the program, exit
$_
```

Scenario Suppose the `svi` mode is changed to make it an executable file, and you have a file called `xyz` in your current directory that you want to edit.

Let's run `svi` and look at the results:

```
$ svi xyz [Return] . . . . . . . . .  Run svi with a specified file
                                      name.
hello . . . . . . . . . . . . . . . . You are in the vi editor, and the
                                      contents of xyz are displayed.
~ . . . . . . . . . . . . . . . . . . Show the rest of the vi screen.
"xyz" 1 Line, 5 characters  . . . . . The vi status line.
$ . . . . . . . . . . . . . . . . . . The prompt is displayed after
                                      you exit vi.
```

What if you do not have the `xyz` file in your current directory? How does the `svi` program work if you don't specify a file name? The `svi` program will work, but it may not be exactly the way you would expect. In the first case, the **cp** command fails to find the `xyz` file in your current directory, and the vi editor is invoked with the `xyz` as a new file. In the second case, the **cp** command fails again, and the vi editor is invoked with no file name. In both cases, you do not see the **cp** command's error messages, as vi is invoked before you have a chance to see them.

The `svi` program needs fixing. It must be able to recognize errors, show the appropriate error messages, and not invoke **cp** or vi if the specified file is not in your directory or the filename is not specified on the command line.

Before modifying the `svi` code to handle these problems, you need to know some other commands and constructs of the shell language.

Terminating Programs: The exit Command

The **exit** command is a built-in shell command that you can use to immediately terminate execution of your shell program. The format of this command is as follows:

```
exit n
```

where *n* is the exit status, which is also called the return code (RC). If no exit value is provided, the exit value of the last command executed by the shell is used. To be consistent with the other UNIX programs (commands) that usually return a status code upon completion, you can program your script to return an exit status to the parent process. As you know, typing **exit** at the **$** prompt terminates your login shell and consequently logs you off.

11.3.4 Conditions and Tests

You can specify that certain commands be executed depending on the results of the other commands' executions. You often need this kind of control when writing shell scripts.

Every command executed in the UNIX system returns a number that you can test to control the flow of your program. A command returns either a 0 (zero), which means success, or some other number, which indicates failure. These true and false conditions are used in the shell programming constructs and are tested to determine the flow of the program. Let's look at these constructs.

The if-then construct

The **if** statement provides a mechanism to check whether the condition of something is true or false. Depending on the results of the **if** test, you may change the sequence of command executions in your program. The following shows the **if** statement (construct) syntax:

```
if [ condition ]
then
  commands
  . . .
  last-command
fi
```

1. *The* **if** *statement ends with the reserved word* **fi** *(* **if** *typed backward).*
2. *The indentation is not necessary, but it certainly makes the code more readable.*
3. *If the condition is true, then all the commands between the* **then** *and* **fi**, *what is called the body of the* **if**, *are executed. If the condition is false, the body of the* **if** *is skipped, and the line after* **fi** *is executed.*

The square brackets around the conditions are necessary and must be surrounded by white spaces, either spaces or tabs.

Modify the `svi` script file to include an **if** statement to match Figure 11.11 (`svi2`). The **if** statement is putting some control over the `svi2` output.

Figure 11.11
Another Version: The `svi2` Script File

```
$ cat -n svi2
   1 # svi2
   2 # save and invoke vi
   3 # A sample program to show the usage of the shell variables
   4 # Version 2: adding the if-then statement
   5 #
   6 DIR=$HOME/keep           # assign keep pathname to DIR
   7 if [ $# = 1 ]            # check for number of command line arguments
   8 then
   9   cp $1 $HOME/keep       # copy specified file to keep
  10 fi                       # end of the if statement
  11 vi $1                    # invoke vi with the specified file name
  12 exit 0                   # end of the program, exit
$_
```

If the count of the command line arguments (*$#*) is one (a filename is specified), then the condition is true and the body of the **if** (the **cp** command) is executed, followed by the vi invocation with the specified file name, exactly as the last `svi` program run.

If the count of the command line arguments (*$#*) is zero (no filename is specified on the command line), then the condition is false, the body of the **if** (the **cp** command) is skipped, and only the vi editor is invoked with no file name.

You can run the svi2 program by typing the following command line:

```
$ svi2 [Return] . . . . . . . Run svi without an argument.
$ svi2 myfirst [Return] . . . Run svi2 with a file name.
```

The if-then-else Construct

By adding the **else** clause to the **if** construct, you can execute certain commands when the test of the condition returns a false status. The syntax of this more complex **if** construct is as follows:

```
if [ condition ]
then
  true-commands
  . . .
  last-true-command
else
  false-commands
  . . .
  last-false-command
fi
```

1. *If the condition is true, then all the commands between the* **then** *and* **else**, *the body of the* **if**, *are executed.*
2. *If the condition is false, the body of the* **if** *is skipped and all the commands between* **else** *and* **fi**, *the body of the* **else**, *are executed.*

Modify the svi2 script file to include the **if** and **else** statements. Figure 11.12 shows this modification, called svi3.

Figure 11.12
Another Version: The svi3 Script File

```
$ cat -n svi3
   1 #
   2 # svi3
   3 # save and invoke vi
   4 # A sample program to show the usage of the shell variables
   5 # Version 3: adding the if-then-else statement
   6 #
   7 # DIR=$HOME/keep      # assign keep pathname to DIR
   8 if  [ $# = 1 ]       # check for the number of command line arguments
   9 then
  10   cp $1 $DIR          # copy specified file to keep
  11   vi $1               # invoke vi with the specified file name
  12 else
  13   echo "You must specify a file name. Try again."  #display error message
  14 fi                    # end of the if statement
  15 exit 0                #end of the program, exit
$_
```

If the count of the command line arguments (*$#*) is one (a filename is specified), then the condition is true and the body of the **if** (the **cp** and **vi** commands) is executed. Next, the body of the **else** is skipped, and the **exit** command is executed, exactly as in the previous `svi2` run.

If the count of the command line arguments (*$#*) is not one, then the condition is false, the body of the **if** (the **cp** and **vi** commands) is skipped, and the body of the **else** (the **echo** command) is executed. The **echo** command shows the error message, and then the **exit** command is executed.

Let's run this new version of the program `svi3`.

```
$ svi3 [Return]   . . . . . . . No command line argument.
You must specify a filename. Try again.
$_ . . . . . . . . . . . . . . Receive prompt.
```

1. *You must specify a filename on the command line; otherwise, you get the error message and the vi editor is not invoked.*
2. *If you do not specify a filename, then the number of arguments (the value of the positional variable* $#*) is zero. Therefore, the* **if** *condition fails and the body of the* **if** *is skipped; the body of the* **else**, *the* **echo** *command, is executed.*

In this version, the `svi3` script does not check for the file's existence. If the specified filename is not in the specified directory, the copy command is going to complain, but the vi editor is invoked with the specified filename as a name for a new file.

The if-then-elif Construct

When you have nested sets of **if** and **else** constructs in a script file, you can use the **elif** (short for **else if**) statement. The **elif** combines the **else** and **if** statements. The complete syntax is as follows:

```
if [condition_1]
then
     commands_1
elif [condition_2]
then
     commands_2
elif [condition_3]
then
     commands_3
...
...
else
     commands_n
fi
```

The script file in Figure 11.13, called `greetings1`, outputs greetings according to the time of the day. It shows the **Good Morning** message before noon, **Good Afternoon** if the time is between 12:00 and 18:00, and so on. The **if-then-elif** construct is used to determine the morning, afternoon, and evening hours.

1. *The* **set** *command is used with* **date** *as its argument to set the positional variables.*
2. *The* **hour:minute:second** *is the fourth field in the date and time string (the output of the* **date** *command), and it is assigned to the positional variable* $4.

Figure 11.13
A Sample Script Using the **if-then-elif** Construct: `greetings1`

```
$ cat -n greetings1
   1 #
   2 # greetings
   3 # greetings program version 1
   4 # A sample program using the if-then-elif construct
   5 # This program displays greetings according to the time of the day
   6 #
   7 # echo
   8 set `date`          # set the positional variables to the date string
   9 hour=$4             # store the part of the date string that shows the hour
  10 if [ "$hour" -lt 12 ]              # check for the morning hours
  11 then
  12    echo "GOOD MORNING"
  13 elif [ "$hour" -lt 18 ]            # check for the afternoon hours
  14 then
  15    echo "GOOD AFTERNOON"
  16 else                               # it must be evening
  17    echo "GOOD EVENING"
  18 fi                                 # end of the if statement
  19 echo
  20 exit                               # end of the program, exit
$_
```

3. *For this version of the* `greetings1` *script to work, you must have the standard shell (*sh*). If your login shell is* ksh *or* bash*, type* **sh** *and [Return] to invoke a copy of the* sh *shell and then execute the* `greetings1` *file. You can type* **exit** *to return to your login shell.*

To run this program, you type the following command line. Notice that it is assumed that you have already made `greetings1` an executable file.

```
$ greetings1 [Return] . . .  Run greetings1.
```

You can write the `greetings` program in many ways depending on the shell commands you know and want to use. For example, instead of using the **set** command and the positional variables, you could use the **date** command capabilities and obtain only the hour field from its output string.

Figure 11.14 shows another version of the `greetings` program, which we will call `greetings2`.

The line **hour=`date +%H`** requires more explanation. The **%H** limits the output of the **date** command to just the part of the date string that shows the hour.

Use the **date** command capabilities:

```
$ date [Return]     . . . . .  Display date and time.
Wed Nov 30  14:00:52 EDT 2005
$ date +%H [Return] . . . .  Show only the hour of the day.
14
$_  . . . . . . . . . . . .  The prompt reappears.
```

Figure 11.14
A Sample Script Using the **if-then-elif** Construct: `greetings2`

```
$ cat -n greetings2
 1 #
 2 # greetings2
 3 # greetings program version 2
 4 # A sample program using the if-then-elif construct
 5 # This program displays greetings according to the time of the day
 6 # Version2: using the date command format control option
 7 #
 8 echo                     # skip a line
 9 hour=`date +%H`          # store the part of the date string that shows the hour
10 if [ "$hour" -le 18 ]   #check for the morning hours
11 then
12    echo "GOOD MORNING"
13 elif [ "$hour" -le 18 ] #check for the afternoon hours
14 then
15    echo"GOOD AFTERNOON"
16 else                     # it must be evening
17    echo"GOOD EVENING"
18 fi
19 echo                     #skip a line
20 exit 0                   # end of the program, exit
$_
```

The **date** command has numerous field descriptors that let you limit or format its output. You type a **+** at the beginning of the argument followed by field descriptors such as **%H**, **%M**, and so on. You can use the **man** command to obtain a full list of the field descriptors. Let's look at more examples.

Use the **date** command field descriptors:

```
$ date '+DATE: %m-%d-%y' [Return] . . . . .   Show only the date separated by hyphens.
DATE: 05-10-99
$ date '+TIME: %H:%M:%S' [Return] . . . . .   Show only the time separated by colons.
TIME: 16:10:52
```

1. *The argument must start with a plus sign. It indicates that the output format is under your control. Each field descriptor is preceded by a percent sign (%).*
2. *If the argument contains white-space characters, then it must be put inside quotation marks.*
3. *You can add the* `greetings2` *program to your* `.profile` *file, so that every time you log in, the appropriate greeting message is displayed.*

True or False: The test Command

The **test** command is a built-in shell command that evaluates the expression given to it as an argument and returns true if the expression is true (if it returns 0) or, if otherwise, false (if it returns nonzero). The expression could be simple, such as testing two numbers for equality, or complex, like testing several commands that are related with logical operators. The **test** command is particularly helpful in writing script files. In fact, the brackets around the condition in the **if** statement are a special form of the **test** command. Figure 11.15 shows a sample script file using the **test** command to test the **if** condition.

Figure 11.15
A Shell Script File Named `check_test`

```
$ cat -n check_test
   1 #
   2 # check_test
   3 # A sample program using the test command
   4 #
   5 echo                                    # skip a line
   6 echo "Are you OK?"
   7 echo "Input Y for yes and N for no:\c"  # prompt user
   8 read answer                             # store user response in answer
   9 if test "$answer" = Y                   # test if user entered Y
  10 then
  11   echo "Glad to hear that!"             # display the message
  12 else                                    # user entered N
  13  echo "Go home!"                        # display the message
  14 fi
  15 echo                                    # skip a line
  16 exit 0                                  # end of the program, exit
$_
```

1. *The* `check_test` *script prompts the user to enter (*Y*)es or (*N*)o, and then reads the user answer.*
2. *If the user response is the letter* Y, *the* **test** *command returns zero, and the* **if** *condition is true, then the body of the* **if** *is executed. The message* **Glad to hear that!** *is displayed, and the body of the* **else** *is skipped.*
3. *Any input from the user other than a single letter* Y *makes the* **if** *condition fail; therefore, the body of the* **if** *is skipped and the* **else** *is executed. The message* **Go home!** *is displayed.*

Invoking test with Brackets The shell offers you another way to invoke the **test** command. You can use square brackets ([and]) instead of the word **test**. So the **if** statement can be written as follows:

```
if test "$variable" = value
```

or

```
if [ "$variable" = value ]
```

11.3.5 Testing Different Categories

Using the **test** command, you can test different categories of things including the following: numeric values, string values, and files. Each of these categories is explained in the following sections.

Numeric Values

You can use the **test** command to test (compare) two integer numbers algebraically. You also can combine expressions comparing numbers with logical operators. The format is as follows:

```
test expression_1 logical operator expression_2
```

The logical operators are as follows:

- Logical *and* operator (**-a**): The **test** command returns 0 (condition code true) if both expressions are true.
- Logical *or* operator (**-o**): The **test** command returns 0 (condition code true) if one or both of the expressions are true.
- Logical *not* operator (**!**): The **test** command returns 0 (condition code true) if the expression is false.

The operators available for comparing variables that hold numeric values are summarized in Table 11.5.

Table 11.5
The **test** Command Numeric Test Operators

Operator	Example	Meaning
-eq	*number1* **-eq***number2*	Is *number1* equal to *number2*?
-ne	*number1* **-ne***number2*	Is *number1* not equal to *number2*?
-gt	*number1* **-gt***number2*	Is *number1* greater than *number2*?
-ge	*number1* **-ge***number2*	Is *number1* greater than or equal to *number2*?
-lt	*number1* **-lt***number2*	Is *number1* less than *number2*?
-le	*number1* **-le***number2*	Is *number1* less than or equal to *number2*?

Scenario Suppose you want to write a script that accepts three numbers as input (command line arguments) and displays the largest of the three. Using the **cat** command, Figure 11.16 shows one way of writing this program, which is named `largest`.

1. *When you run the* `largest` *program, it waits for you to enter three numbers. Then your input is passed through the* **if-then-elif** *construct to find the largest number.*
2. *If neither of the first two numbers you enter is the largest, then the first* **if** *statement fails; next, the* **elif** *also fails, and your program gets to the* **else** *statement. There is no need for more checking. If the largest number is neither the first nor the second number, then it must be the third one.*

Try a sample run of the `largest` program:

```
$ chmod +x largest [Return] . . .  Change it into an executable file.
$ largest [Return]  . . . . . . .  Run the program.
Enter three numbers and I show you the largest of them>>_
100 10 400 [Return] . . . . . . .  Enter three numbers.
```

Figure 11.16
A Shell Program File Named `largest`

```
$ cat -n largest
   1 #
   2 # largest
   3 # A sample program using the test command
   4 # This program accepts three numbers and shows the largest of them
   5 #
   6 echo                     # skip a line
   7 echo "Enter three numbers and I will show you the largest of them>> \c"
   8 read num1 num2 num3
   9 if test "$num1" -gt "$num2" -a "$num1" -gt "$num3"
  10 then
  11    echo "The largest number is: $num1"
  12 elif test "$num2" -gt "$num1" -a "$num2" -gt "$num3"
  13 then
  14    echo "The largest number is:$num2"
  15 else
  16    echo "The largest number is:$num3"
  17 fi
  18 echo "Done! :-)"         # end of the program mesage
  19 echo                     # skip a line
  20 exit 0                   # end of the program, exit
$_
```

```
The largest number is: 400
Done! :-)
$_ . . . . . . . . . . . . . . . . The prompt reappears.
```

Think about how you can improve the `largest` program. For example, could you use fewer lines of code? (e.g., Do you have to repeat the **echo** command in the body of the **if**, **elif**, and **else**?) Could you do error checking? (What if you enter two numbers?)

String Values

You can also compare (test) strings with the **test** command. The **test** command provides a different set of operators for the string comparison. These operators are summarized in Table 11.6. The following examples show the use of the **test** command with string arguments.

Table 11.6
The **test** Command String Test Operators

Operator	Example	Meaning
=	*string1* = *string2*	Does *string1* match *string2*?
!=	*string1* != *string2*	Does *string1* not match *string2*?
-n	**-n** *string*	Does *string* contain characters (nonzero length)?
-z	**-z** *string*	Is *string* an empty string (zero length)?

First experiment with the null variable:

```
$ STRING5 [Return] . . . . . . . Declare a null variable.
$ test -z $STRING [Return] . . Test for zero length.
test: Argument expected. . . . Error message.
$_ . . . . . . . . . . . . . . The prompt reappears.
```

1. *The shell substitutes the* null *string for* $STRING, *thus the error message.*
2. *Enclosing a string variable in quotation marks ensures a proper test, even if the variable contains spaces or tabs.*

Now try the variable *$STRING* in quotation marks:

```
$ test -z "$STRING" [Return]  . . Test for zero length.
```

This time the **test** *command returns zero (true), which means the variable* $STRING *contains a zero length string.*

In another example, the commands are typed directly at the **$** prompt. If the command is not complete, the shell shows the secondary prompt (>) and waits for you to complete the command:

```
$ DATE1= `date` [Return] . . . . . . . . . Initialize DATE1.
$ DATE2= `date` [Return] . . . . . . . . . Initialize DATE2.
$ if test "$DATE1" = "$DATE2" [Return] . . Test for equality.
> then [Return] . . . . . . . . . . . . . . The shell shows the second prompt sign
                                           since the if command is not completed yet.
> echo "STOP! The computer clock is dead!" [Return]
> else [Return]  . . . . . . . . . . . . . Begin the else body.
> echo "Everything is fine." [Return]
> fi [Return]  . . . . . . . . . . . . . . End of if; as soon as you press [Return],
                                           the program is executed.
Everything is fine . . . . . . . . . . . . Get the output of the program.
$_ . . . . . . . . . . . . . . . . . . . . The prompt reappears.
```

The result of this test (**if** *condition) is false. Therefore, the body of the* **else** *is executed, and the message is displayed. Can the values stored in* DATE1 *and* DATE2 *ever be equal?*

Files

You can use the **test** command to test file characteristics, such as file size, file type, and file permissions. More than 10 file attributes can be checked; a few of them are introduced here. Table 11.7 summarizes the file test operators. The following example shows the use of the **test** command in connection with files. The commands are typed directly at the **$** prompt.

Suppose you have a file called `myfile`, with only read access to it. Let's type the following commands and see the output:

```
$ FILE=myfile [Return] . . . . . . . . . . Initialize FILE variable.
$ If test -r "$FILE" [Return]  . . . . . . Check to see if myfile is readable.
> then [Return]  . . . . . . . . . . . . . Secondary prompt.
> echo "READABLE" [Return] . . . . . . . . Show message.
> elif test -w "$FILE" [Return]  . . . . . Check to see if myfile is writable.
> then [Return]
```

Table 11.7
The **test** Command File Test Operators

Operator	Example	Meaning
-r	**-r** *filename*	Does *filename* exist, and is it readable?
-w	**-w** *filename*	Does *filename* exist, and is it writable?
-s	**-s** *filename*	Does *filename* exist, and does it have a nonzero length?
-f	**-f** *filename*	Does *filename* exist, but is it not a directory file?
-d	**-d** *filename*	Does *filename* exist, and is it a directory file?

```
> echo "WRITABLE" [Return] . . . . . . . . Show message.
> else [Return]
> echo "Read and Write Access Denied"
> fi [Return] . . . . . . . . . . . . . . End of the if construct.
READABLE
$_ . . . . . . . . . . . . . . . . . . . . Return back to the primary prompt.
```

1. *As soon as you type* **fi***, the end of the* **if** *construct, the shell executes the commands, produces the output, and shows the* **$** *prompt.*
2. *The first* **if** *tests the read access to* `myfile`*. Since you have read access, the* **test** *command returns 0 (true), and the* **if** *condition is true. So only the body of the first* **if** *is executed, and the message* **READABLE** *is displayed.*
3. *The message* **WRITABLE** *is echoed if you have only write access to* `myfile`.
4. *The message* **Read and Write Access Denied** *is displayed if you do not have write or read access to* `myfile` *and the* **if** *condition and* **elif** *condition fail.*

Figure 11.17 shows yet another version of the `svi` script file. In this version (`svi4`), the existence of the specified file is checked, and only if the file exists will the **cp** and **vi** commands be executed.

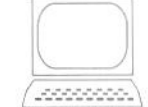

Run this new version of the `svi` program `svi4`:

```
$ svi xyz [Return] . . . . Run it, specifying a nonexistent file xyz.
File not found. Try again.
$ svi [Return] . . . . . . Run it again; no file is specified.
You must specify a filename. Try again.
$_ . . . . . . . . . . . . Return to the prompt.
```

1. *To check the* **if** *conditions, the word* test *is used instead of brackets.*
2. *This program uses nested* **if** *constructs (* **if** *inside another* **if***).*
3. *If the specified file is not found, then it shows the* **File not found. Try again.** *message.*
4. *If the specified file is found, then it copies the file to the* `keep` *directory and invokes the vi editor.*
5. *If no filename is specified on the command line, the first* **if** *condition fails, and it shows the* **You must specify a filename. Try again.** *message.*

Figure 11.17
Another Version of the svi Script File: svi4

```
$ cat -n svi4
   1 #
   2 # svi4
   3 # save and invoke vi
   4 # A sample program to use the test command and file test operators
   5 # Version 4: This version checks for the existence of the file
   6 #
   7 # DIR=$HOME/keep     # assign keep pathname to DIR
   8 # if test $# = 1     # check for number of command line argument
   9 then
  10  if test -f $1      # check for the existence of the file
  11  then
  12    cp $1 $DIR       # copy specified file to keep
  13    vi $1            # invoke vi with the specified filename
  14  else
  15    echo "File not found. Try again"
  16  fi
  17 else
  18   echo " You must specify a filename. Try again."
  19 fi                  # end of the if statement
  20 exit 0              # end of the program, exit
$_
```

11.3.6 Parameter Substitution

The shell provides a parameter substitution facility that lets you test the value of a parameter and change its value according to a specified option. This is useful in shell programming when you want to check if a variable is set to something. For example, when you issue a **read** command in a script file, you want to make sure the user has entered something before taking any action.

The format consists of a dollar sign (**$**), a set of braces (**{** and **}**), a variable, a colon (**:**), an option character, and a word, as follows:

```
${parameter:option character word}
```

An option character determines what you intend to do with the *word.* The four option characters are specified by the **-**, **+**, **?**, and **=** signs. These four options perform differently, depending on whether the variable is an empty variable.

A variable is an *empty variable* (a *null variable*) if its value is an empty string. For example, all of the following variables are set to a null value and are empty variables.

```
EMPTY=   . . . . .  This is an empty variable called EMPTY.
EMPTY="" . . . .    This is an empty variable called EMPTY.
EMPTY='' . . . .    This is an empty variable called EMPTY.
```

To create an empty variable, put no blank characters between the quotation marks.

Table 11.8 summarizes the shell variable substitution (evaluation) options. A brief explanation and example of each type follow.

Table 11.8
The Shell Variable Evaluation Options

Variable Option	Meaning
$*variable*	The value stored in *variable*.
${*variable*}	The value stored in *variable*.
${*variable: -string*}	The value of *variable* if it is set and not empty; otherwise, the value of *string*.
${*variable: +string*}	The value of *string* if *variable* is set and not empty; otherwise, nothing.
${*variable: =string*}	The value of *variable* if it is set and not empty; otherwise, *variable* is set to the value of *string*.
${*variable: ?string*}	The value of *variable* if it is set and not empty; otherwise, print the value of *string* and exit.

${parameter} Placing the variable (parameter) inside the braces prevents any conflict caused by the character that follows the variable name. The following example clarifies this matter.

Suppose you want to change the name of a file called memo, specified in the variable called *FILE*, to memoX.

```
$ echo $FILE [Return] . . . . . . Check what is stored in FILE.
memo
$ mv $FILE $FILEX [Return] . . . . Change memo to memoX.
Usage: mv
$_ . . . . . . . . . . . . . . . . . Return to prompt.
```

This command did not work because the shell considers *$FILEX* the name of the variable, which does not exist.

```
$ mv $FILE ${FILE}X [Return] . . . Change memo to memoX.
$_ . . . . . . . . . . . . . . . . . Job done; the prompt is back.
```

This command worked because the shell considers *$FILE* the variable name and substitutes its value, in this case memo.

${parameter:-*word*} The - (hyphen) option means if the listed variable (parameter) is set and not empty (nonnull), use its value. Otherwise, if the variable is empty (null) or unset, substitute its value with *word*. For example:

```
$ FILE= [Return] . . . . . . . . . . . . . . . . . . . This is an empty variable.
$ echo ${FILE:-/usr/david/xfile} [Return] . . . . . . Display FILE value.
/usr/david/xfile
$ echo"$FILE" [Return] . . . . . . . . . . . . . . . . Check the FILE variable.
 . . . . . . . . . . . . . . . . . . . . . . . . . . . It remains empty.
$_ . . . . . . . . . . . . . . . . . . . . . . . . . . Return to prompt.
```

1. *The shell evaluates the variable* FILE*; it is empty, so the* **-** *option results in substitution of the* `/usr/david/xfile` *string, which is passed to the* **echo** *command to display it.*
2. *The* FILE *variable remains an empty variable.*

${parameter:+*word*} The **+** option is the opposite of the **-** option. It means that if the listed variable (parameter) is set and is not empty (nonnull), then substitute its value with *word.* Otherwise, the variable's value remains the same. For example:

```
$ HELP="wanted" [Return] . . . . . . . . Set the HELP variable
$ echo ${HELP:+"Help is on the way"} [Return]
Help is on the way
$ echo $HELP [Return] . . . . . . . . . Check the HELP variable.
wanted . . . . . . . . . . . . . . . . . It remains the same.
$_ . . . . . . . . . . . . . . . . . . . Return to prompt.
```

1. *The shell evaluates the* HELP *variable; it is set to* wanted. *So the* **+** *option results in the substitution of the* **Help is on the way** *string, which is passed to the* **echo** *command to display it.*
2. *The value stored in the* HELP *variable remains the same.*

${parameter:=*word*} The = option means if the listed variable (parameter) is not set or it is empty (null), then substitute its value with *word.* Otherwise, if the variable is not empty, then its value remains the same. For example:

```
$ MESG= [Return] . . . . . . . . . . . . Empty variable.
$ echo ${MESG:="Hello There!"} [Return] . Display MESG value.
Hello There!
$ echo $MESG [Return] . . . . . . . . . Check the MESG variable.
Hello There! . . . . . . . . . . . . . . It is set to the specified
                                         string.
$_ . . . . . . . . . . . . . . . . . . . Return to prompt.
```

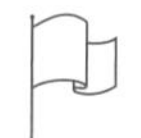

The *word* can be a string with embedded white spaces. It works as long as you place it in quotation marks.

1. *The shell evaluates the variable* MESG. *It is empty. So the* **=** *option results in setting* MESG *to the string* **Hello There!**. *The substitution is done, and the* **echo** *command shows the value stored in* MESG.
2. *The value of the* MESG *variable is changed and it is not an empty variable any more.*

${parameter:?*word*} The **?** option means if the listed variable (parameter) is set and is not empty, then substitute its value. Otherwise, if the variable is empty, print the string *word* and exit the current script. If *word* is omitted, then show the preset message **parameter null or not set**. For example:

```
$ MESG=[Return] . . . . . . . . . . . . MESG is an empty variable.
$ echo ${MESG:? "ERROR!"} [Return] . . Perform the substitution according to the option.
ERROR!
$_ . . . . . . . . . . . . . . . . . . . Return to prompt.
```

The shell evaluates the MESG *variable; it is an empty variable. So the* **?** *option results in the substitution of the ERROR string, which is passed to the* **echo** *command to display it.*

You can use this option to show an error message and terminate your shell script if the user just presses [Return] in response to a **read** command. Figure 11.18 shows an example script file called `name`.

Figure 11.18
A Shell Script File Called `name`

```
$ cat -n name
   1 # name
   2 # save and invoke vi
   3 # A sample program to use the shell parameter substitutions
   4 #
   5 echo                              # skip a line
   6 echo "Enter your name: \c"        # prompts the user
   7 read name
   8 echo ${name:?"You must enter your name"}
   9 echo "Thank you. That's all."
  10 echo                              # skip a line
  11 exit 0                            # end of the program, exit
$_
```

Run the name script:

```
$ name [Return] . . . . . . . . .  Run the name script file.

Enter your name:_ . . . . . . . .  User prompt.
[Return] . . . . . . . . . . . .   Let's say you just press [Return].
You must enter your name  . . .    This is feedback to user.

$_ . . . . . . . . . . . . . . .   The script is terminated, and the
                                   prompt is back.
```

1. *If you press [Return] in response to the* **read** *command, name remains a null variable. The* **?** *option tests the variable; it is a null variable. The specified message is displayed, and the shell program is terminated.*
2. *However, if you enter your name or another word, then the script continues, and the message* **Thank you. That's all.** *is displayed.*

Let's run `name` again:

```
$ name [Return] . . . . . . . . .  Run name.

Enter your name: _ . . . . . . . . User prompt.
Emma [Return] . . . . . . . . . .  Enter your name.
Thank you. That's all.

$_ . . . . . . . . . . . . . . .   Prompt is back.
```

11.4 ARITHMETIC OPERATIONS

The shell does not include a simple, built-in operator for arithmetic operations. For example, let's try to add one to the value of a shell variable. The result is not what you might expect.

```
$ x=10 [Return] . . . . . . Initialize x to 10.
$ echo $x [Return]  . . . . Display the x value.
10
$ x=$x + 1 [Return] . . . . Adds one to the x value.
$ echo $x [Return]  . . . . Display the x value.
10 + 1
$_  . . . . . . . . . . . . Return to prompt.
```

The shell did not add the number 1 to the value of x. *It concatenated the string* +1 *to the string value of* x.

11.4.1 Arithmetic Operations: The expr Command

You can use the **expr** command to evaluate expressions. This command provides arithmetic operations capability and evaluates either numeric or nonnumeric character strings. The **expr** command takes the arguments as expressions, evaluates them, and displays the results on the standard output device.

Arithmetic Operators

The following command sequences introduce the arithmetic operator and demonstrate the use of the **expr** command to evaluate constants.

The following example demonstrates adding and subtracting constants:

```
$ expr 1 + 2 [Return] . . . The + sign is the addition operator.
3
$ expr 15 - 6 [Return]  . . The - sign is the subtraction operator.
9
$ expr 6 - 15 [Return]
-9
$ expr 2.4 - 1[Return]  . . This is a bad argument (2.4); integers only.
expr: nonnumeric argument . UNIX error message.
$ expr 1+1 [Return] . . . . This is a bad argument (1+1).
1+1 . . . . . . . . . . . . It displays string 1+1.
$_
```

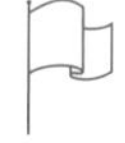

Spaces between the elements of an expression are necessary. The correct command is typed as `expr 1 + 1`. Notice the spaces around the + sign.

This example shows dividing, multiplying, and calculating a division remainder:

```
$ expr 10 / 2 [Return] . . . . . . . The / sign is the division
                                     operator.
5
```

```
$ expr 10 \* 2 [Return] . . . . . . . The * is the multiplication
                                      operator.
20
$ expr 10 \% 3 [Return] . . . . . . . The % is the remainder
                                      operator.
1
$_
```

Because the * (multiplication) and % (remainder) characters have special meaning to the shell, they must be preceded by a backslash [\] for the shell to ignore their special meanings.

The following example demonstrates adding an integer to a shell variable:

```
$ x=10 [Return] . . . . . . . . . . . Initializes x to 10.
$ x=`expr $x + 1` [Return] . . . . . . Adds 1 to x.
$ echo $x [Return] . . . . . . . . . . Displays x.
11
$_ . . . . . . . . . . . . . . . . . . Prompt.
```

The grave accent marks [` and `] surrounding the command are necessary and cause the output of the command (expr) to be substituted.

Relational Operators

The **expr** command provides relational operators that work on both numeric and nonnumeric arguments. If both arguments are numeric, the comparison is numeric. If one or both arguments are nonnumeric, the comparison is nonnumeric and uses the ASCII values. The relational operators are

```
=  . . . . . . . . . . . . . . . . . . equal to
!= . . . . . . . . . . . . . . . . . . not equal to
<  . . . . . . . . . . . . . . . . . . less than
<= . . . . . . . . . . . . . . . . . . less than or equal to
>  . . . . . . . . . . . . . . . . . . greater than
>= . . . . . . . . . . . . . . . . . . greater than or equal to
```

1. *The* **expr** *command displays 1 when the comparison is true.*
2. *The* **expr** *command displays 0 when the comparison is false.*

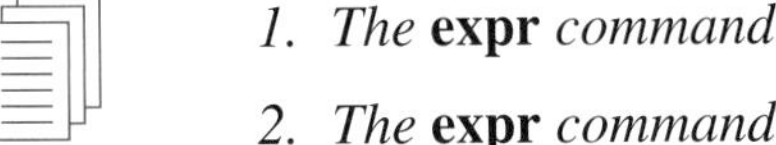

The following command sequences show the use of the **expr** command with the relational operators:

```
$ expr Gabe = Gabe [Return] . . . . Alphabetic comparison
1 . . . . . . . . . . . . . . . . . Display 1, indicating true.
$ expr Gabe = Daniel [Return] . . . Alphabetic comparison.
0 . . . . . . . . . . . . . . . . . Display 0, indicating false.
$ expr 10 \< 20 [Return] . . . . . . Numeric comparison.
1 . . . . . . . . . . . . . . . . . Display 1, indicating true.
$ expr 10 \> 20 [Return] . . . . . . Numeric comparison.
0 . . . . . . . . . . . . . . . . . Display 0, indicating false.
$_ . . . . . . . . . . . . . . . . . Return to prompt.
```

Again, because the > (greater than) and < (less than) characters have special meaning to the shell, they must be preceded by a backslash [\] for the shell to ignore their special meanings.

```
$ expr 6 \> A [Return]  . . . Mixed comparison (numeric and
                              alphabetic).
0  . . . . . . . . . . . . . . Display 0, indicating false.
$_ . . . . . . . . . . . . . . Return to prompt.
```

*The **expr** command treated 6 as an alphabetic argument, and compared the ASCII value of 6 (54) with the ASCII value of letter A (65).*

11.4.2 Arithmetic Operations: The let Command

You can use the **let** command for dealing with integer arithmetic. The **let** command is a simpler alternative to the **expr** command and includes all the basic arithmetic operations, including addition, subtraction, multiplication, and division. For example:

```
$ x=100 [Return]  . . . . . . Initialize x to 100.
$ let x=x+1   . . . . . . . . Add 1 to x.
$ echo $x   . . . . . . . . . Display x value.
101
$ let y=x*2 [Return]  . . . . Use let command for multiplication.
echo $y   . . . . . . . . . . Display y value.
202
$_
```

1. *The **let** command automatically uses the values of the variables. In this example, the value of the variable* x *is obtained by typing* **x** *instead of* **$x**.
2. *The **let** command interprets the * and % signs as the multiplication and remainder operators respectively, and there is no need to use the * or \% to prevent their special meanings.*

11.5 THE LOOP CONSTRUCTS

You use loop constructs in programs when you want to repeat a set of statements or commands. The loop constructs save a lot of time for you, the programmer. Can you imagine writing 100 lines of code to display a simple message 100 times? The shell provides you with three looping constructs: the **for** loop, **while** loop, and **until** loop. These loops enable you to execute commands repeatedly for a certain number of times or until certain conditions are met.

*As was mentioned before, the **cat -n** command is used throughout this chapter to display the source code for the script files. You can create these files in the* `Chapter11/11.5` *directory using the* vi *editor. Creating these files in a specific directory will keep your files organized and is highly recommended.*

11.5.1 The For Loop: The for-in-done Construct

The **for** loop is used to execute a set of commands a specified number of times. Its basic format is as follows:

```
for variable
in list-of-values
do
  commands
  ...
  last-command
done
```

The shell scans the *list-of-values*, stores the first *word* (value) in the *loop variable*, and executes the commands between **do** and **done** (what is called the *body* of the loop). Next, the second *word* is assigned to the loop variable, and again the body of the loop is executed. The commands in the body of the loop are executed for as many values as the *list-of-values* contains.

Figure 11.19 shows the source code for the `for_in_done` script. This program shows the usage of the *for-in-done* loop.

Figure 11.19
The `for_in_done` Script File

```
$ cat -n for_in_done
   1 #
   2 # for_in_done
   3 # A sample program to show the for-in-done construct
   4 #
   5 echo                               # skip a line
   6 echo for count in 1 2 3            # start of the loop
   7 do
   8   echo "In the loop for $count times"
   9 done                               # end of the for loop
  10 echo                               # skip a line
  11 exit 0                             # end of the program
$_
```

The following command sequences show how the **for** loop works. This is a command line version of the `for_in_done` program. Notice that you are typing the program on the command line and the shell prompts you with > until the loop command is completed when you type in **done**. Refer to Figure 11.19 for the source code of this program.

```
$ for count in 1 2 3 [Return] . . . Set the loop header.
> do [Return] . . . . . . . . . . . Program waits for you to
                                    complete the command.
> echo "In the loop for $count times" [Return]
  . . . . . . . . . . . . . . . . . Display message.
> done [Return] . . . . . . . . . . End of the for loop.
In the loop for 1 times
```

```
In the loop for 2 times
In the loop for 3 times
$_ . . . . . . . . . . . . . . . . . Return to prompt.
```

1. *In this example,* count *is the loop variable, the* list-of-values *consists of numbers (1, 2, and 3), and the body of the loop is a single* **echo** *command.*
2. *Because three values are listed in the* list-of-values, *the body of the loop is executed a total of three times.*
3. *The values in the* list-of-values *are assigned one by one to the* count *variable. Each time through the loop, a new value is assigned until the* list-of-values *is exhausted.*

Scenario Suppose you want to save the name of the files you print along with the time you print them in a file. You write a script file called `slp` (super line printer) that prints your files and saves the information about them in a file called `pfile`. Figure 11.20 shows one way to write an `slp` script.

Figure 11.20
The `slp` Script File

```
$ cat -n slp
   1 #
   2 # slp
   3 # super line printer
   4 # A sample program to show the for_in_done construct
   5 #
   6 echo                                        # skip a line
   7 echo "Enter the name of the file(s)>>\c"    # prompt the user
   8 read filename                               # read input
   9 for FILE in $filename                       # start of the for loop
  10 do
  11  echo "\nFile name:$FILE\n Printed:`date`" >> pfile # save in pfile
  12  lp $FILE                            # print the file
  13 done                                 # end of the for loop
  14 echo "\n\07Job done"                 # inform user
  15 echo                                 # skip a line
  16 exit 0                               # end of the program, exit
$_
```

1. *You can enter more than one filename. Each filename from the list of files in variable* filename *is assigned to the loop variable* FILE. *The* **echo** *command takes care of saving the information in* `pfile`, *and the* **lp** *command prints it.*
2. *The* **echo** *command creates* `pfile` *the first time you use this program. On consequent runs, it appends (the >> redirection operator) the filenames of the printed files.*

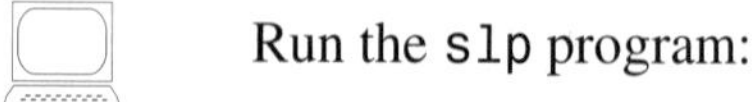

Run the `slp` program:

```
$ slp [Return] . . . . . . . . . . . . . Run it.

Enter the name of the files(s)>>.    . . Waiting for input.
```

```
myfile [Return]  . . . . . . . . . . . . . You enter myfile.
lp: request id is lp 1-9223 (1 file)   . The lp message.
Job done  . . . . . . . . . . . . . . . . Beep and final message.
$_  . . . . . . . . . . . . . . . . . . . Prompt.
$ cat pfile [Return]  . . . . . . . . . . Check pfile.
File name: myfile
Printed: Mon Dec 5   12:01:35 EST 2005

$_  . . . . . . . . . . . . . . . . . . . The shell prompt appears.
```

11.5.2 The While Loop: The while-do-done Construct

The second type of looping construct explained here is the **while** loop. Unlike the **for** loop, whose number of iterations depends on the number of values in the *list-of-values*, the **while** loop continues as long as the loop condition is true. The format is as follows:

```
while [ condition ]
do
  commands
  ...
  last command
done
```

The commands in the body of the loop (between **do** and **done**) are repeatedly executed as long as the loop condition is true (nonzero). The loop condition must eventually return to false (zero); otherwise, your loop is an infinite loop and goes on forever. That is one of those times that you must know the **kill** key on your system so you can terminate the process.

Figure 11.21 shows the source code for the carryon script. This program shows the usage of the **while-do-done** loop.

Figure 11.21
The carryon Script

```
$ cat -n carryon
   1 #
   2 # carryon
   3 # A sample program to show the while-do-done construct
   4 #
   5 echo                                  # skip a line
   6 carryon=Y                             # initialize the carryon variable
   7 while [ $carryon = Y ]                # start the while loop
   8 do
   9  echo "I do the job as long as you type Y:_\b\c"
  10  read carryon                         # read from keyboard
  11 done                                  # end of the while loop
  12 echo "Job Done!"                      # end of the program message
  13 echo                                  # skip a line
  14 exit 0                                # end of the program, exit
$_
```

The following command sequence shows how the **while** loop works. This is a command line version of the carryon program. Notice that you are typing the program on the command line and the shell prompts you with > until the loop command is completed when you type **done**. Refer to Figure 11.21 for the source code of this program.

```
$ carryon=Y [Return] . . . . . . . . . . . . Initial the carryon
                                             variable.
$ while [ $carryon = Y ] [Return] . . . .    Set up the while loop.
> do [Return]
> echo "I do the job as long as you type Y:_\b" [Return]
> read carryon [Return]
> done [Return] . . . . . . . . . . . . . .  Read input.
I do the job as long as you type Y:_ . . .   You are waiting for
                                             input.
Y [Return] . . . . . . . . . . . . . . . .   Press [Y] and
                                             [Return].
I do the job as long as you type Y:_ . . .   You are waiting for
                                             input again.
N [Return] . . . . . . . . . . . . . . . .   Press [N] and
                                             [Return].
$_ . . . . . . . . . . . . . . . . . . . .   The shell prompt
                                             appears.
```

In this example, as long as you press **Y**, *the loop condition is true, and the body of the loop, the* **echo** *and* **read** *commands, is repeated. You stop this program by pressing any character but* **Y**.

Figure 11.22 shows the source code for the script file called counter1. The counter1 program is another example of the **while** loop construct that displays the numbers from 0 to 9. The **expr** command is used to calculate the next number to be displayed. The **cat** command is used to display the source code.

Figure 11.22
The counter1 Script File

```
$ cat -n counter1
   1 #
   2 # counter1
   3 # Counter: Counts from 1-9 using while loop
   4 #
   5 echo                               # skip a line
   6 count=1                            # initialize the count variable
   7 while [ $count -lt 10 ]            # start the while loop
   8 do
   9  echo $count                       # display count value
  10  count=`expr $count + 1`           # increment count
  11 done                               # end of the while loop
  12 echo "End! :-)"                    # end of the program message
  13 echo                               # skip a line
  14 exit 0                             # end of the program, exit
$_
```

Notice that the command **expr $count + 1** *is enclosed with a pair of grave accent marks (` and `). When you run the program, the output of this command is stored in the* count *variable.*

Running the `counter1` program.

```
$ counter1 [Return] . . Execute the counter1 program.

1
2
3
4
5
6
7
8
9
End! :-)

$_ . . . . . . . . . . . The shell prompt appears.
```

Figure 11.23 shows a new version of this script, `counter2`, using the **let** instead of the **expr** command. The **cat** command is used to show the source code.

Figure 11.23
The New Version of the `counter2:` Script File

```
$ cat -n counter2
   1 #
   2 # counter2
   3 # Counter: Counts from 1-9 using while loop and the let command
   4 #
   5 echo                              # skip a line
   6 count=1                           # initialize the count variable
   7 while [ $count -lt 10 ]           # start the while loop
   8 do
   9  echo $count                      # display count value
  10  let count=count+1                # increment count
  11 done                              # end of the while loop
  12 echo "End! :-)"                   # end of the program message
  13 echo                              # skip a line
  14 exit 0                            # end of the program, exit
$
```

The following versions of the `counter` *program work only with the* ksh *shell.*

The let Command Abbreviated

You can abbreviate the **let** command to double parentheses, **(( )).** Figure 11.24 shows yet another version of the program we have been working with, now `counter3`. In this version the **let** command abbreviation is used instead of **expr**.

Figure 11.24
Another Version of the `counter3:` Script File

```
$ cat -n counter3
    1 #
    2 # counter3
    3 # Counter: Counts from 1-9 using while loop and the let command
    4 #          abbreviation (())
    5 #
    6 echo                              # skip a line
    7 count=1                           # initialize the count variable
    8 while (( $count < 10 ))           # start of the while loop
    9 do
   10   echo $count                     # display count value
   11   (( count=count+1 ))             # increment count
   12 done                              # end of the while loop
   13 echo "Done! :-)"                  # end of the program message
   14 echo                              # skip a line
   15 exit 0                            # end of the program, exit
$_
```

11.5.3 The Until Loop: The until-do-done Construct

The third type of the looping construct explained here is the **until** loop. The **until** loop is similar to the **while** loop, except that it continues executing the body of the loop as long as the condition of the loop is false (zero). The **until** loop is useful in writing scripts whose execution depends on other events occurring. The format is as follows:

```
until [ condition ]
do
  commands
  ...
  last-command
done
```

The body of the until loop might never get executed if the loop condition is true (nonzero) the first time it is executed.

Scenario You want to check whether a specified user is on the system or, if not, to be informed as soon as the user logs in. Figure 11.25 shows one way to write the script file uon (User ON).

In the uon script, the **until** loops stops as soon as the loop condition is true. If **grep** (Chapter 8) does not find the specified user ID passed to it by the **who** command in the list of the users, then the loop condition remains false (nonzero) and the **until** loop continues executing the body of the loop, the **sleep** command. As soon as **grep** finds the specified user ID in the users list, the loop condition becomes true (zero) and the loop stops. Next, the command after the loop is executed, which informs you, with two beeps, that the specified user is logged in.

1. *The output of the* **grep** *command is redirected to the null device (never-never land). That means you neither want to see the output nor save it.*
2. *The* **sleep** *command stops your program for half a minute. The net effect is that the* **grep** *command checks the list of users every half minute.*

Figure 11.25
The uon Script File

```
$ cat -n uon
   1 #
   2 # uon
   3 # Let me know if xyz is on the system
   4 #
   5 echo                               # skip a line
   6 until who ¦ grep "$1" > /dev/null# redirect the output of grep
   7 do
   8   sleep 30                         # wait half a minute
   9 done                               # end of the until loop
  10 echo "\07\07$1 is on the system" # inform user
  11 echo                               # skip a line
  10 exit 0                             # end of the program, exit
$
```

There is only one problem left. If you run this program in the foreground and the specified user is not logged in, then you are at the mercy of the user to log in and you cannot use your terminal while the above program is running. A better idea is to run uon in the background.

Run uon in the background:

```
$ uon emma &[Return] . . . . . . . . Run it in the background.
                                     Check to see if emma is
                                     logged in.
4483 . . . . . . . . . . . . . . . . Process ID.
$_ . . . . . . . . . . . . . . . . . The prompt is back.
You can do other work. You will be informed when emma logs in.
emma is on the system . . . . . . . . Beep beep. You are informed.
You can go on with whatever you were doing.
```

11.6 DEBUGGING SHELL PROGRAMS

It is easy to make mistakes when you write long and complex script files. Because you do not compile script files, you do not have the luxury of compiler error checking. Therefore, you have to run the program and try to decipher the error messages displayed on the screen. But do not despair!

11.6.1 The Shell Command

You can use the **sh**, **ksh**, or **bash** command with one of its options to make the debugging of your script files easier. For example, the **-x** option causes the shell to echo each command it executes. This trace of your script execution can help you to find the whereabouts of the bugs in your program.

Shell Options

Table 11.9 summarizes the shell command options, and the following examples show how to use them. In the following examples, **sh** is used, but you can use one of the other shells, **ksh** or **bash**.

Table 11.9
The **sh** Command Options

Option	Operation
-n	Reads commands but does not execute them.
-v	Shows the shell input lines as they are read.
-x	Shows the commands and their arguments as they are executed.

If you want to debug the BOX script file, you type the following:

$ sh -x BOX [Return]

You also can place the **sh** options as commands in your script file. Use the **set** command and type the following line at the beginning of your file, or place it wherever you want the debugging to start:

```
set -x
```

-x Option The **-x** option shows the commands in your script file as they look after the parameter and command substitutions have taken place. Using the **sh** command with the **-x** option, let's execute the BOX script file in different ways and explore the possibilities.

Run BOX with the **-x** option, but do not specify command line arguments:

```
$ sh -x BOX[Return]  . . . . . . . . . Use the -x option.
+ echo The following is the output of the BOX script.
The following is the output of the BOX script.
+ echo Total number of command line arguments: 0
Total number of command line arguments: 0
+ echo The first parameter is:
The first parameter is:
+ echo The second parameter is:
The second parameter is:
+ echo This is the list of all parameters:
This is the list of all parameters:
$_  . . . . . . . . . . . . . . . . . . . Return to prompt.
```

1. *All lines beginning with plus signs (+) are commands that are executed by the shell, and the lines below them show the output of the commands.*
2. *The **echo** commands are displayed after the variable substitutions are done.*

Run BOX using **sh** with the **-x** option, and specify some command line parameters:

```
$ sh -x BOX of candy [Return]  . . . Use the -x option.
+ echo The following is the output of the BOX script.
```

```
The following is the output of the BOX script.
+ echo Total number of command line arguments: 2
Total number of command line arguments: 2
+ echo The first parameter is: of
The first parameter is: of
echo The second parameter is: candy
The second parameter is: candy
+ echo This is the list of all parameters: of candy
This is the list of all parameters: of candy
$_ . . . . . . . . . . . . . . . . Return to prompt.
```

-v Option The **-v** option is similar to the **-x** option. However, it displays the commands before the substitution of the variables and commands is done.

Run the BOX program with the **-v** option, and specify some command line arguments.

```
$ sh -v BOX is full of gold nuggets [Return] . . . Use the -v option.
echo "The following is the output of the $0 script."
The following is the output of the BOX script.
echo "The total number of command line arguments: $#"
Total number of command line arguments: 5
echo "The first parameter is: $1"
The first parameter is: is
echo "The second parameter is: $2"
The second parameter is: full
echo "This is the list of all parameters: $*"
This is the list of all parameters: is full of gold nuggets
$_ . . . . . . . . . . . . . . . . . . . . . . . . . . Return to prompt.
```

1. *One line shows the command that is executed by the shell, and the line below it shows the output of the command.*
2. *The* **echo** *commands are displayed before the variable substitutions are done. (This is different from the* **-x** *option, which displays the commands after the substitutions are done.)*

You can use the **-x** and **-v** options together on a command line. Using both options enables you to look at the commands in your file before and after execution, plus the output they produce. The following command line shows that the **sh** is invoked with both options:

```
$ sh -xv BOX [Return]
```

-n Option The **-n** option is used to detect syntax errors in your script file. Use this option when you want to make sure that you do not have syntax errors in a program before you run it.

For example, suppose you have a script file called check_syntax. Figure 11.26 shows the source code for this program, which has an intentional syntax error.

Run the check_syntax program using the **-n** option, and observe its output:

```
$ sh -n check_syntax [Return] . . . Run it.
check_syntax: syntax error at line 10 'else' unexpected
$_ . . . . . . . . . . . . . . . . . Return to prompt.
```

Figure 11.26
Source Code for the check_syntax Program

```
$ cat -n check_syntax
   1 #
   2 # check_syntax
   3 # A sample program to show the output of the shell -n option
   4 #
   5 echo                     # skip a line
   6 echo "$0: checking the program syntax"
   7 if [ $# -gt 0 ]          # start of the if
   8 # then                   # this line is intentionally commented out
   9     echo "Number of the command line arguments: $#"
  10  else
  11     echo "No command line arguments"
  12 fi                       # end of the if
  13 echo "Bye!"              # end of the program message
  14 exit 0                   # end of the program, exit
$
```

1. *Using the* **-n** *option, none of the commands in your program are executed. This option only locates and recognizes the syntax errors.*
2. *If you use the* **-x** *or* **-v** *option, the program is executed until it reaches the part with the wrong syntax. Then, the error message is displayed and the program is terminated.*

What is wrong at line 10? You have to add the word *then* after the **if** statement to correct the **if-then-else** construct in your program. In our program example, the word *then* is commented out and it is sufficient to remove the # sign before the word *then*.

What is the output of check_syntax after you have fixed the syntax error? The $0 contains the name of the program, in this case check_syntax. The $# contains the number of the command line arguments. If there are command line arguments, the body of the **if** is executed; otherwise, the body of the **else** is executed.

Run check_syntax again:

```
$ check_syntax one two three [Return] . . . Use three command
                                            line arguments.
check_syntax: checking the program syntax
Number of the command line arguments: 3
$ . . . . . . . . . . . . . . . . . . . . . Return to prompt.
$ check_syntax [Return] . . . . . . . . . . No command line
                                            arguments
check_syntax: checking the program syntax
No command line arguments
$_  . . . . . . . . . . . . . . . . . . . . Return to prompt.
```

The **sh** debugging options are mostly useful when you write long and complex shell programs. They will come in handy when you explore the script files presented in Chapter 12.

COMMAND SUMMARY

The following commands and options have been discussed in this chapter.

chmod

This command changes the access permission of a specified file according to the option letters indicating different categories of users. The user categories are **u** (for user/owner), **g** (for group), **o** (for others), and **a** (for all). The access categories are **r** (for read), **w** (for write), and **x** (for executable).

. (dot)

This command lets you run a process in the current shell environment and does not allow the shell to create a child process to run the command.

exit

This command terminates the current shell program whenever it is executed. It can also return a status code (RC) to indicate the success or failure of a program. It also terminates your login shell and logs you off if it is typed at the **$** prompt.

expr

This command is a built-in operator for arithmetic operations. It provides arithmetic and relational operators.

let

This command provides arithmetic operations.

read

This command reads input from the input device and stores the input string in one or more variables specified as the command arguments.

sh, ksh, or bash

This command invokes a new copy of the shell. You can run your script files using this command. Only three of the numerous options are mentioned.

Option	Operation
-n	Reads commands but does not execute them.
-v	Prints the input to the shell as the shell reads it.
-x	Prints command lines and their arguments as they are executed. This option is used mostly for debugging.

test

This command tests the condition of an expression given to it as an argument, and returns true or false depending on the status of the expression. It gives you the capability of testing different types of expressions.

REVIEW EXERCISES

1. How do you execute a shell script file?
2. What is the command to make a file an executable file?
3. When do you use the **.** (dot) command?
4. What is the command to read input from the keyboard?
5. Explain the command line parameters.
6. What are the shell positional variables?
7. How are the positional variables related to the command line parameters?
8. How do you debug your shell script?
9. Name the command that terminates a shell script.
10. What are the constructs in the shell language?
11. What is the use of the loop construct?
12. What is the difference between the **while** and **until** loops?
13. What is the command that performs arithmetic operations?
14. What is the command to add two integer numbers?
15. What is the command to multiply the values of two variables?
16. What is the shell built-in programming language?
17. What is the command to change the file permissions?
18. What is the command to change permission for a file called `xyz` to provide access to all users (user, group, and others)?
19. When do you use **sh** (**ksh**, **bash**) to execute a command?
20. When do we use the **chmod** command?
21. What does the command **sh -n** provide?
22. What does the command **sh -v** provide?
23. What does the command **sh -x** provide?
24. What is the command to read from the keyboard?
25. What is the **test** command?
26. What is the **let** command?
27. How do you write a comment line in a script (program)?
28. How do you compare two strings?
29. How do you compare two numeric values?
30. What is the command to show current date in mm/dd/yy format?

Terminal Session

In this terminal session, you are going to write a few script files and improve on the script file examples presented in this chapter.

1. Create a shell script named `LL` that lists your directory in a long format.
   a. Execute `LL` using the **sh** command.
   b. Change `LL` to an executable file.
   c. Execute `LL` again.
2. Create a script file that performs the following:
   a. Clears the screen.
   b. Skips two lines.
   c. Shows the current date and time.
   d. Shows the number of the users on the system.
   e. Beeps a few times and shows the message **Now at your service**
3. Modify the `largest` script file from this chapter to recognize the number of inputs and display appropriate messages.
4. Write a script file similar to `largest` that calculates the smallest of three integer numbers that are read from the keyboard. Make it able to recognize some input errors.
5. Create a script file for each of the script examples in this chapter that were typed at the **$** prompt. Make the appropriate modifications if necessary and run them. Use **sh** with the **-x** and other options to debug, and investigate the way shell scripts are executed.
6. Write a script file that sums the numbers passed to it as arguments on the command line and displays the results. Use the **for** loop construct in your program. For example, if you name this program `SUM`, and you type
   ```
   $ SUM 10 20 30 [Return]
   the program displays the following:
   10 + 20 + 30 = 60
   ```
7. Rewrite the `SUM` program, this time using the **while** loop.
8. Rewrite the `SUM` program, this time using the **until** loop.
9. Modify the script file `largest` from this chapter to accept numbers from the command line. For example, you type *LARGEST 1 2 3* and the program displays:
   ```
   The largest numbers is: 3
   ```
10. Modify the script file `greetings1` from this chapter and use the **cut** command to calculate the hour of the day.
11. Create a script file called `file_checker` that reads the filename entered and outputs the file properties such as: exists, readable and so on.
12. Write a script file named `d` that shows the current date.
13. Write a script file named `t` that shows the current time.
14. Write a script file called `s` that displays the name of your current shell.
15. Write a script file that checks if you have a `.profile` file in your HOME directory or not and displays an appropriate message.

CHAPTER 12

Shell Scripts: Writing Applications

This chapter builds on the commands and concepts of the previous chapter and discusses additional shell programming commands and techniques. It also presents a simple application program and shows the process of developing programs using the shell language. It introduces new shell commands when they are used in the shell scripts.

In This Chapter

12.1 WRITING APPLICATIONS

Chapter 11 introduced shell programming basics and mentioned using the shell command language to write application programs. This chapter explains the process of developing application programs. The examples are complete programs, and, when necessary, new commands or constructs are introduced and explored.

Table 12.1 lists the shell built-in commands that are covered in this chapter and shows their availability under different shells.

Table 12.1
The Shell Built-in Commands

Command	Built into		
	Bourne Shell	Korn Shell	Bourne Again Shell
trap	sh	ksh	bash
case	sh	ksh	bash

12.1.1 The `lock1` Program

Scenario You want to protect the access to your terminal (say when you are away from it for a few minutes), without logging off and then logging in again. You can write a script file that, when invoked, shows a message on the screen and does not go away until you enter the correct password. But do you really want to lock your terminal? Yes. The goal here is to learn about commands and programming techniques, and apply them to script files, whether you want to lock your terminal or not.

Figure 12.1 shows a script named `lock1` (lock version 1), which does the job by locking your keyboard.

1. *The* **cat -n** *command is used throughout this chapter to display the source code for the script files. The* **-n** *option provides the line numbers that are used to refer to a specific line of code when that line is explained in the text.*
2. *You can create these script files under the* `$HOME/Chapter12/12.1` *directory using the vi editor.*
3. *Creating these files under a specific directory will help keep your files organized and is highly recommended.*

Let's look at the `lock1` program line by line to see how it works.

Lines 1–11: These lines are all started with the # sign and are documentation lines.

Lines 12 and 15: These lines are commands to clear the screen. These lines might not work on your terminal screen. Either find the clear screen code for your terminal or you may use the **tput clear** command instead. The **tput** command is explained later in this chapter.

Line 13: This line prompts you to enter a password. Any sequence of characters, numbers, and embedded white-space characters is accepted. The password you enter has nothing to do with your login password.

See page x for an explanation of the icons used to highlight information in this chapter.

Figure 12.1
The lock1 Script File

```
$ cat -n lock1
   1 #
   2 # lock1 (lock version 1)
   3 # This program locks the keyboard, and you must type the
   4 # specified password to unlock it.
   5 # logic:
   6 #    1- Ask the user to enter a password.
   7 #    2- Lock the keyboard until the correct password is entered.
   8 #
   9 # DO NOT RUN THIS PROGRAM IF YOU DO NOT KNOW WHAT YOU ARE DOING!!
  10 # Please read the first part of chapter 12.
  11 #
  12 echo "\032"                          # can be replaced by tput clear
  13 echo "\n\nENTER YOUR PASSWORD>"      # ask for password
  14 read pword_1                         # read password
  15 echo "\032"                          # clear screen again
  16 echo "\n\n THIS SYSTEM IS LOCKED...."
  17 pword_2=                             # declare an empty variable
  18 until [ "$pword_1" = "$pword_2" ]    # start of the loop
  19 do
  20   read pword_2                       # body of the loop
  21 done                                 # end of the until loop
  22 exit 0                               # end of the program, exit
$_
```

Line 14: This line reads from the keyboard. When you press [Return], your intended password is stored in the variable called *pword_1*.

Line 16: This line shows the appropriate message on the screen. You can change it to any other message you like.

Line 17: This line declares a variable called *pword_2*, which is used to store the input from the keyboard.

Lines 18–21: These lines construct the **until** loop. The condition compares the contents of the *pword_1* variable (your password) with the content of the *pword_2* variable (what is read from the keyboard). The first time through the loop the contents of the two variables are not equal (*pword_1* contains your password; *pword_2* is empty). Therefore, the body of the **until** loop is executed. Remember, the body of the **until** loop is executed when the loop condition is false. The body of the loop is a **read** command, and it waits to read from the keyboard. It doesn't show any prompt; just the cursor is displayed. (After all, you do not want to alert the intruder that your system is waiting for a password.) If you enter anything but the correct password, the **until** loop continues and hangs a read on your keyboard. This **read** command repetition effectively locks your keyboard. It reads, compares, comparison fails, and reads again. When you enter the correct password, the loop condition returns true status (now *pword_1* is equal to *pword_2*). The loop stops, and your keyboard is freed.

Line 22: This line exits the lock1 script with zero status, indicating a normal termination of the program.

Problems with the `lock1` Program

The `lock1` program runs fine, except for some minor problems. There is some potential for improvement. Let's explore it.

`lock1` is not a foolproof program:

- While the keyboard is locked, if you press [Del] (the interrupt key), the `lock1` script is terminated, the **$** prompt is displayed, and the system is ready to accept commands. So much for protecting your system from intruders! To be effective, the `lock1` program must be able to ignore normal interrupt signals.
- The password is displayed. Usually, passwords are not displayed on the screen. On line 10, when the password is requested, whatever you type as your password is echoed on the screen for everyone to see. You must prevent display of the user response to the password prompt.
- The message is hard coded. It always shows **This system is locked**. It would be nice to be able to specify the message to be displayed on the command line.

To solve these problems, a few more commands must be explored.

12.2 UNIX INTERNALS: THE SIGNALS

How do you terminate a process? You terminate a process by generating an interrupt signal. What is a signal? A *signal* is a report to your process about a specific condition. For example, [Del], [Break], and [Ctrl-c] are used to send an interrupt signal to a process to terminate it.

Remember, your process thinks it is working with files and does not know about your terminal. Then how does an interrupt signal that you enter from the keyboard terminate your process? Your interrupt signals go to the UNIX kernel and not to your process. The kernel knows about the devices, and it is notified when you press any of the interrupt keys. Then UNIX sends a signal to your process telling it that an interrupt has occurred. In response to this signal, your process terminates or takes some other action.

There are several different events that make the kernel send a signal to your process. These signals are numbered to specify the specific event they present. Table 12.2 summarizes some of these signals that are usually used in the script files. The signal numbers

Table 12.2
Some of the Shell Signals

Signal Number	Name	Meaning
1	**hang up**	Terminal connection is lost.
2	**interrupt**	One of the **interrupt** keys has been pressed.
3	**quit**	One of the **quit** keys has been pressed.
9	**kill**	The **kill -9** command has been issued.
15	**terminator**	The **kill** command has been issued.

may be different in your system. Ask your system administrator, or look it up in your system reference manual.

The hang-up Signal Signal 1 is used to tell the process that the system has lost the connection to its terminal. This signal is generated when the cord that runs from your terminal to the computer is disconnected, or when the phone line (your modem connection) is lost. Also, on some systems the hang-up signal is generated when you turn off your terminal.

The interrupt Signal Signal 2 is generated when one of the interrupt keys is pressed. This could be [Ctrl-c], [Del], or [Break].

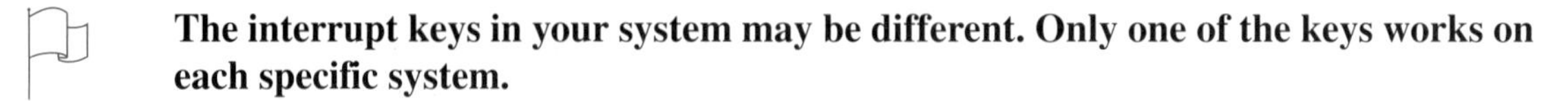

The interrupt keys in your system may be different. Only one of the keys works on each specific system.

The quit Signal Signal 3 is generated from your keyboard when you press [Ctrl-\]. This causes the process to *core dump* before it terminates.

The kill Signals Signals 9 and 15 are generated by the **kill** (see Chapter 8) command. Signal 15 is the default signal, and 9 is generated when the **kill** command option **-9** is used. Both terminate the receiving process.

12.2.1 Trapping the Signals: The trap Command

The default action taken by your process when it receives any of the signals is immediate termination. You can use the **trap** command to change the process default action to whatever you specify. For example, you can instruct your process to ignore interrupt signals or to execute a specified command instead of termination. Let's explore the possibilities. The format of the **trap** command is as follows:

```
trap "optional commands" signal numbers
```

The *commands* part is optional. When it is present, the commands are executed whenever your process receives one of the signals you have specified to be trapped.

The commands that are specified to the trap command must be enclosed in single or double quotation marks.

1. *You can specify more than one signal number to be trapped.*
2. *The signal numbers are the numbers associated with the signals that you want the* **trap** *command to catch.*

The following command sequences show how the **trap** command works:

```
trap "echo I refuse to die!" 15
```

This command executes the **echo** command and displays the **I refuse to die!** message whenever it receives a simple **kill** command. However, your script continues.

```
trap "echo Killed by a signal!; exit" 15
```

If your process receives a **kill** command (signal 15), the **echo** command is executed and shows the **Killed by a signal!** message. Next, the **exit** command is executed, which causes your script to terminate.

```
trap " " 15
```

No command is specified. Now, if your process receives the **kill** signal (signal 15), it ignores it, and the script continues.

The quotation marks must be present even if no command is specified. Without the quotation marks the trap command resets the specified signals.

12.2.2 Resetting the Traps

Issuing a **trap** command in your script changes the default actions of the signals received by your process. Using the **trap** command, without the optional commands part, changes the specified signals to their default actions. This command is useful when you want to trap a certain signal in one part of your script and in another part you want the signals not to be trapped.

For example, if you type the following command in your script file:

```
$ trap " " 2 3 15
```

the **interrupt**, **quit**, and **kill** commands are ignored, and if any of their keys is pressed, your script keeps on running. If you type the following command:

```
$ trap 2 3 15
```

the specified signals are reset. That is to say, the **interrupt**, **quit**, and **kill** keys are restored, and if any of them is pressed, your running script is terminated.

12.2.3 Setting Terminal Options: The stty Command

You use the **stty** command to set and display terminal characteristics. You can control various characteristics of your terminal, such as the baud rate (rate of transmission between the terminal and the computer) and the functions of certain keys (**kill**, **interrupt**, etc.). The **stty** command without arguments shows a selected group of settings. Use the **-a** option to list all your terminal settings.

Figure 12.2 shows an example of the terminal settings. Your system might have different settings.

Figure 12.2
An Example of the Terminal Settings

```
$ stty
speed 9600 baud;-parity
erase = '^h' ;kill = '^u';
echo
$_
```

Terminals have a wide variety of capabilities, and **stty** supports the modification of more than a hundred different settings. Some of these settings change the communication mode of your terminal, some change the value assigned to special keys, and yet another group combines settings.

Table 12.3 lists a very small subset of the available options, the most common ones. Use the **man** command to obtain a more detailed list of options and explanations of their functions.

Table 12.3
A Short List of Terminal Options

Option	Operation
echo [-echo]	Echoes [does not echo] the typed characters; the default is **echo**.
raw [-raw]	Disables [enables] the special meaning of the metacharacters; the default is **-raw**.
intr	Generates an interrupt signal; usually the [Del] key is used.
erase	[Backspace]. Erases the preceding character; usually the # key is used.
kill	Deletes the entire line; usually @ or [Ctrl=u] is used.
eof	Generates the (end-of-file) signal from the terminal; usually [Ctrl-d] is used.
ek	Resets the **erase** and **kill** keys to # and @ respectively.
sane	Sets the terminal characteristics to sensible default values.

1. *The default settings are usually the best for most of the options.*
2. *Some of the options are enabled by typing the option name and are disabled by preceding them with a hyphen.*

Let's look at some examples:

```
$ stty -echo [Return] . . . . Turn echoing off.
$ stty echo [Return]  . . . . Turn echoing back on.
```

The stty echo command is not displayed on your terminal due to the effect of the previous stty -echo command.

Set the **kill** key to [Ctrl-u]:

```
$ stty kill \^u [Return]  . . Now [Ctrl-u] is the the kill key.
$_ . . . . . . . . . . . . .  The prompt reappears.
```

To set a special key, either press three characters ([\], [^], and the specified letter) or type the combination keys directly. In this example, you either type **\^u** *or press [Ctrl-u].*

```
$ stty sane [Return]  . . . . Reset options to reasonable values.
$  . . . . . . . . . . . . .  The prompt reappears.
```

When you have changed the options too many times and lost track of your changes, the **sane** *option comes to your rescue.*

Change the **kill** and **erase** keys to their default values:

```
$ stty ek [Return]  . . . . . Set the kill and erase keys.
$_ . . . . . . . . . . . . .  The prompt reappears.
```

This command sets the **kill** key to @ and the **erase** key to #.

12.3 MORE ABOUT TERMINALS

The UNIX operating system supports numerous types of terminals. Each terminal has its own capabilities and characteristics. These capabilities are documented in the terminal user/technical manual along with a set of escape characters to be used that enables you to manipulate the terminal capabilities. Each terminal type has its own set of escape characters. In Chapter 11, escape character **\032** was used to clear the screen. The **\032** is the clear screen code for vt100-type terminals. You type the following line to clear the screen:

`$ echo "\032" [Return]` . . . Clears screen on the vt100-type terminals.

Terminal capabilities are not limited to the clear screen function. There are dozens of other characteristics, such as boldface text, blinking, underlining, and so on. You use them to make your display more meaningful, better organized, or better looking.

12.3.1 The Terminals Database: The `terminfo` File

Each terminal supported in your system has an entry in the terminal database (file) called *terminfo* (terminal information). The `terminfo` database is a single text file that contains descriptions of many types of terminals. There is a list of capabilities associated with each terminal in the database.

12.3.2 Setting the Terminal Capabilities: The tput Command

The **tput** utility, which is standard on any system with the `terminfo` database, lets you print out the values of any single capability. This makes it possible to use the terminals' capabilities in shell programming. For example, to clear the screen, you type:

```
$ tput clear [Return]
```

This command works regardless of your terminal type, as long as your system includes the `terminfo` *database and your terminal type is in the database.*

Table 12.4 shows some of the terminal capabilities that can be activated using the **tput** command. The **tput** program used with the `terminfo` database lets you choose particular terminal capabilities and print out their values or store them in a shell variable. By default, **tput** assumes that you are using the terminal type specified in the *TERM* shell variable. You can override this by using the **-T** option. For example, you type the following command to specify the terminal type:

```
$ tput -T ansi [Return]
```

1. *Usually when you start a style mode, it remains in effect until you remove it.*
2. *You can store the character sequences in variables and then use the variables.*

You use the sgr0 option to turn off all the user-defined terminal attributes. This is an important option to know, because it is the only way to turn off some attributes such as blink.

Table 12.4
A Short List of the Terminal Capabilities

Option	Operation
bel	Echoes the terminal's bell character.
blink	Makes a blinking display.
bold	Makes a boldface display.
clear	Clears the screen.
cup *r c*	Moves cursor to row *r* and column *c*.
dim	Dims the display.
ed	Clears from the cursor position to the end of the screen.
el	Clears from the cursor position to the end of the line.
smso	Starts stand out mode.
rmso	Ends stand out mode.
smul	Starts underline mode.
rmul	Ends underline mode.
rev	Shows reverse video, black on white display.
sgr0	Turns off all the attributes.

The following command sequences show the use of the **tput** command to change the terminal characteristics.

Using the **tput** command, first clear the screen and then show the message **The terminfo database** in row 10 column 20.

$ tput clear [Return] . . . . The screen is cleared.
$ tput cup 10 20 [Return] . . . . Positions the cursor.
$ echo "The terminfo database" [Return] . . Display the message.

```
The terminfo database
```

You specify the cursor position before the **echo** *command that displays the message at the specified location on the screen.*

These commands can be sequenced together on one line. Commands are separated by colons, as follows:

$ tput clear: tput cup 10 20: echo "The terminfo database" [Return]

Store a character sequence in a variable and then use it to manipulate the screen display.

$ bell=`tput bel` [Return] . . . Store the character sequence for bell in the *bell* variable.

The shell executes the command between the grave accent marks and assigns the command's output, in this case the character sequence for the terminal bell (beep), to the *bell* variable.

$ s_uline=`tput smul` [Return]. . . . . Store code for start underline display.
$ e_uline=`tput rmul` [Return]. . . . . Store code for end underline display.
$ tput clear [Return] . . . . Clear the screen.

```
$ tput cup 10 20 [Return]  . . . . . . . . . . Position the cursor.
$ echo $bell [Return]  . . . . . . . . . . . . Sound the terminal bell.
$ echo $s_uline [Return]  . . . . . . . . . . . Start underline display.
$ echo "The terminfo database" [Return] . . . Display message, underlined.
The terminfo database
$ echo $e_uline [Return]  . . . . . . . . . . . End underline display.
```

The **echo** commands can be combined into one command using the assigned variables as follows:

```
$ echo "$bell${s_uline}The terminfo database$e_uline" [Return]
```

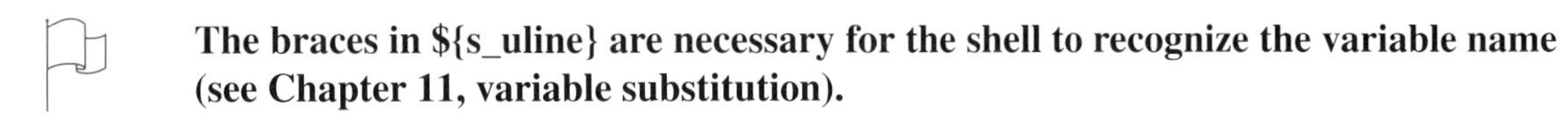

The braces in ${s_uline} are necessary for the shell to recognize the variable name (see Chapter 11, variable substitution).

12.3.3 Solving the `lock1` Program Problems

Let's use the **stty**, **trap**, and **tput** commands, and modify the `lock1` script to fix its problems and create a better user interface. Figure 12.3 shows the new version of the `lock` script, and following is the explanation of the modified or added lines.

Figure 12.3
The `lock2` Script File

```
$ cat -n lock2
   1 #
   2 # lock2 (lock version 2)
   3 # This program locks the keyboard, and you must type the
   4 # specified password to unlock it.
   5 # Logic:
   6 #  1- Ask the user to enter a password.
   7 #  2- Lock the keyboard until the correct password is entered.
   8 #
   9 # DO NOT RUN THIS PROGRAM IF YOU DO NOT KNOW WHAT YOU ARE DOING!!
  10 # Please read the first part of chapter 12.
  11 #
  12 trap " " 2 3                        # ignore the listed signals
  13 stty -echo                          # prohibit echoing the input
  14 tput clear                          # clear the screen
  15 tput cup 5 10; echo "ENTER YOUR PASSWORD> \c" # ask for password
  16 read pword_1                        # read password
  17 tput clear                          # clear screen again
  18 tput cup 10 20 ; echo "THIS SYSTEM IS LOCKED...."
  19 pword_2=                            # declare an empty variable
  20 until [ "$pword_1" = "$pword_2" ]   # start of the loop
  21 do
  22  read pword_2                       # body of the loop
  23 done                                # end of the until loop
  24 stty echo                           # enable echo of input characters
  25 tput clear                          # clear screen
  26 exit 0                              # end of program, exit
$_
```

Line 12: This line traps the signals 2 and 3. This means your script ignores the **interrupt** and **quit** keys when they are pressed and continues running.

Line 13: This line disables the echoing capability of the terminal. Therefore the next characters of input (in this case the password) are not displayed.

Lines 14 and 17 and 25: These lines clear the screen. Instead of using the **echo** command and the clear screen code, the **tput** command is used.

Line 15: This line issues two commands, separated by the semicolon (**;**). The first command (**tput cup 5 10**) positions the cursor at row 5, column 10. The second command echoes the message at the location of the cursor.

Line 18: This line is similar to line 15 and issues two commands separated by the semicolon (**;**). The first command (**tput cup 10 20**) positions the cursor on line 10, column 20. The second command echoes the message at the location of the cursor.

Line 24: This line resets (enables) the echoing capability of your terminal. This happens after you have entered the correct key to unlock the keyboard.

Now, if you change the `lock2` to an executable file and execute it, you can type the following commands:

```
$ chmod +x lock2 [Return]   . . Change the program access mode.
$ lock2 [Return]    . . . . . . Execute it.
```

Then your screen looks like the screen depicted in Figure 12.4, and you are prompted to enter a password.

Figure 12.4
The Prompt from the `lock2` Program

```
ENTER YOUR PASSWORD>_
```

Enter any sequence of characters or numbers (which are not displayed) as your password. You must remember this password to unlock your keyboard. Next, the terminal screen is cleared and your screen looks like the screen depicted in Figure 12.5.

Figure 12.5
The Message from the `lock2` Program

```
THIS SYSTEM IS LOCKED....
```

Now the keyboard is locked, and only your password unlocks it. What happens if you forget your password? You cannot use [Del], [Ctrl-c], or other keys to terminate the `lock2` program. The **trap** command ignores these signals, and the program continues keeping the keyboard locked. You cannot trap the **kill** signals (9 and 15), but you cannot use your keyboard to issue the **kill** command. The solution is simple: log in using another terminal, and issue a **kill** command to terminate the `lock2` from your terminal.

Specifying the Display Message As mentioned before, you can improve the `lock2` program by giving the user freedom to specify the message to be displayed or choose the canned (default) message. The specified message must be passed to the program on the command line. For example, you type

$ lock3 Coffee Break. Will be back in 5 minutes [Return]

You have to modify your `lock2` program to accommodate the new changes in our scenario. With the new version (called `lock3`), you can either enter a message on the command line or execute the program as before without specifying a message. Your program must recognize both cases and show the appropriate message. Figure 12.6 shows the source code for the `lock3` program, and the following line explanations clarify how it works.

Figure 12.6
The `lock` Program, Third Version

```
$ cat -n lock3
   1 #
   2 # lock3 (lock program version 3)
   3 # This program locks the keyboard, and you must type the
   4 # specified password to unlock it.
   5 # Logic:
   6 #   1- Ask the user to enter a password.
   7 #   2- Lock the keyboard until the correct password is entered.
   8 #
   9 # DO NOT RUN THIS PROGRAM IF YOU DO NOT KNOW WHAT YOU ARE DOING!!
  10 # Please read the first part of chapter 12.
  11 #
  12 trap " " 2 3 4                      # ignore the listed signals
  13 stty -echo                          # prohibit echoing the input
  14 if [ $# -gt 0 ]                     # online message is specified
  15 then
  16   MESG="$@"                         # store the specified message
  17 else
  18   MESG="THIS SYSTEM IS LOCKED"      # set to default message
  19 fi
  20 tput clear                          # clear the screen
  21 tput cup 5 10 ; echo "ENTER YOUR PASSWORD>\c"   # ask for password
  22 read pword_1                        # read password
  23 tput clear                          # clear screen again
  24 tput cup 10 20 ; echo "$MESG"
  25 pword_2=                            # declare an empty variable
  26 until [ "$pword_1" = "$pword_2" ] #start of the loop
  27 do
  28    read pword_2                     # body of the loop
  29 done                                # end of the until loop
  30 stty echo                           # enable echo of input characters
  31 tput clear                          # clear screen
  32 exit 0                              # end of program, exit
$_
```

Lines 14–19: These lines make the **if-then-else** construct. The **if** condition checks the value of the positional variable *$#* that contains the number of the line arguments. If *$#* is greater than zero, it indicates that the user has specified a message on the command line, and the body of the **if** is executed. The positional variable *$@* holds the specified message, which is stored in the *MESG* variable.

If *$#* is not greater than zero, then the body of the **else** is executed, and the *MESG* variable holds the canned message.

Line 24: This line displays the contents of the *MESG* variable, either the user-specified message or the canned message.

The remaining lines are the same as the `lock2` version of the program, and they carry out the same functions. As before, you change the access mode of your `lock3` program to executable mode, and then you can run it with or without specifying a message on the command line.

12.4 MORE COMMANDS

In the next section, we are going to put all our shell programming skills together and write an application that consists of more than just one script file. Before presenting the scenario for this application program, however, we need to examine a few more commands.

12.4.1 Multiway Branching: The case Construct

The shell provides you with the **case** construct, which enables you to selectively execute a set of commands from the list of commands. You can use the **if-elif-else** construct to achieve the same result, but when you have to use too many **elif** statements (more than two or three, for example), the **case** construct is preferred. Using the **case** construct rather than multiple **elif** statements is a more elegant way to program.

The syntax of the **case** construct is as follows:

```
case variable in
  pattern_1)
    commands_1 ;;
  pattern_2)
    commands_2 ;;
  ...
  ...
    *)
  default_commands ;;
esac
```

case, **in**, and **esac** (**case** typed backward) are reserved words (keywords). The statements between the **case** and the **esac** are called the *body* of the **case** construct.

When your shell executes the **case** statement, it compares the contents of the variable with each of the patterns until either a match is found or the shell reaches the keyword **esac**. The shell executes the commands associated with the matching pattern. The default **case**, represented by *), must be the last **case** in your program. The end of each **case** is indicated by two semicolons (**;;**).

Let's look at a simple menu program and explore the application of the **case** construct. A menu system provides an easy and simple user interface, and most computer users are familiar with menu systems. It is easy to implement menus in shell scripts.

You usually expect a menu program to perform the following functions:

- Display the possible options.
- Prompt the user to select a function.

- Read the user input.
- Call other programs according to the user input.
- Display an error message if the input is wrong.

The user enters a selection; the menu program must recognize your selection and take action accordingly. This recognition can be easily achieved by using the **case** construct. For example, suppose you have a script called MENU. The source code for your MENU program is depicted in Figure 12.7. The following command sequences show the running and output of this program.

Figure 12.7
Source Code for a Simple MENU Program

```
$ cat -n MENU
   1 #
   2 # MENU
   3 # A sample program to demonstrate the use of the case construct.
   4 #
   5 echo
   6 echo " 0: Exit"
   7 echo " 1: Show Date and Time"
   8 echo " 2: List my HOME directory"
   9 echo " 3: Display Calendar"
  10 echo  "Enter your choice: \c"   # display the prompt
  11 read option                     # read user answer
  12 case $option in                 # beginning of the case construct
  13    0) echo Good bye ;;          # display the message
  14    1) date ;;                   # show date and time
  15    2) ls $HOME ;;               # display the HOMEdirectory
  16    3) cal ;;                    # show the current month calendar
  17    *) echo "Invalid input. Good bye." ;; # show error message
  18 esac                            # end of the case construct
  19 echo
  20 exit 0                          # end of program, exit
$_
```

Run the MENU program:

```
$ MENU [Return] . . . . . . . Execute the menu program.
0: Exit
1: Show Date and Time
2: List my HOME directory
3: Display Calendar
Enter your choice:  . . . . . Wait for input.
```

The menu is displayed, and it prompts you to enter your choice. Your input is stored in the variable called *$option*. The variable *$option* is passed to the **case** construct, which recognizes your selection and executes the appropriate commands.

If you enter 0: The variable *$option* contains 0, which matches the 0 (zero) pattern in the **case** construct. Thus the **echo** command is executed and the message **Good bye** is displayed. No other pattern matches, and the program ends.

If you enter 1: The variable *$option* contains 1, which matches the 1 (one) pattern in the **case** construct. Thus, the command **date** is executed and the date and time of the day are displayed. No other pattern matches, and the program ends.

If you enter 2: The variable *$option* contains 2, which matches the 2 (two) pattern in the **case** construct. Thus, the command **ls** is executed and the contents of your HOME directory are displayed. No other pattern matches, and the program ends.

If you enter 3: Similar to the previous selections. The variable *$option* matches case 3, and the output of the **calendar** command is displayed.

If you enter 5: What if you make a mistake and enter 5 or 7? If your **case** construct has a default case (the * matches all), the default matches when all other cases fail, and its associated commands are executed. In this case, it displays **Invalid input. Good bye**.

12.4.2 Revisiting the `greetings` Program

Let's write another version of the `greetings` program (which was introduced in Chapter 11). In the new version, the **case** construct is used instead of the **if-elif-else**. Figure 12.8 is one way to write this program.

Figure 12.8
The New Version of the `greetings` Program

```
$ cat -n greetings3
   1 #
   2 # greetings3
   3 # greeting program version 3
   4 # This version is using the case construct to check the hour of
   5 # and the day and to display the appropriate greetings.
   6 #
   7 echo
   8 bell=`tput bel`              # store the code for the bell sound
   9 echo $bell$bell              # two beeps
  10 hour=`date +%H`              # obtain hour of the day
  11 case $hour in
  12   0?¦1[0-1] )  echo "Good Morning";;
  13      1[2-7] )  echo "Good Afternoon";;
  14           * )  echo "Good Evening";;
  15 esac                         # end of the case
  16 echo
  17 exit 0                       # end of the program, exit
$_
```

The output of this version is the same as the previous one. It beeps twice, and depending on the hour of the day, it displays appropriate greetings. The hour variable (which contains the hour of the day) is passed to the **case** construct. The value stored in hour is compared to each of the **case** patterns until a match is found. These patterns need more explanation. Lines 11 through 15 make the **case** construct.

The first pattern **0? ¦ 1[0-1]** checks for the morning hours (01 to 11). The **0?** represents any two-digit number that consists of a 0 and another digit (0 to 9). Therefore **0?** matches values from 00 to 09. The **1[0-1]** represents any two-digit number that consists of a 1 and another digit (0 to 1). Therefore, **1[0-1]** matches values 10 and 11. The ¦ (pipe) is the sign for logical **or**; therefore, the whole expression is true if either of the two patterns matches the value in the hour variable. If the value in the hour variable is between 01 and 11, the **echo** command displays **Good Morning**.

The second pattern **1[2-7]** checks for the afternoon hours (12–17). It represents any two-digit number that consists of a 1 and another digit in the range of 2 to 7. Therefore, **1[2-7]** matches values from 12 to 17. If the value in the hour variable is between 12 and 17, the **echo** command displays **Good Afternoon**.

The third pattern * (the default) matches any value. If the value in the *hour* variable is between 18 and 23, the first and second patterns fail, and the **echo** command displays **Good Evening**.

12.5 A MENU-DRIVEN APPLICATION

Scenario Suppose you want to write a menu-driven application that facilitates keeping track of your UNIX books. You want to be able to update the list of your books, to know whether any specified book is in your library (or not), and whether a book has been borrowed, who borrowed it, and when it was borrowed.

Of course this is a typical application for a database, but the goal is to practice some of the UNIX commands and to get a feeling about how you can put the commands and constructs together to create useful programs.

1. *As was mentioned before, the* **cat -n** *command is used throughout this chapter to display the source code for the script files. The* **-n** *option provides the line numbers that are used to refer to a specific line of code when that line is explained in the text.*
2. *Depending on your shell, the* **echo** *command in the programs might not recognize the escape characters. In this case, use the* **echo** *command with the* **-e** *option (***echo -e***). You can also use the* **alias** *command, such as in the following command line:*

   ```
   alias  echo=`echo -e`
   ```

3. *You can create these section script files under the* `$HOME/Chapter12/12.5` *directory using the vi editor.*
4. *Creating these files under a specific directory will help keep your files organized and is highly recommended.*

The Hierarchy Chart

Figure 12.9 shows the hierarchical chart of a menu-driven UNIX library program called `ULIB`. When you invoke the `ULIB`, it shows the main menu and waits for you to enter your choice. Each entry takes you to another level of menus. Like most of the menu-driven user interfaces, the top levels of `ULIB`'s hierarchy chart present the user interface. In this case, the three programs `ULIB`, `EDIT`, and `REPORTS` just show the appropriate menu and wait for the user's selection of the items on the menu.

Figure 12.9
The ULIB Program Hierarchy Chart

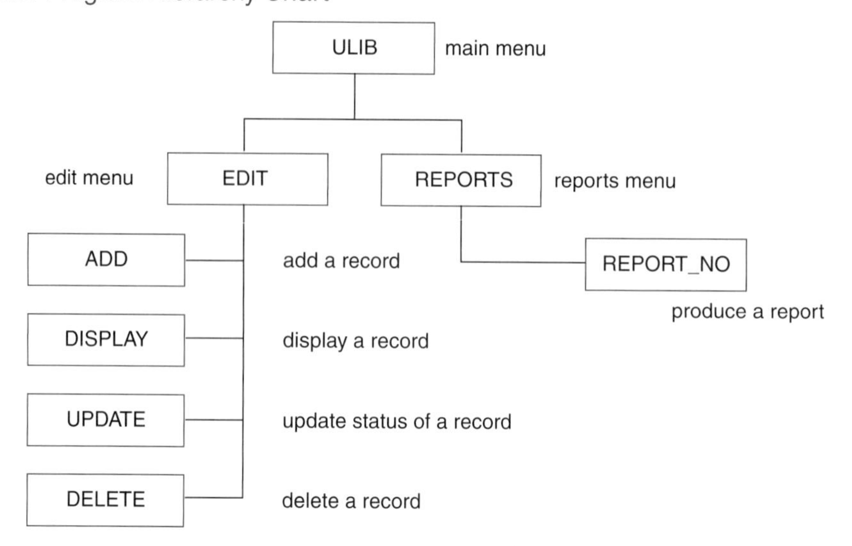

Each box in the hierarchy chart represents a program. Let's start with the first program, the `ULIB` at the top of the chart.

12.5.1 The `ULIB` Program

The `ULIB` program is the driver for our menu-driven library program. This program provides the startup screen (welcoming message) and then displays the main menu. The main menu provides the means to exit the program or to get to the next level of menus, `EDIT` and `REPORTS`.

Figure 12.10 shows one way to write this program. The line numbers are not part of the source code. Whenever necessary, these line numbers are referenced to provide more explanation on the commands or logic of the program.

Lines 1–6: These lines all begin with the # sign and are documentation lines.

Line 7: This line stores the terminal code for the boldface display in a variable called *BOLD*.

Line 8: This line stores the terminal code for the normal display in a variable called *NORMAL*.

Line 9: This line makes the variables *BOLD* and *NORMAL* available to the subshells.

Lines 10–12: These lines all begin with the # sign and are documentation lines.

Line 13: This line clears the screen.

Line 14: This line positions the cursor at row 5, column 15.

Line 15: This line displays the message **Super Duper UNIX Library** on the screen. Using the *BOLD* variable makes the message appear in bold.

Line 16: This line positions the cursor at row 12, column 10.

Figure 12.10
The ULIB Program Source Code

```
$ cat -n ULIB
   1 #
   2 # UNIX library
   3 # ULIB: This program is the main driver for the UNIX library application
   4 #       program. It shows a brief startup message and then displays the main menu.
   5 #       It invokes the appropriate program according to the user selection.
   6 #
   7 BOLD=`tput smso`                  # store code for bold mode in BOLD
   8 NORMAL=`tput rmso`                # store code for end of the bold mode in NORMAL
   9 export BOLD NORMAL                # make them recognized by subshells
  10 #
  11 # show the title and a brief message before showing the main menu
  12 #
  13 tput clear                        # clear screen
  14 tput cup 5 15                     # place the cursor on line 5, column 15
  15 echo "${BOLD}Super Duper UNIX Library"   # show the title in bold
  16 tput cup 12 10                    # place the cursor on line 12, column 10
  17 echo "${NORMAL}This is the UNIX library application"   # the rest of the title
  18 tput cup 14 10 ; echo "Please enter any key to continue..._\b\c"
  19 read answer                       # read user input
  20 error_flag=0                      # initialize the error flag, indicating no error
  21 while true                        # loop forever
  22 do
  23  if [ $error_flag -eq 0 ]         #check for the error
  24  then
  25    tput clear                     # clear screen
  26    tput cup 5 10
  27    echo "UNIX Library - ${BOLD}MAIN MENU${NORMAL}"
  28    tput cup  7 20  ; echo "0: ${BOLD}EXIT${NORMAL} this program"
  29    tput cup  9 20  ; echo "1: ${BOLD}EDIT${NORMAL} Menu"
  30    tput cup 11 20  ; echo "2: ${BOLD}REPORTS${NORMAL} Menu"
  31    error_flag=0                   # reset error flag
  32  fi
  33  tput cup 13 10 ; echo "Enter your choice> _\b\c"
  34  read choice                      # read user choice
  35 #
  36 # case construct for checking the user selection
  37 #
  38   case $choice in                 #check user input
  39        0 )  tput clear ; exit 0 ;;
  40        1 )  EDIT ;;               # call EDIT program
  41        2 )  REPORTS ;;            # call REPORT program
  42        * )  ERROR 20 10           # call ERROR program
  43             tput cup 20 1 ; tput ed # clear the reset of the screen
  44             error_flag=1 ;;       # set error flag to indicate error
  45   esac                            # end of the case construct
  46 done                              # end of the while construct
  47 exit 0                            # exit the program
$_
```

Line 17: This line displays the rest of the message, **This is the UNIX library application**, on the screen. Using the *NORMAL* variable deactivates the bold terminal mode, and the rest of the message appears in normal fashion.

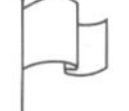

When a terminal mode is set, it remains in effect until it is canceled. Thus, if you use the command tput smso, your terminal displays text in boldface until you cancel the bold mode by typing the command tput rmso.

Line 18: This line consists of two commands. The **tput** command positions the cursor at row 14, column 10, and the **echo** command displays the prompt message.

Line 19: This line waits for input from the keyboard. When you press a key, the program continues.

This is a good point in the program to stop line explanations and to run it to see the output of this part (lines 1 to 19). If you run this program by typing the following command, the `ULIB` program shows the startup screen and waits for your input to continue the program.

```
$ ULIB [Return]
```

Figure 12.11 depicts the startup screen. After displaying the startup screen, the program shows the main menu and waits for the user to specify a selection from the menu. Depending on your selection, the program activates other programs. This process of showing the menu continues until you select the exit function from the menu.

Figure 12.11
The `ULIB` Program Startup Screen

```
                Super Duper UNIX Library

        This is the UNIX library application
        Please enter any key to continue...>_
```

Line 20: This line initializes the **error_flag** to 0 to indicate there is no error yet. The purpose of this flag is to indicate if the user has made a mistake. Depending on the value of this flag, your program shows the whole menu or just the portion of it that is necessary. This prevents clearing of the screen each time the user makes a mistake and causes less distraction for the user.

Lines 21, 22, and 46: These lines make the **while** loop. The **while** loop condition is set to true. The true condition is always true (succeeds); therefore, the **while** loop continues. That is to say, the body of the loop is executed forever, and this is exactly what you want, to show the main menu continuously until the exit function is selected.

Lines 23, 24, and 32: These lines make the **if** construct. The **if** condition checks for the state of the **error_flag**. If the **error_flag** is 0, then there is no error, and the condition is true. Therefore the body of the **if** is executed. If the **error_flag** is not 0, the **if** condition fails, and the body of the **if** is skipped.

Line 25: This line clears the screen so the main menu can be displayed. Remember, the startup screen is displayed before the main menu display.

Lines 27–30: These lines show the main menu on the screen.

Line 31: This line resets the **error_flag** to 0, initializing it for the next loop iteration.

Lines 33 and 34: These lines display the prompt text and wait for the user to enter a selection.

Let's continue the running of this program. Figure 12.12 shows the main menu. The prompt is displayed and the program waits for your input.

Figure 12.12
The ULIB Program Main Menu Screen

```
                UNIX Library - MAIN MENU

                    0:EXIT this program
                    1:EDIT menu
                    2:REPORTS Menu

                Enter your choice>_
```

The bold text on the screen is produced by displaying (using the **echo** *command) the code stored in the* BOLD *variable.*

Your input is saved in the *choice* variable, and *choice* is passed to the **case** construct to determine the next action according to your input.

Lines 35–37: These lines all begin with the # sign and are documentation lines.

Lines 38–45: These lines make the **case** construct.

If you select 0: The value in the *choice* variable is 0; it matches the 0 pattern and executes the associated commands. In this case, it clears the screen and exists the program. Choosing the 0 from the main menu is the normal way to end this program.

If you select 1: The value in the *choice* variable is 1; it matches the 1 pattern and executes the associated commands. In this case, it executes the program called `EDIT`. When the `EDIT` program is terminated, control returns to this program, and the program sequentially executes the next lines. Line 45 indicates the end of the **case** construct. Line 46 indicates the end of the **while** loop; therefore, it goes to line 21 to check the **while** loop condition. The condition is true; hence, the body of the loop is executed and shows the main menu again. This process is repeated until you enter 0 from the main menu.

If you select 2: The value in the choice variable is 2; it matches the 2 pattern and executes the associated commands. In this case, it executes the program called `REPORTS`. Similar to the function 1 selection that was explained earlier, when the `REPORTS` program is finished, control returns to the program menu, and the main menu is displayed, ready for your next selection.

If you make a mistake: If you make a mistake in your selection by inputting any key but the valid 0, 1, and 2 values, the value in the *choice* variable matches the default pattern (the asterisk) and executes the associated commands. In this case, it executes a program called ERROR. The ERROR program shows the error message at the indicated line and column on the screen. The desired line and column are specified on the command line. Therefore, **ERROR 20 10** means the error message will display at row 20, column 10. As before, control returns to the menu program when the ERROR program is finished.

Line 43: This line consists of two **tput** commands. The **tput cup 20 1** places the cursor at row 20, column 1, and the **tput ed** clears the part of the display between the cursor location and the end of the screen. In this case, terminal display lines 20, 21, 22, 23, and 24 are cleared. This erases the error message, but the main menu text remains on your terminal screen.

Line 44: This line sets the **error_flag** to 1, indicating an error has occurred.

Remember **error_flag** *is checked in the* **if** *condition. Because its value is 1, the* **if** *condition fails, and the body of the* **if** *construct is skipped. This means that the main menu text is already on the screen and there is no need to redisplay it.*

Figure 12.13 shows the screen when you make a mistake in your selection. The error message and the prompt text are the output of the ERROR program. When you press a key to signal your intention to continue the program, the error message is erased and you are prompted again to enter your selection from the menu (as was shown in Figure 12.12).

Figure 12.13
The ULIB Program Error Message Screen

```
        UNIX Library - MAIN MENU

            0:EXIT this program
            1:EDIT menu
            2:REPORTS Menu

        Enter your choice>_6
     Wrong Input. Try again.
  Press any key to continue...>_
```

12.5.2 The ERROR Program

The ERROR program shows a *canned* (hard coded) error message any time it is called. It accepts the values indicating the location of the cursor (line and column) from the command line. Figure 12.14 shows the source code for this program. The following explanations provide more detailed information about the code on specific lines in the program.

Line 6: This line places the cursor at a specified location on the screen. The row and column values are stored in the two positional variables *$1* and *$2*. These variables contain the first two parameters passed to the program on the command line. Therefore, if you call this program by typing

```
$ ERROR 10 15 [Return]
```

the positional variables *$1* and *$2* contain the values 10 and 15 respectively, and the error message will be displayed at row 10, column 15. The ULIB program (Figure 12.10) calls

Figure 12.14
The ERROR Program Source Code

```
$ cat -n ERROR
   1 #
   2 # ERROR: This program displays an error message and waits for user
   3 #        input to continue. It displays the message at the specified
   4 #        row and column.
   5 #
   6 tput cup $1 $2                    # place the cursor on the screen
   7 echo "Wrong Input. Try again."    # show the error message
   8 echo "Press any key to continue...> _\b\c"  # display the prompt
   9 read answer                       # read user input
  10 exit 0                            # indicate normal exit
$_
```

the ERROR program when user input is wrong to display an error message at row 20, column 10.

Line 7: This line displays the error message.

Line 8: This line displays the prompt.

Line 9: This line reads the user input. No error checking is performed on the user input; you can enter any key to satisfy the **read** command. Thus, control of the program is in the hands of the user, and the user can continue the program by pressing a key.

12.5.3 The EDIT Program

The EDIT program is activated whenever you select function 1 from the main menu. This program is a driver for the edit menu and is similar to the main menu program. It shows the edit menu, and according to your selection, it activates appropriate programs. It consists of a **while** loop that covers the whole program. The body of the **while** loop consists of the edit menu display code and a **case** structure that determines what commands must be executed to satisfy your selection. Essentially this is the logic for any menu program you write: Show the menu, read the selection, and act according to the selection.

Figure 12.15 shows the source code for the EDIT program. This program is similar to the main (ULIB) program that was explained in detail in the previous section. Figure 12.16 shows the EDIT menu. The EDIT menu is displayed when you choose function 1 from the main menu.

12.5.4 The ADD Program

This program adds a record to your library file and is activated when you select the ADD function from the EDIT menu. It prompts you to enter information, adds the new record to the end of the library file, and asks if you want to add any more records. A **yes** answer continues the program, appending records to the file. A **no** answer terminates the program and returns control to the calling program, in this case the EDIT program, and you are back to the EDIT menu for your next selection. This program assumes that the library file that holds all the information about your UNIX books is called ULIB_FILE;

Figure 12.15
The EDIT Program Source Code

```
$ cat -n EDIT
   1 #
   2 # UNIX Library
   3 # EDIT: This program is the main driver for the EDIT program.
   4 #       It shows the EDIT menu and invokes the appropriate program
   5 #       according to the user selection.
   6 #
   7 error_flag=0                     # initialize the error flag, indicating no error
   8 while true                       # loop forever
   9 do
  10  if [ $error_flag -eq 0 ]        # check for the error
  11  then
  12    tput clear ; tpup cut 5 10    # clear screen and place the cursor
  13    echo "UNIX Library - ${BOLD}EDITMENU${NORMAL}"
  14    tput cup  7 20                # place the cursor
  15    echo "0: ${BOLD}RETURN${NORMAL}To the Main Menu"
  16    tput cup  9 20 ; echo "1:${BOLD}ADD${NORMAL}"
  17    tput cup 11 20 ; echo "2:${BOLD}UPDATESTATUS${NORMAL}"
  18    tput cup 13 20 ; echo "3:${BOLD}DISPLAY${NORMAL}"
  19    tput cup 15 20 ; echo "4:${BOLD}DELETE${NORMAL}"
  20  fi
  21  error_flag=0                    # reset error flag
  22  tput cup 17 10 ; echo "Enter your choice> _\b\c"
  23  read choice                     #read user choice
  24 #
  25 # case construct for checking the user selection
  26 #
  27  case $choice in
  28     0) exit 0 ;;                 # check user input
  29     1) ADD ;;                    # return to the main menu
  30     2) UPDATE;;                  # call UPDATE program
  31     3) DISPLAY ;;                # call DISPLAY program
  32     4) DELETE ;;                 # call DELETE program
  33     *) ERROR 20 10               # call ERROR program
  34        tput cup 20 1 ; tput ed # clear the rest of the screen
  35        error_flag=1 ;;           # set error flag to indicate
  36  esac                            # end of the case construct
  37 done                             # end of the while construct
  38 exit 0                           # end of the program
$_
```

Figure 12.16
The EDIT Menu

```
                UNIX Library - EDIT MENU

                    0: RETURN To The Main Menu
                    1: ADD
                    2: UPDATE STATUS
                    3: DISPLAY
                    4: DELETE

                Enter your choice>_
```

it further assumes that the following information is saved in the record format for each book in the `ULIB_FILE`. Each item is stored as a field in the record, and the following list shows the field names and sample information in each field. Data about this textbook are used as examples to show a possible value for each field.

- Title: UNIX Unbound
- Author: Afzal Amir
- Category: Textbook
  Three valid categories are assumed:
  - System books: abbreviated to *sys*
  - Reference books: abbreviated to *ref*
  - Textbooks: abbreviated to *tb*
- Status: in
  The status indicates whether the book is checked out or is in the library. The book's status is determined by the program, and the status field is set to **in** (checked in) automatically when you add a book. The status is changed to **out** (checked out) when you indicate a book is checked out (borrowed) by someone.
- Borrower's name:
  This field remains empty if the status field indicates the specified book is in the library (status set to **in**) and is set to the name of the person who borrowed the book if the status field shows **out** (checked out).
- Date:
  This field remains empty if the status field indicates the specified book is in the library (status set to **in**) and is set to the date that the book was checked out if the status field indicates **out** (the book is checked out).

1. *The first time a book record is added to the* `ULIB_FILE` *file, the status field is set to* **in**, *and the borrower's name and the date fields remain empty.*
2. *Subsequently, when someone borrows a book, you update your library by selecting the UPDATE function from the main menu. Then, you are asked to enter the name of the person who borrowed the book. The program changes the status field to* **out** *(checked out), and the date field is set to the current date.*

Figure 12.17 shows the source code for this program; line explanations for this program follow.

Lines 1–6: These lines all begin with the # sign and are documentation lines.

Line 7: This line initializes the *answer* variable to the letter **y**, indicating **yes**. As long as this variable indicates **yes**, the **while** loop continues and in each iteration it adds one record to the `ULIB_FILE` file.

Lines 8, 9, and 28: These lines make the **while** loop. The body of this loop is executed as long as the condition of the loop is true—in this case, as long as the value stored in the *answer* variable is the letter **y**.

Line 10: This line clears the screen.

Line 11: This line displays the title of the screen at the indicated cursor position.

Lines 12–14: These lines display the prompts at the specified cursor locations.

Figure 12.17
The ADD Program Source Code

```
$ cat -n ADD
   1 #
   2 # UNIX library
   3 # ADD: This program adds a record to the library file (U_LIB). It asks the
   4 #      title, author, and category of the book. After adding the information to
   5 #      the ULIB_FILE file, it prompts the user for the next record.
   6 #
   7 answer=y                               # initialize the answer to indicate yes
   8 while  [ "$answer" = y ]               # as long as the answer is yes
   9 do
  10   tput clear
  11   tput cup  5 10 ; echo "UNIX Library - ${BOLD}ADDMODE"
  12   echo "${NORMAL}"
  13   tput cup  7 23 ; echo "Title:"
  14   tput cup  9 22 ; echo "Author:"
  15   tput cup 11 20 ; echo "Category:"
  16   tput cup 12 20 ; echo "sys; system, ref: reference, tb: textbook"
  17   tput cup  7 30 ; read title
  18   tput cup  9 30 ; read author
  19   tput cup 11 30 ; read category
  20   status=in                             # set the status to indicate book is in
  21   echo "$title:$author:$category:$status:$bname:$date" >> ULIB_FILE
  22   tput cup  14 10 ; echo "Any more to add? (Y)es or (N)o> _\b\c"
  23   read answer
  24   case $answer in                       # check user answer
  25      [Yy]* ) answer=y ;;                # any word starting with Y or y is yes
  26          * ) answer=n ;;                # any other word indicates no
  27      esac                               # end of the case construct
  28 done                                    # end of the while loop
  29 exit 0                                  # end of the program
$_
```

Line 15: This line is a help message that explains what abbreviations the user can enter in the category field.

Lines 16–19: Each of these lines places the cursor at a specified location on the screen (just after each prompt text), reads the user's answers, and stores them in appropriate variables.

Line 20: This line sets the status variable to **in**, indicating the book is in the library.

At this point all the necessary information is obtained, and the next step is to save it in the ULIB_FILE file. If you select the ADD function from the EDIT menu, the ADD program is activated. Figure 12.18 shows the screen display produced by the ADD program.

The cursor is placed at the Title field. You enter the title of the book followed by [Return], and the cursor moves to the Author field. When you enter the last field, the program saves the information and asks you if you want to add another record. Using this textbook as an example, Figure 12.19 shows the screen after the information has been entered and the record has been saved.

Figure 12.18
The ADD Program Screen Format

```
UNIX Library - ADD MODE

   Title:
  Author:
Category:
sys: system, ref: reference,tb: textbook
```

Figure 12.19
The Example of the Filled in ADD Screen

```
UNIX Library - ADD MODE

   Title: UNIX Unbounded
  Author: Afzal Amir
Category: tb
sys: system, ref: reference,tb: textbook

Any more to add? (Y)es or (N)o>_
```

Line 21: This line saves the information in the `ULIB_FILE` file. By the time the ADD program reaches this line, the variables contain the following values:

- The *title* variable contains **UNIX Unbounded**.
- The *author* variable contains **Afzal Amir**.
- The *category* variable contains **tb**.
- The *status* variable contains the word **in**.
- The *bname* (for borrower's name) variable is empty.
- The *date* (date that the book was checked out) variable is empty.

By default, the **echo** command displays its output on the terminal screen. However, the output of the **echo** command on line 21 is redirected to the `ULIB_FILE` file. Therefore, the information stored in the specified variables is saved in the `ULIB_FILE` file.

1. *When you have just added a record, the variables* bname *(borrower's name) and* date *(checked out date) remain empty.*
2. *The append sign (>>) is used to write information in the* `ULIB_FILE` *file. This is necessary so that each time you add a new record, it is appended to the end of the file and the rest of the records in the file remain intact.*

Delimiter Character

The fields in each record are stored in sequential order. However, this method of saving makes it difficult to retrieve each field when you want to read and display a record. You must designate a field delimiter and separate the fields in the record with your selected character. Keys that UNIX uses as field separators ([Tab], [Return], and [Spacebar]) are not good choices. Remember, the title of a book or name of an author may contain space characters. If you choose the space character as the delimiter, then an author's name with a space between the first and last names is considered as two fields, although you may want your program to see it as one field.

You can choose ~ or ^ as the delimiter key. In the ULIB program, **:** is used as the delimiter character.

The ULIB_FILE file is a text file, and you can use the **cat** command to display its contents or use the vi editor to modify it. If the previous record is the only record in the file, then the content of the ULIB_FILE file looks like the display in Figure 12.20.

Figure 12.20
The Contents of the ULIB_FILE File

```
$ cat ULIB_FILE
  UNIX Unbounded:Afzal Amir:tb:in::
$_
```

1. *Each record in the file is one line, and the fields are separated by colons.*
2. *The two colons at the end of the line are the placeholders for the two variables* bname *(which contains the name of the borrower when not empty) and* date *(which contains the date that the book was borrowed when not empty).*

Lines 22 and 23: The **echo** command displays the prompt and the **read** command waits for the user's answer, which is saved in the *answer* variable.

Lines 24–27: These lines make the **case** construct. The **case** construct checks for the user's *answer* by matching the value in answer to the **case** patterns.

The first pattern is **[Yy]***. This pattern matches all sequences of characters that start with **Y** or **y**. You can type any word that starts with the letter *y* and the program interprets it as the *yes* answer. The *answer* variable is set to the letter *y*, the loop condition (line 8) is true, and the the program continues.

The second pattern is **[*]**. This pattern matches any word that does not start with the letter *y* and interprets it as the no answer. The *answer* variable is set to the letter *n*, the loop condition (line 8) fails, and the program ends.

12.5.5 Record Retrieval

Part of the code in each of the remaining programs in the library application needs to display an existing record on the screen. In order to display a record, you must specify what record you want displayed. For example, if you specify the author's name, the library file is searched to check whether the specified author's name is in the file. If it is found, the record is read (retrieved) and displayed. If the record is not found, an error message is displayed.

12.5.6 The DISPLAY Program

The DISPLAY program displays a specified record from the ULIB_FILE file on the screen, and it is activated when you select the display function from the EDIT menu. It first prompts you to enter the book's title or the author's name. Then, it searches the ULIB_FILE file to find a match for the specified book. If the record is found, it is displayed in a presentable format. If the record is not found, an error message is displayed, and you are prompted for the next input. This process continues until you indicate your intention to end the program. When the DISPLAY program ends, the control is returned to the calling program, in this case the EDIT program, and you are back to the EDIT menu for your next selection.

Figure 12.21 shows the DISPLAY program source code. The following is the line-by-line explanation of this program.

Lines 1–6: These lines all begin with the # sign and are documentation lines.

Line 7: This line copies the value of the environment variable *IFS* to OLD_IFS. You want to save the old value before assigning the new value. Later, you want to be able to restore the *IFS* variable to its original setup.

Line 8: This line initializes the variable *answer* to the letter *y*. Consequently it forces the first iteration of the **while** loop.

Lines 9, 10, and 44: These lines construct the **while** loop. The body of the loop consists of all the lines between lines 10 and 44. As long as the condition of the **while** loop is true (answer is equal to the letter *y*), then the body of the loop is executed.

Line 11: This line consists of three commands. It clears the screen, places the cursor at row 5, column 6, and prompts the user to enter the title or the author of the book to be displayed.

Line 12: This line waits for the user input and saves the input in the *response* variable.

Line 13: This line uses the **grep** command (see chapter 8) to find all the lines in the ULIB_FILE file that contain the pattern stored in the response variable (the user input), and the output is redirected to the TEMP file. If the TEMP file is not empty, it means the specified book record is found, and if TEMP remains an empty file, it means the specified record is not found.

Lines 14, 15, 35, and 37: These lines make an **if-then-else** construct. The **if** condition tests for the existence of a nonzero-length file, in this case the TEMP file. If the condition is true (TEMP exists and there is something in it), then the body of the **if** (lines 16–34) is executed. If the condition is false (TEMP is an empty file), then the body of the **else** (line 36) is executed.

Line 16: This line sets the delimiter to the colon (**:**). In this application, the colon is the symbol that separates the fields in the records.

Line 17: This line reads each field of the specified book record from the TEMP file.

Lines 18 and 19: These lines place the cursor at the specified location and display the record header.

Line 20: This line consists of two commands. It places the cursor at row 7, column 23, and then displays the title of the book. The title variable contains the title of the book as it was read from the TEMP file.

Line 21: This line consists of two commands. It places the cursor at row 8, column 22, and then displays the author of the book. The author variable contains the name of the author as it was read from the TEMP file.

Lines 22–27: These lines make the **case** construct. The **case** construct checks the book's category (stored in the category variable) and changes the category abbreviations to a full word. For example, the category abbreviation *sys* is changed to *system*.

Figure 12.21
The DISPLAY Program Source Code

```
$ cat -n DISPLAY
   1 #
   2 # UNIX library
   3 # DISPLAY: This program displays a specified record from the ULIB_FILE.
   4 #          It asks the Author/Title of the book, and displays the specified
   5 #          book is not found in the file.
   6 #
   7 OLD_IFS="$IFS"                              # save the IFS settings
   8 answer=y                                    # initialize the answer to indicate yes
   9 while [ "$answer" = y ]                     # as long as the answer is yes
  10 do
  11  tput clear ; tput cup 3 5 ; echo "Enter the Author/Title> _\b\c"
  12  read response
  13  grep -i "$response" ULIB_FILE> TEMP        # find the specified book in the library
  14  if [ -s TEMP ]                             # if it is found
  15  then
  16    IFS=":"                                  # set the IFS to colon
  17    read title author category status bname date < TEMP
  18    tput cup 5 10
  19    echo "UNIX Library - ${BOLD}DISPLAYMODE${NORMAL}"
  20    tput cup 7 23 ; echo "Title: $title"
  21    tput cup 8 22 ; echo "Author: $author"
  22    case $category in                        # check the category
  23             [Tt][Bb]) word=textbook ;;
  24         [Ss][Yy][Ss]) word=system ;;
  25         [Rr][Ee][Ff]) word=reference ;;
  26                    *) word=undefined ;;
  27    esac
  28    tput cup 9 20 ; echo "Category: $word" # display the category
  29    tput cup 10 22 ; echo "Status:$status" # display the status
  30    if [ "$status" = "out" ]                 # if it is checked out
  31    then                                     # then show the rest of the information
  32      tput cup 11 14 ; echo "Checked out by: $bname"
  33      tput cup 12 24 ; echo "Date: $date"
  34   fi
  35  else                                       # if book not found
  36    tput cup 7 10 ; echo "$response not found"
  37  fi
  38  tput cup 15 10 ; echo "Any more to look for? (Y)es or (N)o> _\b\c"
  39  read answer                                # read user's answer
  40  case $answer in                            # check user's answer
  41     [Yy]* ) answer=y ;;                     # any word starting with Y or y is yes
  42         * ) answer=n ;;                     # any other word indicates no
  43 esac                                        # end of the case construct
  44 done                                        # end of the while loop
  45 IFS="$OLD_IFS"                              # restore the IFS to its original value
  46 exit 0                                      # exit the program
$_
```

The pattern **[Tt][Bb]** matches any combination of uppercase and lowercase letters representing the category abbreviation *tb*. The string **textbook** is stored in the *word* variable.

The pattern **[Ss][Yy][Ss]** matches any combination of uppercase and lowercase letters representing the category abbreviation *sys*. The string **system** is stored in the *word* variable.

The pattern **[Rr][Ee][Ff]** matches any combination of uppercase and lowercase letters representing the category abbreviation *ref*. The string **Reference** is stored in the *word* variable.

If it is none of the above categories, then it matches the default (asterisk) and the string **undefined** is stored in the *word* variable.

Line 28: This line places the cursor at row 9, column 20, and displays the specified book category field (contents of the *word* variable).

Line 29: This line places the cursor at row 10, column 22, and displays the specified book status field (contents of the *status* variable).

Lines 30 and 34: These lines make an **if-then** construct. If the **if** condition is true (the book is checked out), then the body of the **if** (lines 31–33) is executed. Otherwise, if the status shows the book is **in**, then the body of the **if** is skipped.

Line 32: This line places the cursor at row 11, column 14, and displays the name of the person who has borrowed the book (contents of the *bname* variable).

Line 33: This line places the cursor at row 12, column 24, and displays the date (contents of the *date* variable) the book was checked out.

Line 36: This line is executed when the `TEMP` file is empty (the specified book is not found). It places the cursor at row 7, column 10, and displays the error message.

Similar to the `ADD` program, the rest of the code is to check whether the user's intention is to display another record or to return to the `EDIT` menu.

If you run the `ULIB` program and select the DISPLAY function from the `EDIT` menu, a prompt is displayed. Figure 12.22 shows the screen at this point.

Figure 12.22
The Prompt Displayed by the `DISPLAY` Program

```
Enter the Author/Title>_
```

You can enter the author's name or the book's title to specify the book record you want to be displayed. Let's say you enter **UNIX Unbounded**. Figure 12.23 shows the screen that displays your specified record. This screen is similar to the ADD screen depicted in Figure 12.19. The program prompts you to enter your intention to look at more records or to end the `DISPLAY` program. When this program ends, control returns to the `EDIT` program, the `EDIT` menu is displayed again, and you can select your next function.

If your specified record is not found, the error message is displayed. For example, if you do not have a book titled *XYZ*, Figure 12.24 shows the error message and the prompt following it.

Figure 12.23
Sample of the DISPLAY Program Screen

```
UNIX Library - DISPLAY MODE

   Title : UNIX Unbounded
  Author : Afzal Amir
Category : texbook
  Status : in

Any more to look for?(Y)es or (N)o>_
```

Figure 12.24
Sample Screen Showing the Error Message

```
Enter the Author/Title>XYZ


      XYZ not found

      Any more to look for?(Y)es or (N)o>_
```

12.5.7 The UPDATE Program

The UPDATE program changes the status of a specified record in the ULIB_FILE file, and it is activated when you select the UPDATE STATUS function from the EDIT menu. You choose this option when a book is checked out or when a checked-out book is checked in. The UPDATE program first prompts you to enter the title or the author of the book whose status you want to change. Then it searches the ULIB_FILE file to find a match for the specified book. If the record is found, it is displayed in a presentable format. If the record is not found, an error message is displayed, and you are prompted for the next input. This process continues until you indicate your intention to end the program. When the UPDATE program ends, control is returned to the calling program, in this case the EDIT program, and you are back to the EDIT menu for your next selection.

Figure 12.25 shows the UPDATE program source code, and a line-by-line explanation of this program follows. The source code for the UPDATE program is longer than

Figure 12.25
Source Code for the UPDATE Program

```
$ cat -n UPDATE
   1 #
   2 # UNIX library
   3 # UPDATE:This program updates the status of a specified book. It asks the
   4 #        Author/Title of the book, and changes the status of the specified
   5 #        book from in (checked in) to out (checked out), or from out to in.
   6 #        If the book is not found in the file, an error message is displayed.
   7 #
   8 OLD_IFS="$IFS"                        # save the IFS settings
   9 answer=y                              # initialize the answer to indicate yes
```

Figure 12.25
(continued)

```
10 while [ "$answer" = y ]
11 do
12  new_status= ; new_bname= ; new_date= # declare empty variables
13  tput clear                           # clear screen
14  tput clear ; tput cup 3 5 ; echo "Enter the Author/Title > _\b\c"
15  read response
16  grep -i "$response" ULIB_FILE > TEMP # find the specified book
17  if [ -s TEMP ]                       # if it is found
18  then                                 # then
19    IFS=":"                            # set the IFS to colon
20    read title author category status bname date < TTEMP
21    tput cup 5 10
22    echo "UNIX Library - ${BOLD}UPDATESTATUSMODE${NORMAL}"
23    tput cup 7 23 ; echo "Title: $title"
24    tput cup 8 22 ; echo "Author: $author"
25    case $category in                        # check the category
26            [Tt][Bb] ) word=textbook ;;
27        [Ss][Yy][Ss] ) word=system ;;
28        [Rr][Ee][Ff] ) word=reference ;;
29                   * ) word=undefined ;;
30    esac
31    tput cup  9 20 ; echo "Category: $word" # display the category
32    tput cup 10 22 ; echo "Status: $status" # display the status
33    if [ "$status" = "in" ]                 # if it is checked in
34    then                                    # then show the rest of the information
35      new_status=out                        # indicate the new status
36      tput cup 11 18 ; echo "New status: $new_status"
37      tput cup 12 14 ; echo "Checked out by: _\b\c"
38      read new_bname
39      new_date=`date +%D`
40      tput cup 13 24 ; echo "Date: $new_date"
41    else
42      new_status=in
43      tput cup 11 14 ; echo "Checked out by: $bname"
44      tput cup 12 24 ; echo "Date:$date"
45      tput cup 15 18 ; echo "New status:$new_status"
46    fi
47    grep -iv "$title:$author:$category:$status:$bname:$date" ULIB_FILE> TEMP
48    cp TEMP ULIB_FILE
49    echo "$title:$author:$category:$new_status:$new_bname:$new_date">> ULIB_FILE
50  else                                      # if book not found
51    tput cup 7 10 ; echo "response not found"
52  fi
53  tput cup 16 10; echo "Any more to update? (Y)es or (N)o> _\b\c"
54  read answer                               # read user answer
55  case $answer in                           # check user answer
56         [Yy]* ) answer=y ;;                # any word starting with Y or y is yes
57             * ) answer=n ;;                # any other word indicates no
58  esac                                      # end of the case construct
59 done                                       # end of the while loop
60 IFS="$OLD_IFS"                             # restore the IFS to its original value
61 rm TEMP TTEMP                              # delete files
62 exit 0                                     # exit the program
$_
```

that of the previous programs, but most of the lines are copied from the DISPLAY program. The lines of code at the beginning and end of the DISPLAY, UPDATE, and DELETE programs are all the same. All these programs prompt you to specify the author or title of the desired book, and the program either finds the record and displays it or does not find the record and displays an error message. At the end of the program, you are prompted to indicate your intention to continue the program or to end it. Therefore, only the new or changed lines are explained here.

Line 12: This line declares three empty variables. These variables store new values when the status of the book is changed from **in** to **out** (checked out) and initialize them to be empty when the status of a book is changed from **out** to **in** (checked in).

- The *new_status* variable holds the new status, the word *in* or *out*.
- The *new_bname* variable holds the name of the borrower or is empty.
- The *new_date* variable holds the current date or is empty.

Lines 33–46: These lines make the **if-then-else** construct. If the **if** condition is true (the *status* variable contains the word *in* indicating the book is in the library), then the body of the **if** (lines 35–40) is executed. If the **if** condition is false (the status variable contains the word *out*, indicating the book is checked out), then the body of the **else** (lines 42–45) is executed.

Lines 35–40: These lines are the body of the **if** and change the status of the record from **in** to **out**. The only item of information needed is the name of the person who checked out the book.

Line 35: This line stores the word *out* in the *new_status* variable, indicating the the book is checked out.

Line 36: This line places the cursor at the specified row and column and displays the new status. In this case, the new status is **out**.

Line 37: This line prompts the user to enter the name of the person who checked out the book.

Line 38: This line reads the user's response and saves it in the *new_bname* variable.

Line 39: This line stores the current date in the *new_date* variable. The **%D** option indicates only the date and not the time in the date string (see Chapter 11).

Lines 42–45: These lines are the body of the **else** and change the status of the record from **out** to **in**.

Line 42: This line stores the word *in* in the *new_status* variable, indicating the the book is checked in.

Line 43: This line places the cursor on the specified line and column, and it displays the contents of the *bname* variable, the borrower's name.

Line 44: This line places the cursor at the specified row and column and displays the contents of the *date* variable, the date on which the book was checked out.

Line 45: This line places the cursor at the specified row and column and displays the contents of the *new_status* variable, in this case the word *in*, indicating that the book is checked in.

Line 47: This line saves (copies) every record in the ULIB_FILE file to the TEMP file, except for the specified record on the display. It does that by using the **grep** command with the **i** and **v** options (see Chapter 8) and redirecting the output of the **grep** from the terminal display to the TEMP file. The **i** option causes the **grep** to ignore the difference between uppercase and lowercase letters. The **v** option causes the **grep** to pick all the lines that do not contain a match for the specified pattern.

Line 48: This line uses the **cp** command (see Chapter 5) to copy the TEMP file to the ULIB_FILE file. Now the ULIB_FILE file contains every record except the record whose status has just been changed.

Line 49: This line appends the changed record to the ULIB_FILE file. Now the ULIB_FILE file contains the specified record with its updated status.

Let's continue running the program. The terminal display helps you associate the line explanations with the actual output on the screen.

Figure 12.26 shows the UPDATE program display, assuming the UPDATE STATUS function is selected from the EDIT menu and the specified book is **UNIX Unbounded**.

Figure 12.26
The UPDATE Program Display

```
UNIX Library - UPDATE STATUS MODE

          Title: UNIX Unbounded
         Author: Afzal Amir
       Category: textbook
         Status: in
     New status: out
 Checked out by: Steve Fraser
           Date: 12/12/05

Any more to update?(Y)es or (N)o>_
```

1. *The status change is reflected on the screen by showing the old status and the new status fields.*
2. *You are prompted to enter the name of the person who checked out the specified book. It is assumed that the name* **Steve Fraser** *is entered in response to the* **Checked out by:** _ *prompt.*

Figure 12.27 shows the ULIB_FILE after the change of status. If you compare this display to the display in Figure 12.20, you will notice the changes in the fields of the specified record.

Figure 12.27
The Contents of the ULIB_FILE File

```
$ cat ULIB_FILE file
   UNIX Unbounded:Afzal Amir:tb:out:Steve Fraser:12/12/05
$_
```

Figure 12.28 shows the display produced by the DISPLAY program, showing the same record.

Figure 12.28
Sample Record Display

```
UNIX Library - DISPLAY MODE

           Title: UNIX Unbounded
          Author: Afzal Amir
        Category: textbook
          Status: out
  Checked out by: Steve Fraser
            Date: 12/12/05
Any more to look for?(Y)es or (N)o>_
```

Compare this display (where the status indicates the book is checked out) with the display depicted in Figure 12.23 (where the status indicates the book is checked in).

12.5.8 The DELETE Program

The DELETE program deletes a specified record from the ULIB_FILE file, and it is activated when you select the DELETE function from the EDIT menu. It first prompts you to enter the book's title or the author's name. Then it searches the ULIB_FILE file to find a match for the specified book. If the record is found, it is displayed in a presentable format, and the confirmation prompt is displayed. It lets you decide whether the displayed record is the record you want to delete. If the record is not found, an error message is displayed, and you are prompted for the next input. This process continues until you indicate your intention to end the program. When the DELETE program ends, control is returned to the calling program, in this case the EDIT program, and you are back to the EDIT menu for your next selection.

Figure 12.29 shows the source code for the DELETE program. The length of the program should not worry you. The beginning and the end parts of the DELETE program are similar to those of the DISPLAY and UPDATE programs. The major differences are the code for the confirmation check before deletion of the record and the code for actual removal of the specified record from the ULIB_FILE file.

After the record is found and displayed, the confirmation prompt is displayed. At this point, the program is on line 36.

Line 36: This line places the cursor at the specified row and column and displays the prompt.

Figure 12.29
Source Code for the DELETE Program

```
$ cat -n DELETE
     1 #
     2 # UNIX library
     3 # Delete: This program deletes a specified record from the ULIB_FILE.
     4 #         It asks the Author/Title of the book, and displays the specified
```

Figure 12.29
(continued)

```
 5 #        book, and deletes it after confirmation, or shows an error  message
 6 #        if the book is not found in the file.
 7 #
 8 OLD_IFS="$IFS"                       #save the IFS setting
 9 answer=y                             # initialize the answer to indicate yes
10 while [ "$answer" = y ]              # as long as the answer is yes
11 do
12  tput clear ; tput cup 3 5 ; echo "Enter the Author/Title> _\b\c"
13  read response
14  grep -i "$response" ULIB_FILE> TEMP  # find the specified book in the library
15  if [ -s TEMP ]                       # if it is found
16  then                                 # then
17    IFS=":"                            # set the IFS to colon
18    read title author category status bname date < TEMP
19    tput cup 5 10
20    echo "UNIX Library - ${BOLD}DELETEMODE${NORMAL}"
21    tput cup 7 23 ; echo "Title: $title"
22    tput cup 8 22 ; echo "Author: $author"
23    case $category in                        # check the category
24             [Tt][Bb] ) word=textbook ;;
25         [Ss][Yy][Ss] ) word=system ;;
26         [Rr][Ee][Ff] ) word=reference ;;
27                    * ) word=undefined ;;
28     esac
29     tput cup  9 20 ; echo "Category: $word"  # display the category
30     tput cup 10 22 ; echo "Status: $status"  # display the status
31     if [ "$status" = "out" ]                 # if it is checked out
32     then                                     # then show the rest of the information
33     tput cup 11 14 ; echo "Checked out by: $bname"
34     tput cup 12 24 ; echo "Date: $date"
35     fi
36     tput cup 13 20 ; echo "Delete this book?(Y)es or (N)o> _\b\c"
37     read answer
38     if [ $answer = y -o $answer = Y ]        # test for Y or y
39     then
40     grep -iv "$title:$author:$category:$status:$bname:$date" ULIB_FILE> TEMP
41     mv TEMP ULIB_FILE
42   fi
43  else                                        # if book not found
44   tput cup 7 10 ; echo "$response not found"
45  fi
46  tput cup 15 10; echo "Any more to delete? (Y)es or (N)o> _\b\c"
47  read annswer                                # read user answer
48  case $answer in                             # check user answer
49        [Yy]* ) answer=y ;;                   # any word starting with Y or y is yes
50            * ) answer=n ;;                   # any other word indicates no
51 esac
52 done                                         # end of the while loop
53 IFS="$OLD_IFS"                               # restore the IFS to its original value
54 rm TEMP                                      # delete the TEMP file
55 exit 0                                       # exit the program
```

Line 37: This line reads the user's response and stores it in the *answer* variable.

Lines 38–42: These lines make the **if** construct. If the **if** condition is true, then the body of the **if** (lines 40 and 41) is executed. The **if** condition tests for *Y* or *y*, and the record is deleted. Any other letter fails the test, the body of the **if** is skipped, and the record is not deleted.

Line 40: This line saves (copies) every record in the ULIB_FILE file to the TEMP file, except the specified record that is currently displayed. It does that by using the **grep** command with the **i** and **v** options and redirecting the output of the **grep** from the terminal display to the TEMP file.

Line 41: This line uses the **mv** command to rename the TEMP file to the ULIB_FILE file. Now the ULIB_FILE file contains every record except the deleted record.

Figure 12.30 shows a sample of the DELETE program display, assuming the DELETE function is selected from the EDIT menu and the specified book exists in the ULIB_FILE file.

Figure 12.30
The DELETE Program Display

```
UNIX Library - DELETE MODE

          Title: UNIX Unbounded
         Author: Afzal Amir
       Category: textbook
         Status: out
 Checked out by: Steve Fraser
           Date: 12/12/05

Delete this book? (Y)es or (N)o>Y

Any more to delete? (Y)es or (N)o>_
```

12.5.9 The REPORTS Program

The other branch of the hierarchy chart (see Figure 12.9) handles the reports produced by using the information stored in the ULIB_FILE file. The REPORTS program is activated whenever you select function 2 from the main menu. This program is a driver for the REPORTS menu and is similar to the EDIT program. It shows the REPORTS menu, and according to your selection, it activates appropriate programs to produce the desired report. It consists of a **while** loop that covers the whole program. The body of the **while** loop consists of the code for the REPORTS menu and a **case** construct that determines what commands must be executed to satisfy your selection.

Figure 12.31 shows the REPORTS program source code. There is no need for a line-by-line explanation. You know it all! However, there is another program involved. The program called REPORT_NO is called to produce the desired reports. Depending on the report selection each time, a different report number is specified on the command line. This report number (1, 2, or 3) is stored in the *$1* positional variable and is accessed in the REPORT_NO program to identify the desired report.

Figure 12.32 shows the REPORTS menu, assuming the report function is selected from the main menu.

Figure 12.31
Source Code for the REPORTS Program

```
$ cat -n REPORTS
   1 #
   2 # UNIX Library
   3 # Reports:  This program is the main driver for the REPORTS menu.
   4 #           It shows the reports menu and invokes the appropriate
   5 #           program according to the user selection.
   6 #
   7 error_flag=0                    # initialize the error flag, indicating no error
   8 while  true                     # loop forever
   9 do
  10  if [ $error_flag -eq 0 ]       # check for the error
  11  then
  12   tput clear ; tput cup 5 10 # clear screen and place the cursor
  13   echo "UNIX Library - ${BOLD}REPORTSMENU${NORMAL}"
  14   tput cup  7 20                # place the cursor
  15   echo "0: ${BOLD}RETURN${NORMAL} To The Main Menu"
  16   tput cup  9 20 ; echo "1: Sorted by ${BOLD}TITLES ${NORMAL}"
  17   tput cup  11 20 ; echo "2: Sorted by ${BOLD}AUTHOR ${NORMAL}"
  18   tput cup 13 20 ; echo "3: Sorted by ${BOLD}CATEGORY ${NORMAL}"
  19  fi
  20  error_flag=0                   # reset error flag
  21  tput cup 17 10 ; echo "Enter your choice> _\b\c"
  22 read choice                     # read user choice
  23 #
  24 # case construct for checking the user selection
  25 #
  26  case $choice in                # check user input
  27    0 ) exit 0 ;;                # return to the main menu
  28    1 ) REPORT_NO 1 ;;           # call REPORT_NO program, passing 1
  29    2 ) REPORT_NO 2 ;;           # call REPORT_NO program, passing 2
  30    3 ) REPORT_NO 3 ;;           # call REPORT_NO program, passing 3
  31    * ) ERROR 20 10              # call ERROR program
  32       tput cup 20 1 ; tput ed # clear the rest of the screen
  33       error_flag=1 ;;           # set error flag to indicate error
  34  esac                           # end of the case construct
  35 done                            # end of the while construct
  36 exit 0                          # end of the program
$_
```

Figure 12.32
The REPORTS Menu

```
                UNIX Library - REPORTS MENU

                     0: RETURN To The Main Menu
                     1: Sorted by TITLE
                     2: Sorted by AUTHOR
                     3: Sorted by CATEGORY
                Enter your choice>_
```

12.5.10 The REPORT_NO Program

The REPORT_NO program is activated whenever you select a report number from the reports menu. This program checks the value in the *$1* positional variable and sorts the ULIB_FILE according to this value. If the *$1* value is 1, then the ULIB_FILE is sorted on the title field, the first field in the record. If the *$1* value is 2, then the ULIB_FILE file is sorted on the author field, the second field in the record, and so on.

The sorted records are saved in the file called TEMP, and it is this file that can be passed to the **pg** or **pr** (see Chapter 7) command to be displayed. However, the report is not nicely arranged and fields in each record are separated by semicolons, similar to Figure 12.27. To make the output more presentable, records are read from the TEMP file, formatted, and then stored in PTEMP. It is the PTEMP file that is passed to the **pg** command to be displayed.

Figure 12.33 shows the REPORT_NO program source code. The line-by-line explanation of some new lines of code in the program follows.

Lines 9, 10, and 11: These lines are the body of the **case** construct. Each of these lines executes a **sort** (see Chapter 8) command to sort ULIB_FILE according to a specified field. The output of the **sort** command is stored in the TEMP file. Let's look at the **sort** options used in each command:

- The **-f** option considers all lowercase letters to be uppercase letters.
- The **-d** option ignores all blanks or nonalphanumeric characters.
- The **-t** option indicates which character is to be used as the field separator. In our program, the colon (:) symbol is specified as the field separator.
- The number option (**+1** and **+2**) sorts the file on the next field specified by the field number. In ULIB_FILE, field 1 is the *title* field, field 2 is the *author* field, and field 3 is the *category* field. Those are the three fields of interest in the REPORT_NO program. If no field position number is specified, the default is field 1. If **+1** specified, it menas skip field 1 (that is, sort on field 2). If **+2** is specified, it means skip field 1 and 2 (that is, sort on field 3) and so on.

Line 9: The field position is not specified, thus the default is field 1 and the file is sorted on the *title* field.

Line 10: The field position is specified as **+1**, thus the file is sorted on field 2, which is the *author* field.

Line 11: The field position is specified as **+2**, thus the file is sorted on field 3, which is the *category* field.

The **-k** option can be used unstead of field position. In this case, the number after the **-k** indicates the actual field number. For example, you can replace lines 9 through 11 in the REPORT_NO program with the following lines using the **-k** option:

```
 9 1) sort -f -d -t : -k 1 ULIB_FILE>TEMP;; # sort on title field
10 2) sort -f -d -t : -k 1 ULIB_FILE>TEMP;; # sort on author field
11 3) sort -f -d -t : -k 1 ULIB_FILE>TEMP;; # sort on category field
```

Lines 17–36: These lines make the **while** loop that continues until all the records in the TEMP file have been read. The return status of the **read** command is tested, which is 0 (zero) as long as there are records in the TEMP file and is 1 (one) when it reaches the end

Figure 12.33
Source Code for the REPORT_NO Program

```
$ cat -n REPORT_NO
   1 #
   2 # UNIX library
   3 # REPORT_NO:  This program produces report from the ULIB_FILE file.
   4 #             It checks for the report number passed to it on the
   5 #             command line, sorts and produces reports accordingly.
   6 #
   7 IFS=:"                                          # set delimiter to ;
   8 case $1 in                                      # check the contents of the $1
   9    1 ) sort -f -d -t : ULIB_FILE > TEMP ;;     # sort on title field
  10    2 ) sort -f -d -t : +1 ULIB_FILE > TEMP ;;# sort on author field
  11    3 ) sort -f -d -t :+2 ULIB_FILE > TEMP ;; # sort on category field
  12 esac                                            # end of the case
  13 #
  14 # read records from the sorted file TEMP. Format and store them in
  15 # PTEMP.
  16 #
  17 while read title author category status bname date  # read a record
  18 do
  19  echo "   Title:$title" >> PTEMP                      # format title
  20  echo "       Author: $author" >> PTEMP               # format author
  21  case $category in                                   # check the category
  22          [Tt][Bb] ) word=textbook ;;
  23      [Ss][Yy][Ss] ) word=system ;;
  24      [Rr][Ed][Ff] ) word=reference ;;
  25                 * ) word=undefined ;;
  26  esac
  27  echo "   Category: $word" >> PTEMP                   # format category
  28  echo "   Status: $status\n" >> PTEMP                 # format status
  29  if [ "$status" = "out"                              # if it is checked out
  30     then                                             # then
  31      echo " Checked out by:$bname" >> PTEMP          # format bname
  32      echo "     Date:$date\n" >> PTEMP               # format date
  33  fi
  34  echo >> PTEMP
  35 done < TEMP
  36 #
  37 # ready to display the formatted records in the PTEMP
  38 #
  39 pg -c -p "Page %d:" PTEMP                       # display PTEMP page by page
  40 rm TEMP PTEMP                                   #remove files
  41 exit 0                                          #end of the program
$_
```

of the file. The body of the loop is very similar to the DISPLAY program, but the output of the **echo** commands is redirected from the terminal display to the PTEMP file.

Line 36: This line is the end of the **while** loop, and the **< TEMP** redirects, input from the standard input to the TEMP file. This is called *loop redirection*, and the shell runs a redirected loop in a subshell. When the loop is ended, the PTEMP contains the sorted and formatted report that can be displayed using any number of commands. Here the **pg** command is used.

Line 40: This line displays the PTEMP file. The **pg** command shows the report one screen at a time, and using the **pg** (see Chapter 7) options you can scan your report.

Let's look at the options used in this command:

- The **-c** option clears the screen before displaying each page.
- The **-p** option places the specified string at the bottom of the screen instead of the default colon prompt. Here, the specified string is **Page %d:**. The character sequence *% d* stands for the current screen number.

Line 41: The two files TEMP and PTEMP are deleted.

Figure 12.34 shows a sample of a report produced by selecting option 2 from the REPORTS menu. The report is sorted on the second field, the author of the books.

Figure 12.34
A Sample Report Produced by REPORT_NO

```
          Title: UNIX Unbounded
         Author: Afzal Amir
       Category: textbook
         Status: out
 Checked out by: Steve Fraser
           Date: 1/12/05

          Title: UNIX For All
         Author: Brown David
         Author: reference
         Status: in

          Title: A Brief UNIX Guide
         Author: Redd Emma
         Author: reference
         Status: out
 Checked out by: Steve Fraser
           Date:1/12/05
 Page 1:
```

The prompt at the bottom of the screen shows the page number. You can press [Return] to display the next page or type any other **pg** *action command to see the desired part of the report.*

COMMAND SUMMARY

The following commands have been introduced in this chapter.

stty

This command sets options that control the capabilities of your terminal. There are more than a hundred different settings, and the following table lists only some of the options.

Option	Operation
echo [-echo]	Echoes [does not echo] the typed characters; the default is **echo**.
raw [-raw]	Disables [enables] the special meaning of the metacharacters; the default is **-raw**.
intr	Generates an interrupt signal; usually the [Del] key is used.
erase	[Backspace]. Erases the preceding character; usually the # key is used.
kill	Deletes the entire line; usually @ or [Ctrl=u] is used.
eof	Generates the (end-of-file) signal from the terminal; usually [Ctrl-d] is used.
ek	Resets the **erase** and **kill** keys to # and @ respectively.
sane	Sets the terminal characteristics to sensible default values.

tput

This command is used with the `terminfo` database, which contains codes for terminal characteristics and facilitates the manipulation of your terminal characteristics such as boldface text, clear screen, and so on.

Option	Operation
bel	Echoes the terminal's bell character.
blink	Makes a blinking display.
bold	Makes a boldface display.
clear	Clears the screen.
cup *r c*	Moves cursor to row *r* and column *c*.
dim	Dims the display.
ed	Clears from the cursor position to the end of the screen.
el	Clears from the cursor position to the end of the line.
smso	Starts stand out mode.
rmso	Ends stand out mode.
smul	Starts underline mode.
rmul	Ends underline mode.
rev	Shows reverse video, black on white display.
sgr0	Turns off all the attributes.

trap

This command sets and resets the interrupt signals. The following table shows some of the signals you can use to control the termination of your program.

Signal Number	Name	Meaning
1	**hang up**	Terminal connection is lost.
2	**interrupt**	One of the **interrupt** keys has been pressed.
3	**quit**	One of the **quit** keys has been pressed.
9	**kill**	The **kill -9** command has been issued.
15	**terminator**	The **kill** command has been issued.

REVIEW EXERCISES

1. What is the **trap** command for?
2. How do you terminate a process?
3. Where do you use the **trap** command?
4. What is the command to display your terminal settings?
5. What is the `terminfo` database? What information is stored in it?
6. What is the **kill** key for your system?
7. What is the **erase** key for your system?
8. What is the command to set the terminal characters to default values?
9. What is the command to set the **erase** key to # and the **kill** key to @?
10. What is the **case** construct used for?

Terminal Session

The following terminal session gives you the opportunity to practice some of the commands on your terminal and write/modify the programs in this chapter.

1. What type is your terminal?
2. Display a partial listing of your terminal settings.
3. Display a full listing of your terminal settings.
4. Change the setting for the **kill** key. Check to see if it is changed.
5. Change the setting for the **erase** key. Check to see if it is changed.
6. Reset the **erase** and **kill** keys to # and @, respectively.
7. Check to see if you have the **tput** utility on your system.
8. Use the **tput** command to clear the screen.

9. Write a script file named `CLS` that clears the screen when it is invoked.
10. Write a script file similar to the `MENU` program. Make your menu show some of the commands you often use but have trouble remembering the exact syntax of, or those that are long and you are tired of typing.
11. Change the UNIX library program in this chapter to store more information about each book; for example, the price of the book, date of publication, and so on.
12. Make the `REPORTS` program more sophisticated. For example, give the user a choice of reports to be printed or displayed on the terminal.
13. There is a lot of room for improvement in the `ULIB` program. For example, the problem of having more than one book with the same author, or the same book title with different authors is not addressed. Can you fix it?
14. The UNIX library program can be used as a prototype for other similar programs. For example, it can be adapted to store names, addresses, and phone numbers of your friends (creating a personal phone directory), or to make a database to keep track of your music CDs and records. Use your imagination and write the code to create a useful application similar to the UNIX library program that you can use in your environment.

CHAPTER 13

Farewell to UNIX

Now that we have covered the UNIX basics and you know about shell programming, it is time to add a few flourishes. The commands discussed in this chapter include disk commands, file manipulation commands, and spelling commands. You have already learned quite a few file manipulation commands, and the new commands in this chapter supplement your previous knowledge. The chapter concludes with a few security and system administration commands to help you understand more about the UNIX system and give you a more comfortable feeling about your system.

In This Chapter

13.1 DISK SPACE

There is a limit to the number of files you can store on a disk or in a file system. That limit depends on two things:

- The total amount of storage space available
- The amount of space set aside for i-nodes

An i-node number (discussed fully in Chapter 7) is assigned to each file in the system; these numbers are kept in the i-node list. An i-node contains specific file information, such as its location on the disk, its size, and so on.

13.1.1 Finding Available Disk Space: The df Command

You can use the **df** (disk free) command to find the total amount of disk space or the space available on a specified file system. If you do not specify a particular file system on the command line, then the **df** command reports the free space for all file systems.

Find the total amount of disk space available:

```
$ df [Return] . . . . . . . . . . No name is specified.
/      (/dev/dsk/c0d0s0).     14534 blocks     2965 i-nodes
/usr   (/dev/dsk/c0d0s2).     203028 blocks   51007 i-nodes
$_ . . . . . . . . . . . . . . . . . Ready for the next command.
```

The output shows that this system has two file systems. The first value is the number of free blocks, and the second is the number of free i-nodes. Each block is usually a 512-byte block; some systems use 1024-byte blocks for this report.

-t Option Using the **-t** option makes the **df** include the total number of blocks in the file system in the output.

Invoke the **df** command with the **-t** option:

```
$ df -t [Return] . . . . . . . . . Use the -t option.
/       (/dev/dsk/c0d0s0).       14534 blocks   2965 i-nodes
                         total:  31552 blocks   3936 i-nodes
/usr    (/dev/dsk/c0d0s2).      203028 blocks  51007 i-nodes
                         total: 539136 blocks  65488 i-nodes
$_ . . . . . . . . . . . . . . . . . Prompt returns.
```

1. *Use the* **man** *command to obtain the list of available options for the* **df** *command.*
2. *For Linux users, use the* **--help** *option to see the help page.*
3. *You may want to place the* **df** *command in your* `.profile` *file so you get a report as soon as you log in.*

See page x for an explanation of the icons used to highlight information in this chapter.

13.1.2 Summarizing Disk Usage: The du Command

You can use the **du** (disk usage) command to obtain a report that includes each directory in the file system, the number of blocks used by the files in that directory, and its subdirectories. This command is useful when you want to know how the space on a file system is being used.

Figure 13.1 shows the directory structure that is used to demonstrate the **du** command examples.

Figure 13.1
Directory Structure for the **du** Command Examples

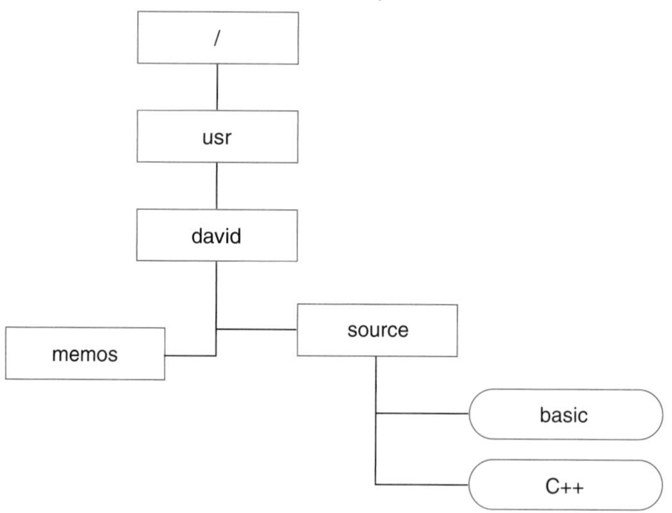

Obtain a disk usage report on the current directory and its subdirectories, assuming that your current directory is `david`.

```
$ du
4       ./memos
12      ./source
20      .
$_
```

The current directory is indicated by the dot (`.`). Here `./memos` occupies 4 blocks, `./source` occupies 12 blocks, and dot (`.`) is 20 blocks.

The **du** *command can be used to obtain reports on any directory structure.*

du Options

Table 13.1 summarizes the **du** command options. Use the **man** command to obtain a detailed list of the available options for the **du** command.

Table 13.1
The **du** Command Options

Option		Operation
UNIX	Linux Alternative	
-a	**--all**	Display the size of directories and files.
-b	**--bytes**	Displays the size of directories and files in bytes.
-s	**--summarize**	Displays only the total blocks for the specified directory; subdirectories are not listed.
	--help	Displays a usage message.
	--version	Displays the version information.

-a Option The **-a** option displays the space used by each file in the specified directory, as well as the directory.

Assuming your current directory is `david`, the following command line shows the usage of the **du** command with **-a** option:

```
$ du -a
4       ./memos
4       ./source/basic
4       ./source/c++
12      ./source
20      .
$_
```

Here, the size of directories (`./memos, ./source, .`) are shown, plus the size of the files (`./source/basic` and `./source/c++`).

-b Option The **-b** option displays the space used by each file in bytes instead of the default, which is in blocks.

Assuming your current directory is `david`, the following command line shows the usage of the **du** command with **-b** option.

```
$ du -b
4096       ./memos
12288      ./source
20480      .
$_
```

Here, the size of directories (`./memos, ./source, .`) are shown in bytes.

-s Option The **-s** option displays the total size of the directories or files in blocks. For example, the following command line shows the size of the current directory in blocks:

```
$ du -s [Return] . . . . .  Show the current directory size in blocks.
20 . . . . . . . . . . . .  The current directory (.) is 20 blocks.
$_ . . . . . . . . . . . .  Prompt is back.
```

Assuming your current directory is david, the following command line shows the usage of the **du** command with **-s** and **-b** options:

```
$ du -bs ./source
12288       ./source
$_
```

Here, the size of the source directory (./source) is shown in bytes.

Practicing Linux Alternative Command Options

Use the Linux alternative options for the **du** command, as suggested in the following command lines:

```
$ du --all [Return] . . . . . .     Same as du -a command.
$ du --byte [Return]  . . . . .     Same as du -b command.
$ du --summarize [Return] . . .     Same as du -s command.
$ du --help [Return]  . . . . .     Display the help page.
$ du --version [Return] . . . .     Display the version information.
```

Read the help page and familiarize yourself with other options available for the **du** *command.*

13.2 MORE UNIX COMMANDS

This section adds to your command repertoire and introduces new commands that help you to manipulate certain aspects of the UNIX environment, create banners, and obtain necessary information about the status of your programs.

1. *As was suggested many times in other chapters and sections of this book, use the* **man** *command to learn more about each command.*
2. *For Linux users, use the* **--help** *to get a help page and use the* **--version** *to get the version information for a specified command.*

13.2.1 Displaying Banners: The banner Command

You can use the **banner** command to produce output in large letters. It displays its arguments (10 characters or less) on the standard output, one argument per line. This is useful for creating banners, signs, report titles, and so on.

Make a happy birthday banner.

```
$ banner happy birthday ¦ lp [Return]. . . .  Make a banner and send it to the printer.
$_ . . . . . . . . . . . . . . . . . . . . .  Return to the prompt.
```

The pipe (¦) routes the output to the printer. Each word (argument) is printed on a separate line. You can use quotation marks to make two or more words a single argument.

Make a GO HOME banner, with both words on the same line.

```
$ banner "GO HOME" ¦ lp [Return]
$_ . . . . . . . .  Return to the prompt.
```

13.2.2 Running Commands at a Later Time: The at Command

You use the **at** command to run a command or list of commands at a later time. This is useful if you want to run your programs when the computer is less busy, or when you want to send mail on a certain date. You can specify the time and date part of the command in various formats, and the syntax is quite flexible.

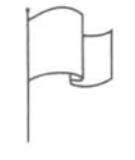

On any given UNIX system, there may be restrictions on who can use the at command. The system administrator can limit the access to the at command to only a few users. You can try it; if you are not authorized, the following message is displayed:

```
at: You are not authorized to use at. Sorry.
```

1. *You specify on the command line the time and date when you want your command to be carried out.*
2. *You do not have to be logged in when the commands are scheduled to run.*

Specifying the Time

If the time part of the **at** command is one or two digits (HH format), then it is interpreted as the time in hours. Thus, 04 and 4 both mean four o'clock in the morning. If the time part is four digits (HHMM format), then it is interpreted as time and minutes. Thus 0811 is 08:11. You can also specify the time by typing the word *noon*, *midnight*, or *now* in the time part of the command.

UNIX assumes a 24-hour clock, unless you specify that the a.m. or p.m. suffix should be used. Accordingly, 2011 is 08:11 p.m.

Specifying the Date

The date part of the **at** command can be a day of the week such as Wednesday or Wed (three-letter abbreviation). It can be in month, day, and year format such as Aug 10, 2005. It can also consist of the special word *today* or *tomorrow*. Here are some examples of legal **at** time and date formats:

```
$ at 1345 Wed [Return] . . . . . . . . . . . .  Run a job on Wednesday at 1:45 p.m.
$ at 0145 pm Wed [Return] . . . . . . . . . .  Run a job on Wednesday at 1:45 p.m.
$ at 0925 am Sep 18 [Return] . . . . . . . .  Run a job on September 18 at 9:25 a.m.
$ at 11:00 pm tomorrow [Return] . . . . . . .  Run a job tomorrow at 11 p.m.
```

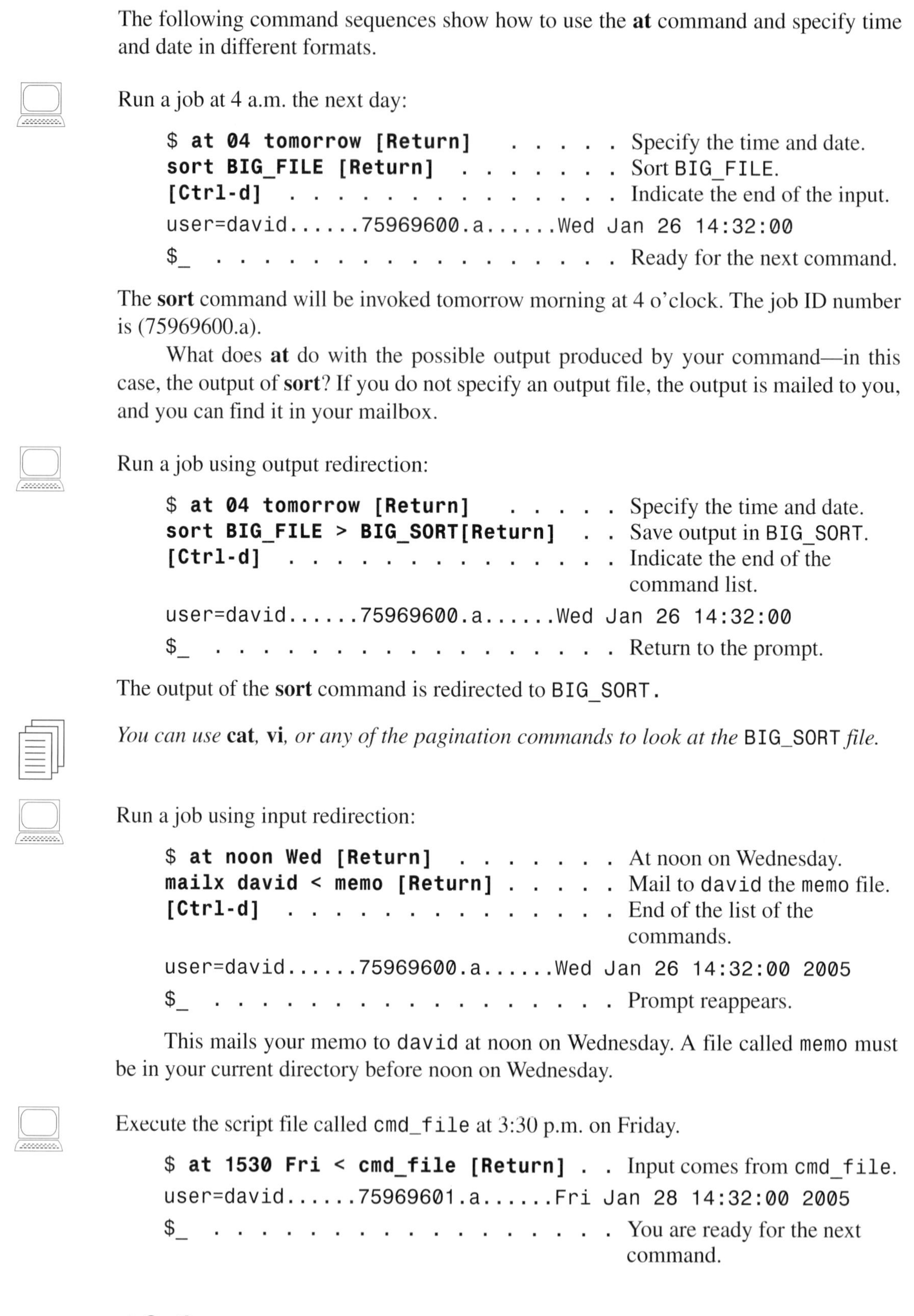

The following command sequences show how to use the **at** command and specify time and date in different formats.

Run a job at 4 a.m. the next day:

```
$ at 04 tomorrow [Return]  . . . . .  Specify the time and date.
sort BIG_FILE [Return] . . . . . . .  Sort BIG_FILE.
[Ctrl-d] . . . . . . . . . . . . . .  Indicate the end of the input.
user=david......75969600.a......Wed Jan 26 14:32:00
$_ . . . . . . . . . . . . . . . . .  Ready for the next command.
```

The **sort** command will be invoked tomorrow morning at 4 o'clock. The job ID number is (75969600.a).

What does **at** do with the possible output produced by your command—in this case, the output of **sort**? If you do not specify an output file, the output is mailed to you, and you can find it in your mailbox.

Run a job using output redirection:

```
$ at 04 tomorrow [Return]  . . . . .  Specify the time and date.
sort BIG_FILE > BIG_SORT[Return] . .  Save output in BIG_SORT.
[Ctrl-d] . . . . . . . . . . . . . .  Indicate the end of the
                                      command list.
user=david......75969600.a......Wed Jan 26 14:32:00
$_ . . . . . . . . . . . . . . . . .  Return to the prompt.
```

The output of the **sort** command is redirected to BIG_SORT.

You can use **cat**, **vi**, *or any of the pagination commands to look at the* BIG_SORT *file.*

Run a job using input redirection:

```
$ at noon Wed [Return] . . . . . . .  At noon on Wednesday.
mailx david < memo [Return] . . . . . Mail to david the memo file.
[Ctrl-d] . . . . . . . . . . . . . .  End of the list of the
                                      commands.
user=david......75969600.a......Wed Jan 26 14:32:00 2005
$_ . . . . . . . . . . . . . . . . .  Prompt reappears.
```

This mails your memo to david at noon on Wednesday. A file called memo must be in your current directory before noon on Wednesday.

Execute the script file called cmd_file at 3:30 p.m. on Friday.

```
$ at 1530 Fri < cmd_file [Return] . . Input comes from cmd_file.
user=david......75969601.a......Fri Jan 28 14:32:00 2005
$_ . . . . . . . . . . . . . . . . .  You are ready for the next
                                      command.
```

at Options

Table 13.2 summarizes the **at** command options.

Table 13.2
The **at** Command Options

Option	Operation
-l	Lists all jobs that are submitted with **at**.
-m	Mails you a short message of confirmation at completion of the job.
-r	Removes the specified job numbers from the queue of jobs scheduled by **at**.

Show the list of your jobs that are scheduled to run later (submitted to **at**):

```
$ at -l [Return] . . . . . . . . List all jobs.
75969601.a...........at Fri Jan 26 14:32:00 2005
$_ . . . . . . . . . . . . . . . Prompt reappears.
```

Remove a job from the **at** job queue:

```
$ at -r 75969601.a [Return]  . . Remove the job.
$ at -l [Return] . . . . . . . . Check the at queue.
$_  . . . . . . . . . . . . . .  Nothing is in the queue.
```

You must specify the job process ID you intend to remove from the queue. You can remove more than one job from the queue by specifying the jobs' process IDs on the command line, separated by a space.

13.2.3 Revealing the Command Type: The type Command

The **type** command is useful when you want to know more about a command. It shows whether the specified command is a shell program or a shell built-in command. The *built-in commands* are part of the shell, and no child process is created when any of them is invoked.

Let's look at some examples:

```
$ type pwd [Return]  . . . . . . Find more about the pwd command.
pwd is a shell built-in
$ type ls [Return] . . . . . . . And the ls command.
ls is /bin/ls
$ type cd [Return] . . . . . . . The cd command.
cd is a shell built-in
$_ . . . . . . . . . . . . . . . Prompt reappears.
```

13.2.4 Timing Programs: The time Command

You can use the **time** command to obtain information about the computer time your command uses. It reports the real time, user time, and system time. You type **time** followed by the command name you want to time.

Real time is the actual time (*elapsed time*) from the moment you enter the command until the command is finished. This includes I/O time, time spent waiting for other users, and so on. The real time can be several times greater than the total CPU time.

User time is the CPU time dedicated to the execution of your command.

System time is the time spent executing UNIX kernel routines to service your command.

CPU time is the time in seconds and fractions of seconds that the CPU spends to execute your command.

Report the time it takes to sort BIG_FILE:

```
$ time sort BIG_FILE > BIG_FILE.SORT [Return]
60.8 real     11.4 user     4.6 sys
$ . . . . . . . . . . . . . . You are ready for the next command.
```

The reports show that the program will run for 60.8 seconds of real time, using 11.4 seconds of user time, and 4.6 seconds of system time, for a total of 16 seconds of the CPU time.

13.2.5 Reminder Service: The calendar Command

You can use the **calendar** command to remind yourself of your appointments and other things you want to do. To use this service, you must create a file called calendar in your HOME directory or the current directory. It displays those lines in the calendar file that contain today's or tomorrow's date. If your system is set up to run the **calendar** command automatically, then it sends you e-mail that contains appropriate lines from your calendar file. You can also run the **calendar** command at the $ prompt to display lines from your calendar file.

The date in each line can be expressed in various formats. Figure 13.2 shows an example of a calendar file, with each line containing a different format of the date string.

Figure 13.2
Example of a calendar File

```
$ cat calendar
1/20 Call David
Meet with your advisor on January 22.
1/22 You have a dental appointment.
The BIG MEETING is on Jan. 23.
3/21 Time to clean up your desk.
$_
```

Run the **calendar** command, assuming today is January 22:

```
$ calendar [Return] . . . . . Run the command.
1/22 You have a dental appointment.
Meet with your advisor on January 22.
The BIG MEETING is on Jan. 23.
$_ . . . . . . . . . . . . . . Prompt reappears.
```

Any line in the calendar file that contains the date 1/22 or 1/23 is displayed.

1. *You can place the* **calendar** *command in your* `.profile` *(startup file) so you are informed of your schedule as soon as you log in.*
2. *This command is not available on all systems. Check its availability in your system before placing the command in the* `.profile` *file.*

13.2.6 Detailed Information on Users: The finger command

The **who** command (introduced in Chapter 3) is used to obtain information about people on the system. You can use the **finger** command to obtain more detailed and informative reports about the users who are logged on. By default, the **finger** command displays information in a multicolumn format. Figure 13.3 is an example of output displayed by the **finger** command.

- The *Login* column shows the login name of the users.
- The *Name* column shows the full name of the users.
- The *TTY* column shows the device number of the users' terminals. The (*) before the terminal name (such as `*console`) indicates that sending messages to that terminal is blocked (see **mesg**, Chapter 9).
- The *Idle* column shows the elapsed time since each user last typed on the keyboard.
- The *When* column shows the time that each user logged in.
- The *Where* column shows the addresses of the users' terminals.

Figure 13.3
The **finger** Command

```
$ finger
Login       Name          TTY      Idle    When        Where
dan     Daniel Knee    *console  5:08   Mon 14:14
gabe    Gabriel Smart  *p0       1:33   Mon 17:15   205.130.70
emma    Emma Good      p2        8:21   Mon 11:02   205.130.71
$_
```

If **finger** is issued with a user's name as an argument, the information about the specified user is displayed regardless of whether he or she is logged in. Figure 13.4 shows information about a user called `gabe`.

Figure 13.4
The **finger** Command with an Argument

```
$ finger gabe
 Login name: gabe   (message off)      In real life: Gabriel Smart
 Directory: /home/students/gabe        shell:/bin/ksh
 On since July 30 18:45:38 On p0
 Mail last read Tue Jul 23 09:08:06 2005
 No Plan.
$_
```

The .plan and .project Files

The only curious thing here is the last line of output: "No Plan." The **finger** command displays the contents of two hidden files called `.plan` and `.project` in the HOME directory of the specified person. In this example, "No Plan." indicates that gabe has not prepared a `.plan` file. Assuming the following `.project` and `.plan` files exist in Gabe's HOME directory, Figure 13.5 shows the output of the **finger** command.

Figure 13.5
The **finger** Command Output Displaying `.project` and `.plan` Files

```
$ finger gabe
Login name: gabe          (message off)        In real life: Gabriel Smart
Directory: /home/students/gabe                 shell:/bin/ksh
On since July 30 18:45:38 On p0
Mail last read Tue Jul 23 09:08:06 2005
Project: C11 Application Project
Plan:
Hi,
I am all smiles these days!
Gabe :-)
$_
```

The `.project` file in Gabe's home directory:

```
C++ Application Project
This project is assigned to the Gold Team.
For more information about this Project contact Gabe Smart.
```

The `.plan` file in Gabe's home directory:

```
Hi,
I am all smiles these days!
Gabe:-)
```

The **finger** *command displays the first line of the* `.project` *file, if it exists in the user's HOME directory* (`$HOME/.project`), *and the whole content of the* `.plan` *file, if it exists in the user's HOME directory* (`$HOME/.plan`).

finger Options

The **finger** command options mostly manipulate the format of the output. Table 13.3 shows some of these options.

The long format option displays a format similar to the single user report (Figure 13.4) for all the users.

13.2.7 Saving and Distributing Files: The tar Command

You can use the **tar** (*t*ape *ar*chiver) command to copy a set of files into a single file called `tarfile`. A `tarfile` is usually saved to a magnetic tape but it can be on any other media, such as a floppy disk. The **tar** command packs multiple files into a single

Table 13.3
The **finger** Command Options

Option	Operation
-b	Suppresses displaying the user's HOME directory and shell in a long form at display.
-f	Suppresses displaying the header in a nonlong format output.
-h	Suppresses displaying the `.project` file in a long format output.
-l	Forces long format output.
-p	Suppresses displaying the `.plan` file in a long format output.
-s	Forces short format output.

file in `tar-format` that can be later unpacked by **tar**. For example, to archive all the files in your current directory and subdirectories to a single `tarfile`, you type

```
$ tar -cvf /dev/ctape1 . [Return]
```

In this example, **tar** packs files from your current directory, which is represented by the dot (.) at the end of the command line, into a single `tarfile` and sends it to the device `/dev/ctape1`.

tar Options

Table 13.4 summarizes the common options for **tar**.

Table 13.4
The **tar** Command Options

Option		Operation
UNIX	Linux Alternative	
-c	**--create**	(**create**) Creates a new `tarfile`. Writing begins at the beginning of the `tarfile`.
-f	**--file**	(**file**) Uses the next argument as a place where the archive is to be placed.
-r	**--concatenate**	(**replace**) Writes a new archive at the end of the `tarfile`.
-t	**-- list**	(**table of contents**) Lists the name of the files in the `tarfile`.
-x	**--extract** or **--get**	(**extract**) Extracts files from the `tarfile`.
-v	**--verbose**	(**verbose**) Provides additional information about the `tarfile` entries.
	--help	Displays a usage message.
	--version	Displays the version information.

The following command sequences show examples using **tar** to archive files.

Assume your current directory is called `projects` and that it contains two files named `head` and `tail` and a subdirectory called `tarfiles`. The following command archives the current directory in a `tarfile` called `projects.tar` in the `tarfiles` directory.

$ tar -cvf ./tarfiles/projects.tar . [Return] . . . Archive the current directory.

The option **c** is for creating a new `tarfile`, **v** is for listing everything as **tar** goes along, and **f** indicates the next argument is where the `tarfile` is placed.

```
$ tar -cvf ./tarfiles/projects.tar .
  a ./projects/                   0K
  a ./projects/head               3K
  a ./projects/tail               6k
$_
```

The **v** option provides additional information about the `tarfiles` entries. The output consists of three fields. The letter **a** is a function letter indicating archiving, followed by the pathname of each file and then its size.

The name of the `tarfile` *can be any valid UNIX filename. Here, the extension* `.tar` *is added for clarity and is not a necessity.*

The following screen shows the **tar** output without the **v** option. Notice that no information is provided.

```
$ tar -cf ./tarfiles/projects.tar .
$_
```

The following command sequences show examples using **tar** to retrieve files from a `tarfile`.

Assume you have a `tarfile` called `projects.tar` in a directory called `tarfiles`. The following command displays the table of contents of the `projects.tar` file:

$ tar -tvf ./tarfiles/projects.tar [Return] . . . List the table of contents in `projects.tar` file.

The option **t** is for listing a table of contents, **v** is for listing everything as **tar** goes along, and **f** indicates the next argument is where the `tarfile` is placed:

```
$ tar -tvf ./tarfiles/projects.tar
  drwx r-- r-- 2029/1005 0    Sep 3 10:33 2005 ./tarfiles/project/
  -rw- r-- r-- 2029/1005 2350 Sep 3 10:33 2005 ./tarfiles/project/head
  -rw- r-- r-- 2029/1005 5374 Sep 3 10:33 2005 ./tarfiles/project/tail
$_
```

The output is similar to the format produced by the **-l** *option of the* **ls** *command.*

The columns have the following meanings:

Column 1	Column 2	Column 3	Column 4	Column 5
-rw- r-- r--	2029/1005	2350	Sep 3 10:33 2005	./tarfiles/project/head
Access permission to head	User ID/ Group ID of head	Size of head in bytes	Modification date of head	Name of the file, head

Now, if you want to extract (retrieve) all files from a tarfile, you type

$ tar -xvf ./tarfiles/projects.tar [Return] . . . . . Extract all files from projects.tar into current directory.

The option **x** indicates you want **tar** to extract files from the projects.tar file. The **v** option causes a verbose output:

```
$ tar -xvf ./tarfiles/projects.tar .
  x ./projects, 0 bytes, 0 tape blocks
  x ./projects/head, 2350 bytes, 5 tape blocks
  x ./projects/tail, 5374 bytes, 11 tape blocks
$_
```

Notice it provides the size of each file in bytes and tape blocks.

You also can retrieve certain files from the backup medium by specifying the file names on the command line, such as

$ tar -xvf ./tarfiles/projects.tar tail [Return] . . . Retrieve tail from projects.tar into current directory.

```
$ tar -xvf ./tarfiles/projects.tar tail
  x ./projects/tail, 5374 bytes, 11 tape blocks
$_
```

Practicing Linux Alternative Command Options

Use the Linux alternative options for the **tar** command, as suggested in the following command lines:

$ tar --create --verbose --file ./tarfiles/projects.tar . [Return] . Same as **tar -cvf** command.

$ tar --create --file ./tarfiles/projects.tar . [Return] . . . . . . Same as **tar -cf** command.

$ tar --list --verbose --file ./tarfiles/projects.tar . [Return] . . Same as **tvf** command.

$ tar --help [Return]. . . . . Display the help page.

$ tar --version [Return] . . . . Display the version information.

Read the help page and familiarize yourself with other options available for the **tar** *command.*

13.3 SPELLING ERROR CORRECTION

You can use the **spell** command to check the spelling of the words in your documents. The **spell** command compares the words in a specified file against a dictionary file. It displays the words that are not found in the dictionary file. You can specify more than one file, but when no file is specified, **spell** gets its input from the default input device, your keyboard. Let's look at some examples.

Run the **spell** command without specifying any filename:

```
$ spell [Return] . . . . . . . . . . . . No argument is specified.
lookin goood [Return]  . . . . . . . . . Input is from the keyboard.
[Ctrl-d] . . . . . . . . . . . . . . . . This is the end of input.
lookin
goood
$_ . . . . . . . . . . . . . . . . . . . And the prompt appears.
```

You signal the end of your input by pressing [Ctrl-d] at the beginning of a blank line. The **spell** command does not suggest correct spelling; it just displays the suspected words. The output is one word per line.

The **spell** command is case sensitive: it is happy with *David* but complains about *david.*

Assuming you have a file called my_doc, check its spelling and save the output in another file:

```
$ spell my_doc > bad_words [Return]   . Spell check my_doc.
$_ . . . . . . . . . . . . . . . . . . . Prompt.
```

The **spell** command output is redirected to the bad_words file. You can use the **cat** command to look at this file.

You can specify more than one file as the **spell** command argument. The specified file names are separated by at least one space.

Assuming you are in the vi editor, invoke the spelling checker:

```
: !spell [Return] . Invoke the spelling checker.
pervious  . . . . . Type the word whose spelling you are in doubt about.
[Ctrl-d]  . . . . . Indicate end of the input.
pervious  . . . . . Misspelled word.
[Hit return to continue]
```

You can invoke any command from within vi by pressing **!** *at the colon prompt followed by the name of the command (see Chapter 11).*

13.3.1 spell Options

Table 13.5 shows the **spell** command options. Explanations and examples for each option follow.

Table 13.5
The **spell** Command Options

Option	Operation
-b	Checks British spelling.
-v	Displays words that are not in the spelling list and their derivation.
-x	Displays plausible stems for each word being checked.

-b Option This option makes the **spell** command check your file with British spelling. Words like *colour*, *centre*, and *programme* are accepted.

-v Option This option shows all the words that are not literally in the spelling list and their plausible derivations. The derivations are indicated by the plus sign.

Run the **spell** command with the **-v** option. The input is from the keyboard:

```
$ spell -v [Return]    . . . . Input is expected from the keyboard.
appointment looking worked preprogrammed [Return]
[Ctrl-d] . . . . . . . . . . Signal the end of your input.
+ ment          appointment
+ ing           looking
+ ed            worked
+ pre           preprogrammed
$_ . . . . . . . . . . . . . Return to the prompt.
```

-x Option This option displays plausible stems of each word until a matching word is found, or the list is exhausted. The stems are prefixed by the equal sign.

Run the **spell** command with the **-x** option and input from the keyboard:

```
$ spell -x [Return]    . . . . Input is expected from the keyboard.
appointment looking worked preprogrammed [Return]
[Ctrl-d] . . . . . . . . . . Signal the end of your input.
= appointment
= appoint
= looking
= looke
= look
= preprogrammed
= programmed
= worked
= worke
= work
$_ . . . . . . . . . . . . . Return to the prompt.
```

13.3.2 Creating Your Own Spelling List

On most UNIX systems, you can create your own dictionary file that supplements the standard dictionary with additional words. For example, you can create a file that contains the correct spellings of the special words and terms that are specific to your site or project. Using the plus sign option, you specify your dictionary file on the command line. The **spell** command first checks its own dictionary and then checks the words against your file. You can use the vi editor to create your own spelling list.

1. *Each word in your dictionary file must be typed on one line.*
2. *Your dictionary file must be sorted alphabetically.*

Assuming you have created your own dictionary file called `my_own`, specify that file on the command line using the plus sign option:

```
$ spell +my_own BIG_FILE > MISSPELLED_WORDS [RETURN]
$_ . . . . . . . . . . . . . . . . . . . Prompt reappears.
```

The `+my_own` *indicates you want to use your own dictionary file called* `my_own`. *The* **spell** *command lists the words that are not found in either of the two dictionary files, its own and yours. The output is redirected into the* `MISSPELLED_WORDS` *file.*

Scenario The default spelling list (dictionary) does not contain the shell commands or other specific UNIX words. Therefore, if you have words such as **grep** and **mkdir** in your text, they are displayed as suspicious words.

You can create a file similar to the one depicted in Figure 13.6, which contains the shell commands or UNIX words with correct spelling.

Figure 13.6
Example of a Dictionary File

```
$ cat U_DICTIONARY
grep
i-node
ls
mkdir
pwd
```

Run the **spell** command with and without your spelling list:

```
$ spell +U_DICTIONARY [Return] . . . Run with your spelling list.
grep pwd mkdir is [Return]. . . . This is your input.
$_ . . . . . . . . . . . . . . . . . . . No spelling errors; the prompt
                                         is back.
$ spell [Return] . . . . . . . . . . Run without your spelling list.
grep pwd mkdir ls [Return]. . . . This is your input, same as
                                         before.
grep
ls
mkdir
pwd
$_ . . . . . . . . . . . . . . . . . . . The prompt is back.
```

13.4 UNIX SECURITY

Information and computer time are valuable resources that require protection. System security is a very important part of multiuser systems. There are various aspects of system security to consider:

- Keeping unauthorized people from gaining access to the system
- Keeping an authorized user from tampering with the system files or other users' files
- Granting some users certain privileges

Security on the UNIX system is implemented by using simple commands, and the system security can be as lax or tight as you desire. Let's examine some of the available means to secure your system.

13.4.1 Password Security

All the information the system needs to know about each user is saved in a file called `/etc/passwd`. This file includes each user's password; however, it is encrypted, using an encoding method that makes the deciphering of the passwords a very difficult task. (Encryption is discussed in more detail later in this section.)

The `passwd` file contains one entry for each user. Each entry is a line that consists of seven fields, which are separated by colons. The following line shows the format of each line in the `passwd` file, followed by the explanation of each field in the line.

```
login-name:password:user-ID:group-ID:user-info:directory:program
```

login-name: This is your login name, the name you enter in response to the login prompt.

password: This is your encrypted password. In System V Release 4, the encrypted password is not stored in the `passwd` file; instead, it is stored in a file called `/etc/shadow` and the letter *x* is used as the placeholder for the `password` field in the `passwd` file.

You can print the `/etc/passwd` *file, but the* `/etc/shadow` *file is not readable by ordinary users.*

user-ID: This field contains the user ID number. The user ID is a unique number assigned to each user, and user ID 0 indicates the *superuser*.

group-ID: This field contains the group ID. The group ID identifies the user as a member of a group.

user-info: This field is used to further identify the user. Usually it contains the user's name.

directory: This field contains the absolute pathname of the HOME directory assigned to the user.

program: This field contains the program that is executed after the user logs in. Usually it is the shell program. If the program is not identified, `/usr/bin/sh` is assumed. You can change your login shell to `/usr/bin/chs` to log in to the C shell, or change it to any other program you wish to log in to.

Figure 13.7
Sample of the passwd File

```
$ cat /etc/passwd
root:x:0:1:admin:/:/usr/bin/sh
david:x:110:255:David Brown:/home/david:/usr/bin/sh
emma:x:120:255:Emma Redd:/home/emma:/usr/bin/sh
steve:x:130:255:Steve Fraser:/home/steve:/usr/bin/csh
$_
```

Figure 13.7 shows some sample entries in the passwd file. This file can be printed out by any user on the system.

13.4.2 File Security

File security limits access to the files. The UNIX system provides you with the commands to specify who can access a file and, once accessed, what type of operation can be done on the file.

The **chmod** command was discussed in Chapter 11; here we will explore the other way of setting file permission. When you use the **ls** command with the **-l** option (Chapter 5), you get your directory listing in full detail. Part of this detailed information is the pattern of **rwx** on each entry, which represents the file permission. The **chmod** command is used to change the file access mode (permission). For example, if you want to give execution permission on myfile to all users, you type the following:

$ chmod a=x myfile [Return]

However, you can specify the new mode as a three-digit number that is computed by adding together the numeric equivalents of the desired permission. Table 13.6 shows the numeric value assigned to each permission letter.

Table 13.6
The File Permission Numeric Equivalents

owner	group	others
r w x	r w x	r w x
4 2 1	4 2 1	4 2 1

Assuming you have a file called mayflies in your current directory, the following examples use the **chmod** command to change the file permission. You can use the **ls -l** command to verify the changes.

Change mayflies permission to allow **read**, **write**, and **execute** permission (access) to all users.

$ chmod 777 mayflies [Return] . . Change the mayflies access mode.
$_ . . . . Done; the prompt is back.

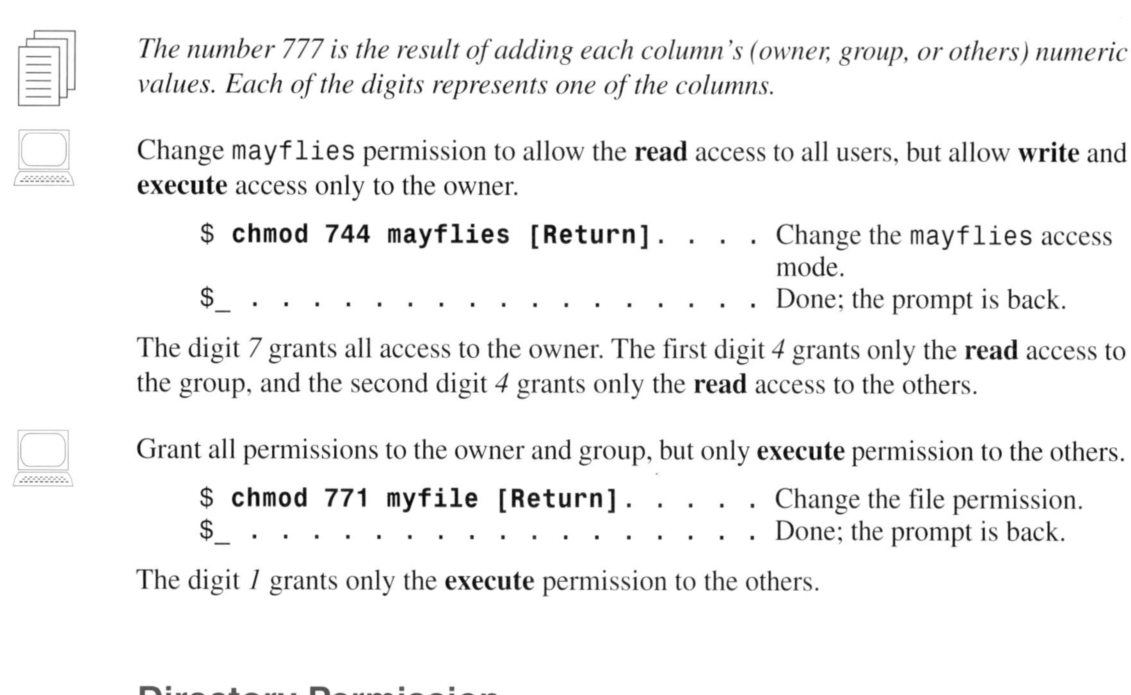

The number 777 is the result of adding each column's (owner, group, or others) numeric values. Each of the digits represents one of the columns.

Change `mayflies` permission to allow the **read** access to all users, but allow **write** and **execute** access only to the owner.

$ chmod 744 mayflies [Return]. . . . Change the `mayflies` access mode.
$_ . . . . Done; the prompt is back.

The digit *7* grants all access to the owner. The first digit *4* grants only the **read** access to the group, and the second digit *4* grants only the **read** access to the others.

Grant all permissions to the owner and group, but only **execute** permission to the others.

$ chmod 771 myfile [Return]. . . . . Change the file permission.
$_ . . . . Done; the prompt is back.

The digit *1* grants only the **execute** permission to the others.

13.4.3 Directory Permission

Directories have permission modes that work in a similar manner to file permission modes. However, the directory access permissions have different meanings:

read: The read (**r**) permission in a directory means you can use the **ls** command to list the filenames.

write: The write (**w**) permission in a directory means that you can add and remove files from that directory.

execute: The execute (**x**) permission in a directory means you can use the **cd** command to change to that directory or use the directory name as part of a pathname.

Grant access permission to all users on the directory called `mybin`:

$ chmod 777 mybin [Return] . . . . . Change the access mode of a directory.
$_ . . . . Done; prompt is back.

As with file permissions, each digit stands for one of the user groups (owner, group, and others). The digit *7* means granting write, read, and execute access to a particular group of users.

13.4.4 The Superuser

It is time to know who the *superuser* is, as some of the commands explained in this chapter might be available only to a superuser on your system. The *superuser* is a user who has privileged authority and is not restricted by the file permissions. The system administrator must be a superuser in order to perform administrative tasks such as establishing new accounts, changing passwords, and so on. Usually, the superuser logs in as **root** and can read, write to, or remove any file in the system, or even shut the system down. Superuser status is usually granted to the system administration personnel.

13.4.5 File Encryption: The crypt Command

The superuser can access any file regardless of file permissions. How can you protect your sensitive files from others (superuser or not)? UNIX provides the **crypt** command, which encrypts your file and makes it unreadable to others. The **crypt** command changes each character in your file in a reversible way, so you can obtain the original file later. The encoding mechanism relies on simple substitution. For example, the letter *A* in your file is changed to the symbol ~. The **crypt** command uses a key to scramble its standard input into an unreadable text that is sent to the standard output.

The **crypt** command is used for both encryption and decryption. In fact, for decryption of a file, you must provide the same key that you specified to its encryption. Let's look at some examples.

Encrypt a file called `names` in your current directory:

```
$ cat names [Return] . . . . . . . . . . . . . . .  Display the content of names.
David    Emma    Daniel    Gabriel    Susan    Maria
$ crypt xyz < names > names.crypt [Return] . . .    Use xyz as the encryption key.
$ cat names.crypt [Return] . . . . . . . . . . .    Display the content of
                                                    names.crypt, the encrypted file.
<<güé^>>ZQX_∂å£ëÁÒ7ÂpgŒYñÅÍ[ÎrÂÆ
$_
$ rm names [Return] . . . . . . . . . . . . . . .   Remove names.
$ crypt xyz < names.crypt > names [Return] . . .    Decode names.crypt.
$ cat names [Return] . . . . . . . . . . . . . . .  Check the contents of names.
David    Emma    Daniel    Gabriel    Susan    Marie
$_ . . . . . . . . . . . . . . . . . . . . . . . .  The prompt is back.
```

The **xyz** is the key used to encrypt the file. The key is actually a password that you use to uncode your file later.

Using input/output redirection, the input to the command is `names` and the output is stored in `names.crypt`. When the file is encrypted, the input to the command is `names.crypt` and the output is stored in `names` when the file is decrypted by issuing the **crypt** command again.

Usually, you remove the original copy of a file after file encryption and leave only the encrypted version, making the information in the file accessible only to someone who knows the encryption key.

Encrypt the `names` file without specifying the encryption key on the command line:

```
$ crypt < names > names.crypt [Return] . . . . .    No encryption key is specified.
Key . . . . . . . . . . . . . . . . . . . . . . .   The prompt for entering the key
                                                    appears.
$_ . . . . . . . . . . . . . . . . . . . . . . . .  Return to the prompt.
```

If you do not specify the encryption key on the command line, then **crypt** *prompts you to enter one. The key that you enter is not echoed, and this method is preferred over typing the key on the command line.*

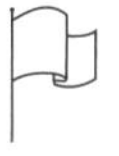

If you type the wrong code while entering the password (key) the first time when you encrypt a file, you will not be able to decrypt the file later. As a safety check, you may want to try decrypting the encrypted file before you remove the original.

13.5 USING FTP

The *File Transfer Protocol* (FTP) is one of the most frequently used services available on your system. You can use the FTP commands to transfer files from one system to another. Files of any type can be transferred, although you may have to specify whether the file is an ASCII or a binary file.

1. *FTP (file transfer protocol) is not just the name of the protocol; it is also the name of a program or command.*
2. *FTP is a popular way to share information over the Internet.*

You issue the FTP command by typing **ftp**, followed by the address of another site, and then pressing the [Return] key.

`$ ftp server2 [Return]` . . Issue the FTP command

You are prompted to enter the user name and password. That means you must have the login permission on the site called *server2.*

`Username: daniel` . . . . . Enter user name, in this case `daniel`.
`Password:` . . . . Enter password (not echoed)

If you have a login name on *server2*, you can transfer files from your system to *server2*. FTP provides a set of commands that allows you to transfer files from one system to another, list directories, and copy files in either direction.

13.5.1 FTP Basics

FTP works as a client/server process. You give the command **ftp** using a remote address, such as the following:

`$ ftp server2 [Return]` . . . Issue the **ftp** command to contact *server2*.

In this case, *server2* is the name of the remote site you want to contact. The **ftp** command immediately attempts to establish a connection to the FTP server that you specified on the command line. In this case, it is the FTP server on *server2*.

When you issue the **ftp**, the FTP running on your system is a client to an FTP process that acts as a server on *server2*. You issue commands to the FTP process at *server2*, and it responds appropriately.

In an FTP session, your system is referred to as **local-host**, *and the system you are connected to is referred to as* **remote-host**.

You can use the FTP interactive mode to connect to the remote site. If you do not specify the name of the FTP server, **ftp** enters its interactive mode and prompts you for further instructions. For example

`$ ftp [Return]` . . The remote site to be contacted is not specified.
`$ ftp>` . . . . . . **ftp** enters into interactive mode and displays its prompt.

To see a list of all **ftp** commands, type **?** and press [Return] during an FTP session. For example

`ftp>` . . . . Prompt indicating an FTP session is in progress.
`ftp> ?[Return]` . . List the commands.

To terminate the current FTP session and return to the prompt, type **bye** as follows:

ftp> bye[Return] . . . . . Terminating and aborting FTP session.

Or you can use the **quit** command as follows:

ftp> quit[Return] . . . . . Terminating and aborting FTP session.

In both cases, the end of the session message is displayed:

```
221 Goodbye.
```

Figure 13.8 shows the list of commands that is displayed when the **?** command is used in a **FTP** session.

Figure 13.8
List of the **ftp** Commands

```
ftp> ?

Commands may be abbreviated. Commands are:
   !         cr           macdef     proxy        send
   $         delete       mdelete    sendport     status
account      debug        mdir       put          struct
append       dir          mget       pwd          sunique
ascii        disconnect   mkdir      quit         tenex
bell         form         mls        quote        trace
binary       get          mode       recv         type
bye          glob         mput       remotehelp   user
case         hash         nmap       rename       verbose
cd           help         ntrans     reset          ?
cdup         lcd          open       rmdir
close        ls           prompt     runique

ftp>
```

You probably need only a few of these commands for most of your FTP work. The following tables list and categorize most of the **ftp** commands.

FTP Connection Commands

The FTP connection related commands are used to establish connection to the remote site and terminate the connection at the end of the FTP session. Table 13.7 summarizes these commands.

FTP File Transfer Commands

The FTP file transfer-related commands are used to transfer files between the local and remote hosts or vice versa. The two most common modes of transfer are ASCII and binary. By default, the transfer mode is set to **ascii**. You can transfer any text (ASCII) file using **ascii** mode. However, binary files such as compiled and executable files must be transferred using the **binary** mode. Table 13.8 summarizes these commands.

FTP File and Directory Commands

The FTP file and directory-related commands are used to manage and manipulate files in an FTP session. These commands are similar to the UNIX file and directory commands. Table 13.9 summarizes these commands.

Table 13.7
FTP Access Commands

Command	Description
open *remote-host name*	Opens a connection to the FTP server on the specified host. It prompts you to enter the user name and password to log in on the remote host.
close	Closes current open connection and returns to the local FTP command. At this point you may issue the **open** command for a different remote host.
quit (bye)	Closes the current FTP session with the remote server and exits **ftp**. That is, it returns to UNIX shell level.

Table 13.8
FTP File Transfer Commands

Command	Description
ascii	Sets the file transfer mode to ASCII. This is the default file type.
binary	Sets the file transfer mode to binary.
bell	Sounds a bell when file transfer is completed.
get *remote-filename [local-filename]*	Copies a single file from the remote to the local host. If no local filename is specified, the copy has the same name on the local host.
mget *remote_filenames*	Copies multiple files from the remote to the local host.
put *local-filename [remote-filename]*	Copies a single file from the local to the remote host. If no remote filename is specified, the copy has the same name on the remote host.
mput *local-filenames*	Copies multiple files from the local to the remote host.

Table 13.9
FTP Files and Directory Commands

Command	Description
cd *remote-directory-name*	Changes the current directory on the remote host to the specified directory.
lcd *local-directory-name*	Changes the currect directory on the local host to the specified directory.
dir	Lists the current directory on the remote host.
pwd	Prints the name of the current directory on the remote host.
mkdir *remote-directory-name*	Makes a new directory on the remote host. Typically, you must have permission to do this.
delete *remote-filename*	Deletes a single specified file on the remote host.
mdelete *remote-filenames*	Deletes multiple files on the remote host.

Miscellaneous Commands

The (**?**) command provides help about **ftp** commands. The (**!**) is the shell escape command. It is used to perform commands on the local host. The **hash** command provides feedback messages in file transfer. This command is particularly useful when you are transferring large files. Table 13.10 summarizes these commands.

Table 13.10
Miscellaneous FTP Commands

Command	Description
? or **help**	Displays an informative message about the meaning of the specified command. If no argument is given, a list of the known commands is displayed.
!	Switches to escape shell mode.
hash	Displays hash sign (#) as feedback for each data block transferred. The size of a data block is 8192 bytes.

The following FTP sessions are examples to show the use of the **ftp** commands.

Opening an FTP Session

The **ftp open** command is used to begin a session with a remote host. The following command sequences shows how to begin an FTP session with remote host called *server2*.

```
$ ftp [Return] . . . . . . . . . . . . . . Issue the ftp command.
ftp> . . . . . . . . . . . . . . . . . . . ftp prompt is displayed.
ftp> open [Return] . . . . . . . . . . . . Issue the open command.
(to) . . . . . . . . . . . . . . . . . . . ftp open prompt is displayed.
(to) server2 [Return]  . . . . . . . . . . Enter the remote host name.
Responses from FTP server on the remote host (server2)
Connected to server2
220 server2 FTP server (UNIX(r) System V Release 4.0) ready.
Name  (server2:gabe):  . . . . . . . . . . ftp prompt is displayed.
Name  (server2:gabe): gabe [Return]  . . . Enter your login name on the remote host.
331 password requires for gabe
password:  . . . . . . . . . . . . . . . . ftp password prompt is displayed.
password: [Return] . . . . . . . . . . . . Enter the password (not displayed).
230 User gabe logged in.
ftp> bye [Return]  . . . . . . . . . . . . End of the FTP session.
221 Goodbye  . . . . . . . . . . . . . . . ftp end of session message.
```

You can also specify the remote host name on the command line. For example, to begin an FTP session with the remote host called *server2*, you type

```
$ ftp server2 [Return] . . . . . . . . . . Remote host name is specified on the
                                           command line.
Connected to server2.
220 server2 FTP server (UNIX(r) System V Release 4.0) ready.
```

Similar to the previous FTP session examples, **ftp** prompts you to enter your login name and password on the remote host before the FTP session is established.

The following command sequences show how to use the help and a few other **ftp** commands. It is assumed you have started an FTP session with a remote host called *server2*.

Using the bell Command

```
ftp> help bell [Return] . . . Display description of the bell command.
bell    beep when command completed
ftp>_ . . . . . . . . . . . . . ftp prompt is displayed.
ftp> bell [Return]  . . . . . Display the bell mode.
Bell mode on.
ftp> bell [Return]  . . . . . Turn the bell off.
Bell mode off.
ftp>_ . . . . . . . . . . . . . ftp prompt is displayed.
```

Using the hash Command

The **hash** command is used to get feedback when a block of file is transferred. The block size is 8192 bytes. Hash signs, such as those shown on the following line, will be displayed at completion of each block transferred.

```
#####
```

The hash mode is usually set before transferring large files:

```
ftp> help hash [Return] . . . Display description of the hash command.
hash    toggle printing '#' for each buffer transferred
ftp> _  . . . . . . . . . . . ftp prompt is displayed.
ftp> hash [Return]  . . . . . Set the hash mode.
Hash mark printing on (8192 bytes/hash mark).hash
ftp> hash [Return]  . . . . . Turn the hash off.
Hash mark printing off.
ftp> _  . . . . . . . . . . . Display the bell mode.
```

Using the File Transfer Mode Commands

```
ftp> help bin [Return]  . . . Display description of the bin command.
binary    set binary transfer type
ftp> bin[Return]  . . . . . . Change to the file transfer binary mode.
200 Type set to I.
ftp> ascii [Return] . . . . . Change to the file transfer ASCII mode.
200 Type set to A.
ftp>_ . . . . . . . . . . . . . ftp prompt is displayed.
```

Getting a File from Another System

Suppose you have started an FTP session with a remote host called *server2* and the remote host has a file named `Notes` in a directory named `Memos`. The following command sequences show how to retrieve Notes from the remote host. As you issue the commands, you receive some feedback from the server, indicating that it is carrying out your orders.

The **cd** command is used to change to the directory named Memos on the remote host:

```
ftp> cd Memos [Return]   . . . . . . Change to the Memos directory.
ftp>
```

The **dir** command is used to list the current directory on the remote host (*server2*), to see if the file is there:

```
ftp> dir [Return] . . . . . . . . . . List the filenames
-rwx --x --x    1 tuser other  3346 Aug     9 11:51 Notes
ftp>
```

Or you can use the **ls** command:

```
ftp> ls [Return]   . . . . . . . . . . List the filenames.
ftp> get [Return]  . . . . . . . . . . Issue a get command.
(remote) file  . . . . . . . . . . . . Prompt to enter the filename
                                       is displayed.
(remote-file) Notes [Return] . . . . . Enter the filename on the
                                       remote host.
(local-file) . . . . . . . . . . . . . Prompt to enter the filename
                                       is displayed.
(local-file) Notes [Return]  . . . . . Enter the filename on the local
                                       host.
```

The remote host transfers the file to your current directory on the local host, and you receive feedback about how many bytes were sent and the rate of transfer:

```
200 PORT command successful.
150 ASCII data connection for Notes
(192.246.52.166,37175).
226 Transfer complete.
ftp>_
```

Miscellaneous Commands

As was mentioned, the **?** command is used to get the list of **ftp** commands. You can also use the **?** for obtaining help for an **ftp** command by specifying a command as the first argument to the **?**. For example:

```
ftp> ? close [Return] . . . . . . . . Help for the close command.
Close    Terminate FTP session.
```

The shell escape command (**!**) is used to start a subshell on the local host. This is very useful if you want to perform some operations on the local host while you are connected to a remote FTP server. After you are finished working on the local host, exit the subshell to return to your FTP session. The following command sequences demonstrate the **!** usage.

Assuming you are in an FTP session, the following command invokes a subshell:

```
ftp> ![Return]   . . . . . . . . . . . Issue a shell escape command.
$_  . . . . . . . . . . . . . . . . . Shell prompt issued by the
                                       local host.
```

You are now in your current directory on the local host and any commands you use are invoked on the local host. You can use any of the available UNIX commands, and the commands are applied to the files on the local host. For example:

$ **cd [Return]** . . . . Change to the HOME directory on the local host.
$_ . . . . Shell prompt issued by the local host.

This changes your current directory on the local host to be your HOME directory. You type **exit** to terminate the subshell and return to your FTP session:

$ **exit [Return]** . . . . Terminate the shell and return to the FTP session.
ftp>_ . . . . Back to the FTP session.

13.5.2 Anonymous FTP

Anonymous FTP uses the `ftp` command and a special restricted account named *anonymous* as the remote host. Using anonymous FTP, you can access and share many files such as documents, source code, and executable files on the Internet. There is a special login for **ftp** that allows you to anonymously access files on a remote host. The following steps describes the process that is similar to using any FTP:

1. You need to know the address or host name of a site that allows anonymous FTP.
2. You use *anonymous* as the user name.
3. You enter any nonempty string for the password. Some systems do password validation on anonymous logins. In this case, you give your e-mail address as your password.

Once you are logged in as *anonymous*, you are granted limited access to the anonymous `ftp` subdirectory on the remote host. All of the commands described in this section can be used.

Usually, FTP sites place publicly accessible files in the directory `/public`. Some sites have a directory called `incoming` somewhere in the `ftp` subdirectory where you place your files.

Anonymous FTP Session

The following command sequence is an example of an FTP anonymous session. It shows the FTP commands and the FTP feedback messages.

Let us assume the anonymous FTP site is `ftp.xyz.net`, and the file you want to get is `pub/services/example.tar`:

$ftp **.xyz.net [Return]** . . . . Start an FTP session with the remote site connected to `ftp.xyz.net`
220 ftp.xyz.net FTP server (Version 1.1 Nov 17 2005) ready.
Name (ftp.xyz.net: xyz): anonymous [Return] . . Specify *anonymous* as the user name.
331 Guest login ok, send ident as password.
Password: [Return] . . . . Give your e-mail address as the password (not displayed).
230 Guest login ok, access restrictions apply.

```
ftb> cd/pub/services [Return]  . . . . . . . . . Change to desired directory.
250 CWD command successful.
ftp> get examples.tar [Return] . . . . . . . . . Get the file.
200 PORT command successful.
266 Transfer complete.
```

1. *Note that many anonymous FTP sites contain a file named* README, *which contains information about available files. To read that file or any text file without retrieving it, you type*

   ftp> **get README/pg [Return]**

2. *Once you have successfully logged in as* anonymous, *you can use all of the FTP commands described in this section.*

The **help** command is used to obtain a one-line description of the **get** command:

```
ftp> help get [Return]  . . . . . . . Help for the get command.
get     receive file
ftp>
```

The **get** command is used to retrieve a file named `Notes` from the remote host:

```
ftp> get [Return]  . . . . . . . . . . Issue the get command.
(remote-file)  . . . . . . . . . . . . Prompt to enter the filename.
(remote-file) Notes [Return]  . . . .  Enter the filename on the
                                       remote host.
(local-file)  . . . . . . . . . . . .  Prompt to enter the filename.
(local-file) Notes.txt [Return] . . .  Enter the filename on the
                                       local host.
```

The remote host transfers the file to your current directory on the local host. Information about how many bytes transferred and the rate of transfer is displayed:

```
200 PORT command successful.
150 ASCII data connection for Notes.txt (192.246.52.166,
37176) (0 bytes).
226 ASCII Transfer complete.
```

You can enter the filenames on the command line such as

```
ftp> get Notes Notes.txt [Return] . . Retrieve Notes.
```

Again, the remote host transfers `Notes` to your current directory on the local host. The name of the file on the local host is `Notes.txt`.

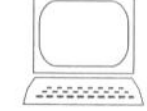

Transferring a File to Another System

Transferring a file from your system (local host) to another system (remote host) is done with the ftp command **put**. Suppose you have started an FTP session with a remote host called *server2* and you want to transfer a file named `november.rpt` from the directory `Reports`, a subdirectory of your current directory. The following command sequences show how to transfer `november.rpt` from the local host. As you issue the commands, you receive some responses from the server, indicating that it is carrying out your orders.

To transfer a file named november.rpt from the directory Reports, a subdirectory of your current directory, start an FTP session on the remote system and follow these steps.

The **lcd** command is used to change the directory on your system (local host):

```
ftp> lcd reports [Return] . . Change directory.
```

The **help** command is used to obtain a one-line description of the **put** command:

```
ftp> help put [Return]  . . . Help for put command.
put     send one file
```

The **put** command is used to transfer the file november.rpt from your system to the remote host:

```
ftp> put [Return]  . . . . . . . . . . . . . . . Issue the put command.
(local - file) . . . . . . . . . . . . . . . . . . Prompt to enter the filename.
(local - file) november.rpt [Return] . . . . . Enter the filename on the local host.
(remote - file). . . . . . . . . . . . . . . . . . Prompt to enter the filename.
(remote - file) november.rpt [Return]. . . . . Enter the filename on the remote host.
```

The local host transfers the file called november.rpt to your current directory on the remote host and the following information is displayed:

```
200 PORT command successful.
150 ASCII data connection for november.rpt (137.161.110.77,1386).
226 Transfer complete.
ftp: 19456 bytes sent in 0.00Seconds 19456000.00Kbytes/sec.
ftp>_
```

13.6 WORKING WITH COMPRESSED FILES

Large files and software packages are often compressed to save space at anonymous FTP archives. These files are usually made available in *tar-format* with file extension such as .tar.Z., and must be retrieved in binary mode. You must first uncompress the file at your site and then use **tar** to recover the original file.

13.6.1 The compress and uncompressed Commands

You can use the **compress** command to reduce the size of the file to save space. When you run a **compress** command on a file, your original file is replaced by a file with the same name but with the file extension .Z. The **compress** command is also used with the encryption command as security measures. In this case, the file is first compressed and then encrypted by the **crypt** command. The following command sequences show the use of the **compress** command.

Assume you have a file named important in your current directory:

```
$ compress important [Return]  . Issue the command.
$ ls Important* [Return] . . . . List the files.
important.z  . . . . . . . . . . The file is compressed.
$_
```

You can use the **-v** option to make **compress** report the efficiency of its compression:

```
$compress -v important [Return] . . . . . Use the -v option.
important: compression: 49.18%—replaced with important.Z
$_
```

You use the **uncompress** command to recover the original file. This uncompresses the compressed file and removes the compressed file. Continuing with the previous example, assume you have the compressed file named `important.z` in your current directory:

```
$ uncompress important [Return]   . . . . Uncompress the
                                          important.z file.
$_ . . . . . . . . . . . . . . . . . . . . Prompt is back.
```

COMMAND SUMMARY

The following commands have been discussed and explored in this chapter.

at

This command runs another command or a list of commands at a later time.

Option	Operation
-l	Lists all jobs that are submitted with **at**.
-m	Mails you a short message of confirmation at completion of the job.
-r	Removes the specified job numbers from the queue of jobs scheduled by **at**.

banner

This command displays its argument, the specified string, in large letters.

calendar

This command is a reminder service and reads your schedule from the `calendar` file in the current directory.

compress

This command is used to compress the specified file, thus reducing the size of the file and saving space. The uncompress command is used to recover the original file and remove the compressed file.

crypt

This command is used to encrypt and decrypt a file. The command changes each character in your file in a reversible way, so you can obtain the original file later.

df

This command reports the total amount of the disk space or the space available on a specified file system.

du

This command summarizes the total space occupied by any directory, its subdirectories, or each file.

Option		Operation
UNIX	Linux Alternative	
-a	--all	Display the size of directories and files.
-b	--bytes	Displays the size of directories and files in bytes.
-s	--summarize	Displays only the total blocks for the specified directory; subdirectories are not listed.
	--help	Displays a usage message.
	--version	Displays the version information.

finger

This command displays detailed information on users.

Option	Operation
-b	Suppresses displaying of the user's HOME directory and shell in a long format display.
-f	Suppresses displaying the header in a nonlong format output.
-h	Suppresses displaying the `.project` file in a long format output.
-l	Forces long format output.
-p	Suppresses displaying the `.plan` file in a long format output.
-s	Forces short format output.

FTP

This command (utility) is used to transfer files from one system to another. Files of any type can be transferred, and you can specify whether the file is an ASCII or a binary file. You type **ftp** to start an FTP session.

FTP access commands	
Command	**Description**
open *remote-hostname*	Opens a connection to the FTP server on the specified host. It prompts you to enter the user name and password to log in on the remote host.
close	Closes current open connection and returns to the local FTP command. At this point you may issue the **open** command for a different remote host.
quit (bye)	Closes the current FTP session with the remote server and exits **ftp**. That is, it returns to UNIX shell level.

FTP file and directory commands	
Command	**Description**
cd *remote-directory-name*	Changes the current directory on the remote host to the specified directory.
lcd *local-directory-name*	Changes the current directory on the local host to the specified directory.
dir	Lists the current directory on the remote host.
pwd	Prints the name of the current directory on the remote host.
mkdir *remote-directory-name*	Makes a new directory on the remote host. Typically, you must have permission to do this.
delete *remote-filename*	Deletes a single specified file on the remote host.
mdelete *remote-filenames*	Deletes multiple files on the remote host.

FTP miscellaneous commands	
Command	**Description**
? or **help**	Displays an informative message about the meaning of the specified command. If no argument is given, a list of the known commands is displayed.
!	Switches to escape shell mode.
hash	Displays hash sign (#) as feedback for each data block transferred. The size of a data block is 8192 bytes.

spell

This command checks the spelling of a specified document or words entered from the keyboard. It only displays the words not found in the spelling list and does not suggest a correct spelling.

Option	Operation
-b	Checks with British spelling.
-v	Displays the words that are not in the spelling list and their derivation.
-x	Displays plausible stems for each word being checked.

tar

This command is used to copy a set of files into a single file, called a `tarfile`. A `tarfile` is usually saved on a magnetic tape but it can be on any other media such as a floppy disk. It packs multiple files into a single file (in `tar-format`) that can be unpacked later by **tar**.

Option		Operation
UNIX	Linux Alternative	
-c	**--create**	(**create**) Creates a new `tarfile`. Writing begins at the beginning of the `tarfile`.
-f	**--file**	(**file**) Uses the next argument as a place where the archive is to be placed.
-r	**--concatenate**	(**replace**) Writes a new archive at the end of the `tarfile`.
-t	**-- list**	(**table of contents**) Lists the name of the files in the `tarfile`.
-x	**--extract** or **--get**	(**extract**) Extracts files from the `tarfile`.
-v	**--verbose**	(**verbose**) Provides additional information about the `tarfile` entries.
	--help	Displays a usage message.
	--version	Displays the version information.

time

This command provides information about the computer time your command uses. It reports the real time, user time, and system time required by a specified command.

type

This command gives more information about another command, such as whether the specified cmmand is a shell command or a shell built-in command.

REVIEW EXERCISES

1. Explain the UNIX system security. What are the ways to secure your file system?
2. Can the superuser delete your files?
3. Can the superuser read your encrypted files?
4. Is the **cd** command a built-in command? How do you find out?
5. Explain *elapsed time*, *user time*, and *system time*.
6. What is the command that reports the disk space?
7. Can you invoke the **spell** command within the vi editor? If so, how?
8. What is the command to use if you want to execute a program at a later time?
9. Do the access modes for files and directories mean the same thing?
10. What is the **tar** command used for?
11. What is a `tarfile`?
12. Can you use the **tar** command to archive files on media other than tapes?
13. What is the command to list the files in a `tarfile` called `save.tar`?
14. What is the **compress** command used for?
15. What does FTP stand for?
16. What is the command to open an FTP session?
17. What is the command to close an FTP session?
18. What is the command to reduce the file size?
19. What is the command to protect your files from others?
20. What is the command to check your spelling?

Terminal Session

Practice the following commands on your system.

1. Sort a file at 13:00 tomorrow.
2. Send mail to another user at 6 p.m. on Wednesday.
3. Use the **time** command on a few commands such as **sort**, **spell**, and so on, and observe the time reports.
4. Change the access mode of your directory to owner only **read**, **write**, and **execute**.
5. Check the spelling in one of your text files.
6. Create your own dictionary file and make the **spell** command use your dictionary file in addition to its own file.
7. Find out the available space on your disk.
8. Find out how many blocks are occupied by your HOME directory, its subdirectories, and the files in them.
9. Place the **du** and **df** commands in your `.profile` file. Observe the reports when you log in.

10. Encrypt a file (if you are authorized). Display the encrypted file on your terminal. Decrypt the file and display it again.
11. Make a banner that shows your initials on the screen.
12. Send your initials banner to the printer.
13. Can you make the system show your initials as soon as you log in?
14. Modify the `greetings` program (from Chapter 12) to show the greetings in large letters.
15. Make a `calendar` file in your HOME directory and type your schedule.
16. Use the **calendar** command to display your current schedule.
17. Place the **calendar** command in your `.profile` file. Observe the report when you log in.
18. Use the **tar** command to archive files in your current directory in another directory called `my_tar_files`.
19. Use the **tar** options, such as the **-t** and **-v** options, and observe the output.
20. Use the **tar x** option to extract a specified file from the `tarfile`.
21. Use the **compress** command to compress a large file. Observe the feedback message that displays the compression percentage. List the file and check the file extension.
22. Uncompress the file. Check that the compressed file is removed.
23. Open an FTP session, from your PC to the university computer (you usually can use the university computer from your PC). Do the following commands when connection is established:
    a. List the available commands.
    b. Get help for a specific command.
    c. Set **bell** and **verbose** to on mode.
    d. Set file transfer mode to ASCII mode.
    e. Make a new directory on the remote host.
    f. Delete a file from the remote host.
    g. Transfer a file from your PC to your account on the remote host.
    h. Transfer a file from the remote host to your PC.

Appendixes

Appendix A: Command Index

Appendix B: Command Index by Category

Appendix C: Command Summary (Including Command Options)

Appendix D: Summary of vi Editor Commands

Appendix E: The ASCII Table

APPENDIX A

Command Index

This appendix is a quick index to the commands covered in this book. The commands are in alphabetical order. Numbers after each command indicate the page numbers.

Command	Description	Page
alias	creates aliases for commands	251
at	executes commands at a later time	415
banner	displays banner (in large characters)	414
cal	provides calendar service	42
calendar	provides reminder service	418
cancel	removes (cancels) the printing requests	113
cat	concatenates/displays file(s)	109
cd	changes the current directory	93
chmod	changes file/directory permissions	317
compress	reduces the size of the file for storage	439
cp	copies files	175
crypt	encrypts/decrypts files	430
cut	selects the specified fields/columns from files	196
date	displays date and time	38
df	shows total amount of free disk space	411
.dot	runs processes in the current shell environment	323
du	provides disk usage report	412
echo	displays (echoes) its argument	221
ed	UNIX line-oriented editor	55
emacs	the emacs editor	55
ex	standard UNIX line-oriented editor	55
exit	terminates the current shell program	322
export	exports variables to other shells	247
expr	provides arithmetic operations	346

fc	lists commands from `history` file	254
find	finds and acts on a specified file	190
finger	displays information on users	419
ftp	transfers files from one system to another	431
grep	searches in file(s) for a pattern	240
head	displays the first part of a specified file	194
help	invokes menu-driven help utility	43
history	keeps a list of all the entered commands	252
kill	terminates processes	237
learn	invokes courses/lessons utility	43
let	provides arithmetic operations	348
ln	links files	181
lp	prints file(s) on the line printer	110
lpr	prints specified file	113
lpstat	provides the status of the printing requests	114
ls	lists contents of a directory	98
mailx	provides an electronic mail processing system (e-mail)	291
man	finds information from the electronic manual	44
mesg	permits/denies messages from the **write** command	272
mkdir	creates a directory	106
more	displays files one screen at a time	200
mv	moves/renames files	179
news	provides access to the local system news	273
nohup	keeps commands running after you log off	237
passwd	changes your login password	32
paste	joins files together line by line	198
pg	displays file one screen at a time	163
pr	formats file before printing	171
ps	provides the process status report	235
pwd	displays the current/working directory pathname	92
r (redo)	repeats commands from the `history` file	254
read	reads input from the input device	324
rm	removes (deletes) file(s)	115
rmdir	removes (deletes) empty directories	98
set	sets/displays the values of the shell variables	226
sh	invokes a new copy of the shell	316
sleep	makes the process wait for a specified time in seconds	235
sort	sorts file(s) in specified order	242
spell	provides spelling checker	424

stty sets terminal options 369
tail displays the last part of a specified file 195
talk provides terminal-to-terminal communication 275
tar archives a set of files into a `tarfile` on a tape................ 420
tee splits output 238
test tests an expression for true/false conditions 336
time times a command 417
tput provides access to the `terminfo` database 371
trap sets/resets the interrupt signals 368
type displays the type of the specified command 417
unset removes a shell variable 226
vi invokes standard UNIX screen editor 52
view invokes read only vi editor 163
wall writes to all currently logged in terminals (*write all*) 274
wc counts lines, words, or characters in a specified file(s) 184
who shows who is on the system 38
write provides terminal-to-terminal communication 271

APPENDIX B

Command Index by Category

This appendix is a quick index to the commands covered in this book. The commands are organized according to their functions and in alphabetical order. Numbers after each command indicate the page numbers.

File and Directory Commands

<> << >> redirection operators 165
| pipe operator 234
cat concatenates/displays file(s) 109
cd changes the current directory 93
chmod changes file/directory permissions 317
compress reduces the size of the file for storage. 439
cp copies files 175
cut selects the specified fields/columns from files 196
ln links files 181
ls lists contents of a directory 98
mkdir makes (creates) a directory 106
mv renames/moves files 179
paste joins files together line by line 198
pwd displays the current/working directory pathname 92
rm removes (deletes) file(s) or directories 115
rmdir removes (deletes) empty directories 98
tar archives a set of files into a `tarfile` on a tape 420

Communication Commands

mailx provides e-mail services 291
mesg permits/denies messages from the **write** command 272
news provides access to the local system news 273

talk	provides terminal-to-terminal communication	275
wall	writes to all currently logged in terminals (*write all*)	274
write	provides terminal-to-terminal communication	271

Help Commands

help	menu-driven help utility	43
learn	invokes courses/lessons utility	43
man	finds information from the electronic manual	44

Process Control Commands

alias	creates aliases for commands	251
at	executes commands at a later time	415
fc	lists commands from `history` file	254
kill	terminates processes	237
nohup	keeps commands running after you log off	237
ps	provides the process status report	235
r (redo)	repeats commands from the `history` file	254
sleep	makes the process wait for a specified amount of time in seconds	235

Line Printer Commands

cancel	removes (cancels) the printing requests	113
lp	prints file(s) on the line printer	110
lpstat	provides the status of the printing requests	114
pr	formats files before printing	171

Information Handling Commands

df	shows the total amount of free disk space	411
du	provides disk usage report	412
expr	provides arithmetic operations	346
find	finds and acts on specified files	190
finger	displays information on users	419
grep	searches files for a specified pattern	240
head	displays the first part of a specified file	194
history	keeps a list of all the entered commands	252
let	provides arithmetic operations	348
more	displays files one screen at a time	200
ps	provides process status report	235
pwd	displays the current/working directory pathname	92
set	sets/displays the values of the shell variables	226
sort	sorts file(s) in a specified order	242

spell provides spelling checker . . . 424
tail displays the last part of a specified file . . . 195
time times a command . . . 417
type shows the specified command type . . . 417
unset removes/unsets a shell variable . . . 226
wc counts lines, words, or characters in specified files . . . 184
who shows who is on the system . . . 38

Terminal Commands

more displays files one screen at a time . . . 200
pg displays file one screen at a time . . . 163
stty sets terminal options . . . 369
tput provides access to the `terminfo` database . . . 371

Security Commands

chmod changes file/directory permissions . . . 317
crypt encrypts/decrypts files . . . 430
passwd changes your login password . . . 32

Starting/Ending Sessions

[Ctrl-d] ends a session . . . 34
exit ends a session (log off) . . . 34
login sign-on prompt . . . 31
passwd changes login password . . . 32

UNIX Editors

ed UNIX line-oriented editor . . . 55
emacs the emacs editor . . . 55
ex standard UNIX line-oriented editor . . . 55
vi standard UNIX screen-oriented editor . . . 52
view vi in read only mode . . . 163

APPENDIX C

Command Summary

The following is a list of the UNIX commands (utilities) in alphabetic order. The numbers after the commands are the page numbers. To refresh your memory, the command line format is shown again in Figure C.1.

Figure C.1
The Command Line Format

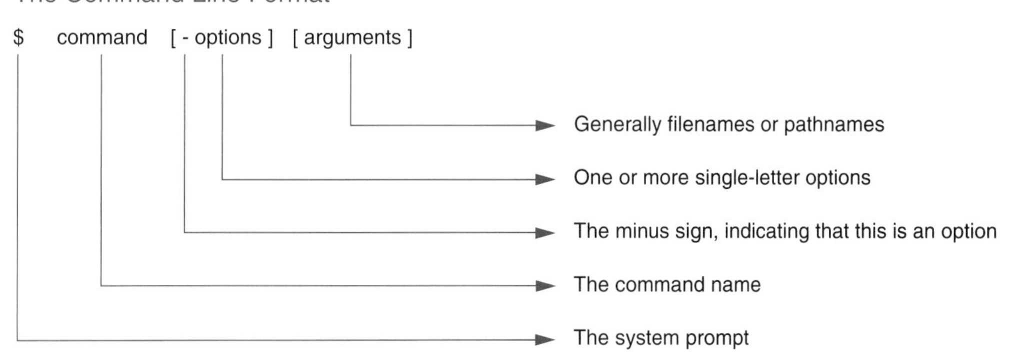

alias 251
This creates aliases for commands.

at. 415
This command runs another command or a list of commands at a later time.

Option	Operation
-l	Lists all jobs that are submitted with **at**.
-m	Mails you a short message of confirmation at completion of the job.
-r	Removes the specified job numbers from the queue of jobs scheduled by **at**.

banner 414
This command displays its argument, the specified string, in large letters.

cal 42
Displays the calendar for a specified year or month of a year.

calendar 418
This command is a reminder service and reads your schedule from the `calendar` file in the current directory.

cancel (cancel print requests) 113
You can use this command to cancel print requests that are in queue waiting to be printed or are currently being printed.

cat (concatenate). 109
This concatenates/displays file(s).

cd (change directory) 93
This command changes your current directory to another directory.

chmod. 317
This command changes the access permission of a specified file according to the option letters indicating different categories of users. The user categories are **u** (for user/owner), **g** (for group), **o** (for others), and **a** (for all). The access categories are **r** (for read), **w** (for write), and **x** (for executable).

compress 439
This command is used to compress the specified file, thus reducing the size of the file and saving space. The uncompress command is used to recover the original file and remove the compressed file.

cp 175
This command copies file(s) within the current directory or from one directory to another.

Option		Operation
UNIX	Linux Alternative	
-b	**--backup**	Makes a backup of the specified file if file already exists.
-i	**--interactive**	Asks for confirmation if the target file already exists.
-r	**--recursive**	Copies directories to a new directory.
	--verbose	Explain what is being done.
	--help	Displays the help page and exits.

crypt 430

This command is used to encrypt and decrypt a file. The command changes each character in your file in a reversible way, so you can obtain the original file later.

cut 196

This command is used to "cut out" specific columns or fields from a file.

Option		Operation
UNIX	Linux Alternative	
-f	**--fields**	Specifies the field position.
-c	**--characters**	Specifies the character position.
-d	**--delimiter**	Specifies the field separator (delimiter) character.
	--help	Displays the help page and exits.
	--version	Displays the version information and exits.

date 38

Displays the day of the week, month, date, and time.

df 411

This command reports the total amount of the disk space or the space available on a specified file system.

. (dot) 323

This command lets you run a process in the current shell environment and does not allow the shell to create a child process to run the command.

du 412

This command summarizes the total space occupied by any directory, its subdirectories, or each file.

Option		Operation
UNIX	Linux Alternative	
-a	**--all**	Display the size of directories and files.
-b	**--bytes**	Displays the size of directories and files in bytes.
-s	**-- summarize**	Displays only the total blocks for the specified directory; subdirectories are not listed.
	--help	Displays a usage message.
	--version	Displays the version information.

echo 221

This command displays (echoes) its arguments on the output device.

Escape Character	Meaning
\a	Audible alert (bell)
\b	Backspace
\c	Inhibit the terminating newline
\f	Form feed
\n	Carriage return and a line feed (newline)
\r	Carriage return without the line feed
\t	Horizontal tab
\v	Vertical tab

exit 322

This command terminates the current shell program whenever it is executed. It can also return a status code (RC) to indicate the success or failure of a program. It also terminates your login shell and logs you off if it is typed at the **$** prompt.

export 247

This command makes a specified list of variables available to other shells.

expr 346

This command is a built-in operator for arithmetic operations. It provides arithmetic and relational operators.

find 190

This command locates files that match a given criterion in a hierarchy of directories. With the action options, you can instruct UNIX about what to do with the files once they are found.

Search Option	Description
-name *filename*	Finds files with the given *filename*.
size +*n*	Finds files with the size *n*.
-type *file type*	Finds files with the specified access mode.
-atime +*n*	Finds files that were accessed *n* days ago.
-mtime +*n*	Finds files that were modified *n* days ago.
-newer *filename*	Finds files that were modified more recently than *filename*.
Action Option	**Description**
-print	Prints the pathname for each file found.
-exec *command* \;	Lets you give *commands* to be applied to the files.
-ok *command* \;	Asks for confirmation before applying the command.

finger 419

This command displays detailed information on users.

Option	Operation
-b	Suppresses displaying of the user's HOME directory and shell in a long format display.
-f	Suppresses displaying the header in a nonlong format output.
-h	Suppresses displaying the `.project` file in a long format output.
-l	Forces long format output.
-p	Suppresses displaying the `.plan` file in a long format output.
-s	Forces short format output.

FTP 431

This command (utility) is used to transfer files from one system to another. Files of any type can be transferred, and you can specify whether the file is an ASCII or a binary file. You type **ftp** to start an FTP session.

FTP access commands 432	
Command	**Description**
open *remote-hostname*	Opens a connection to the FTP server on the specified host. It prompts you to enter the user name and password to log in on the remote host.
close	Closes current open connection and returns to the local FTP command. At this point you may issue the **open** command for a different remote host.
quit (bye)	Closes the current FTP session with the remote server and exits **ftp**. That is, it returns to UNIX shell level.

FTP file and directory commands 432	
Command	**Description**
cd *remote-directory-name*	Changes the current directory on the remote host to the specified directory.
lcd *local-directory-name*	Changes the current directory on the local host to the specified directory.
dir	Lists the current directory on the remote host.
pwd	Prints the name of the current directory on the remote host.
mkdir *remote-directory-name*	Makes a new directory on the remote host. Typically, you must have permission to do this.
delete *remote-filename*	Deletes a single specified file on the remote host.
mdelete *remote-filenames*	Deletes multiple files on the remote host.

FTP miscellaneous commands 434	
Command	**Description**
? or **help**	Displays an informative message about the meaning of the specified command. If no argument is given, a list of the known commands is displayed.
!	Switches to escape shell mode.
hash	Displays hash sign (#) as feedback for each data block transferred. The size of a data block is 8192 bytes.

grep (Global Regular Expression Print) 240

Searches for a specified pattern in file(s). If the specified pattern is found, the line containing the pattern is displayed on your terminal.

Option		Operation
UNIX	Linux Alternative	
-c	--count	Displays only the count of the matching lines in each file that contains the match.
-i	--ignore-case	Ignores the distinction between lower and uppercase letters in the search pattern.
-l	--files-with-matches	Displays the names of the files with one or more matching lines, not the lines themselves.
-n	--line-number	Displays a line number before each output line.
-v	--revert-match	Displays only those lines that do not match the pattern.
	--help	Displays help page and exits.
	--version	Displays version information and exits.

head 194

This command displays the first part of a specified file. This is a quick way to check the contents of a file. The number of lines to be displayed is an option, and more than one file can be specifed on the command line.

Option		Operation
UNIX	Linux Alternative	
-l	--lines	Counts by lines. This is the default option.
-c	--chars=num	Counts by characters.
	--help	Displays the help page and exits.
	--version	Displays the version information and exits.

help. 43

Brings to the display a series of menus and questions that lead you to the descriptions of the most commonly used UNIX commands.

history (ksh). 252

This command is a Korn and Bourne Again shell feature that keeps a list of all the commands you enter during your sessions.

kill 237

Terminates an unwanted or unruly process. You have to specify the process ID number. The process ID0 kills all programs associated with your terminal.

learn 43

A computer-aided instruction program that is arranged in a series of courses and lessons. It displays the menu of courses and lets you select your desired course and lesson.

let 348

This command provides arithmetic operations.

ln 181

Creates links between an existing file and another filename or directory. It lets you have more than one name for a file.

lp (line printer) 110

This prints (provides hard copy of) the specified file.

Option	Operation
-d	Prints on a specific printer.
-m	Sends mail to the user mailbox after completion of the print request.
-n	Prints specified number of copies of the file.
-s	Suppresses feedback messages.

lpr (line printer) 113

This command prints the specified file. **lpr** reads from the standard input if no filename is specified.

Option	Operation
-p	Prints on a specific (named) printer.
-#	Prints specified number of copies of the file.
-T	Prints a specified title on the banner page of the output.
-m	Sends mail to the user mailbox after completion of the print request.

lpstat (line printer status) 114

This provides information about your printing request jobs, including printing a request ID number that you can use to cancel a printing request.

Option	Operation
-d	Prints the name of the system default printer for print requests.

ls (list). 98

This command lists the contents of your current directory, or any directory you specify.

Option		Operation
UNIX	Linux Alternative	
-a	**--all**	Lists all files, including the hidden files.
-C	**-- format=vertical** **-- format=horizontal**	Lists files in multicolumn format. Entries are sorted down the columns.
-F	**--classify**	Places a forward slash (/) after each filename if that file is a directory, and an asterisk (*) if it is an executable file.
-l	**--format=single-column**	Lists files in a long format, showing detailed information about the files.
-m	**--format=commas**	Lists files across the page, separated by commas.
-p		Places a forward slash (/) after each filename if that file is a directory.
-r	**--reverse**	Lists files in reverse alphabetical order.
-R	**--recursive**	Recursively lists the contents of the directory.
-s	**--size**	Show size of each file in blocks.
-x	**--format=horizontal** **--format=across**	Lists files in multicolumn format. Entries are sorted across the line.
	--help	Displays a usage message.

mailx 291

This utility provides the electronic mail system for the users. You can send messages to other users on the system, regardless of whether they are logged on or not.

Option	Operation
-f *filename*	Reads mail from the specified *filename* instead of the system *mailbox*. If no file is specified, it reads from `mbox`.
-H	Displays a list of the message headers.
-s *subject*	Sets the subject field to the string *subject*.

mailx command mode. 291

When you invoke **mailx** to read your mail, it places itself in command mode. The prompt for this mode is the question mark (**?**).

Command	Operation
!	Lets you execute the shell commands (the shell escape).
cd *directory*	Changes to the specified directory, or to the HOME directory if none is specified.
d	Deletes the specified messages.
f	Displays the headlines of the current message.
q	Exits **mailx** and removes the messages from the system mailbox.
h	Displays active message headers.
m *users*	Sends mail to specified *users*.
R *messages*	Replies to the sender of the *messages*.
r *messages*	Replies to the sender of the *messages* and all the other recipients of the same *messages*.
s *filename*	Saves (appends) the indicated messages to the file named *filename*.
t *messages*	Displays (types) the specified *messages*.
u *messages*	Undeletes the specified *messages*.
x	Exits **mailx** does not remove messages from the system mailbox.

mailx tilde escape commands 278

When you invoke **mailx** to send mail to others, it places itself in input mode, ready for you to compose your message. The commands in this mode start with a tilde (~) and are called tilde escape commands.

Command	Operation
~?	Displays a list of all the tilde escape commands.
~! *command*	Lets you invoke the specified shell *command* while composing your message.
~e	Invokes an editor for editing your message. The editor to be used is defined in the mail variable called *EDITOR*.
~q	Quits input mode. Saves your partially composed message is saved in the file called `dead.letter`.
~r *filename*	Reads the file named *filename* and adds its contents to your message.
~< *filename*	Reads the file named *filename* (using the redirection operator) and adds its contents to your message.
~<! *command*	Executes the specified command and places its output into your message.
~v	Invokes the default visual editor, the vi editor, or uses the value of the mail variable *VISUAL*, which can be set up for other editors.
~w *filename*	Writes the message currently being composed to the specified *filename*.

man 44

This command shows pages from the online system documentation.

mesg 272

This command is set to **n** to prohibit unwanted **write** messages. It is set to **y** to receive messages.

mkdir (make directory) 106

This command creates a new directory in your working directory, or in any other directory you specify.

Option	Operation
-p	Lets you create levels of directories in a single command line.

more 200

Displays files one screen at a time. This is useful for reading large files.

Option		Operation
UNIX	Linux Alternative	
-lines	**-num** *lines*	Displays the indicated number of *lines* per screen.
+*line-number*		Starts up at *line-number.*
+/*pattern*		Starts two lines above the line containing the *pattern.*
-c	**-p**	Clears the screen before displaying each page instead of scrolling. This is sometimes faster.
-d		Displays the prompt **[Hit space to continue, Del to abort]**.
	-- help	Displays the help page and exits.

mv 179

This command renames files or moves files from one location to another.

Option		Operation
UNIX	Linux Alternative	
-b	**--backup**	Makes backup of the specified file if file already exists.
-i	**--interactive**	Asks for confirmation if the target file already exists.
-f	**--force**	Removes target file if file already exists and does not ask for confirmation.
-v	**--verbose**	Explains what is being done.
	--help	Displays the help page and exits.
	--version	Displays version information and exits.

news 273

This command is used to look at the latest news in the system. It is used by the system administrator to inform others of the events happening.

Option	Operation
-a	Displays all the news items, whether they are in old or new files.
-n	Lists only the names of the news files (headers).
-s	Displays the number of the current news items.

nohup 237

This prevents the termination of the background process when you log out.

passwd 32

This command changes your login password.

paste 198

This command is used to join files together line by line, or to create new files by pasting together fields from two or more files.

Option		Operation
UNIX	Linux Alternative	
-d	**--delimiters**	Specifies the field separator (delimiter) character.
	--help	Displays the help page and exits.
	--version	Displays the version information and exits.

pg 163

Displays files one screen at a time. You can enter the options or other commands when **pg** shows the prompt sign.

Option	Operation
-n	Does not require [Return] to complete the single-letter commands.
-s	Displays messages and prompts in reverse video.
-*num*	Sets the number of lines per screen to the integer *num*. The default value is 23 lines.
-p*str*	Changes the prompt **:** (colon) to the string specified as *str*.
+*line-num*	Starts displaying the file from the line specified in *line–num*.
+/*pattern*	Starts viewing at the line containing the first occurrence of the specified *pattern*.

The pg Command Key Operators 164

These keys are used when **pg** displays the prompt sign.

Key	Operation
+*n*	Advances *n* screens, where *n* is an integer number.
-*n*	Backs up *n* screens, where *n* is an integer number.
+*n***l**	Advances *n* lines, where *n* is an integer number.
-*n***l**	Backs up *n* screens, where *n* is an integer number.
n	Goes to screen *n*, where *n* is an integer number.

pr 171

Your file is formatted before printing or viewing it on the screen.

Option		Operation
UNIX	Linux Alternative	
+*page*	**--pages=***page*	Starts displaying from the specified *page*. The default is page 1.
-*columns*	**--columns=***columns*	Displays output in the specified number of *columns*. The default is one column.
-a	**--across**	Displays output in columns across (rather than down) the page, one line per column.
-d	**--double-space**	Displays output double spaced.
-h*string*	**--header=***string*	Replaces the filename in the header with the specified *string*.
-l*number*	**--length=***number*	Sets the page length to the specified *number* of lines. The default is 66 lines.
-m	**--merge**	Displays all the specified files in multiple columns.
-p		Pauses at the end of each page and sounds the bell.
-*character*	**--separator=***character*	Separates columns with a single specified *character*. If character is not specified, then [Tab] is used.
-t	**--omit-header**	Suppresses the five-line header and five-line trailer.
-w*number*	**--width=***number*	Sets line width to the specified number of characters. The default is 72.
	--help	Displays help page and exits.
	--version	Displays version information and exits.

ps (process status) 235
This command displays the process ID of the programs associated with your terminal.

Option	Operation
-a	Displays the status of all the active processes, not just the user's.
-f	Displays a full list of information, including the full command line.

pwd (print working directory). 92
This command displays the pathname of your working directory or any other directory you specify.

r (redo) 254
This is a Korn shell command that repeats the last command or commands from the history file.

read. 324
This command reads input from the input device and stores the input string in one or more variables specified as the command arguments.

rm (remove) 115
This command removes (deletes) files from your current directory, or any other directory you specify.

Option		Operation
UNIX	Linux Alternative	
-i	**--interactive**	Asks for confirmation before deleting any file.
-r	**--recursive**	Deletes the specified directory and every file and subdirectory in it.
	--help	Displays a usage message.

rmdir (remove directory). 98
This command deletes the specified directory. The directory must be empty.

set. 226
Displays the environmental/shell variables on the output device. The command **unset** removes the unwanted variables.

sh, ksh, or bash 316
This command invokes a new copy of the shell. You can run your script files using this command. Only three of the numerous options are mentioned.

Option	Operation
-n	Reads commands but does not execute them.
-v	Prints the input to the shell as the shell reads it.
-x	Prints command lines and their arguments as they are executed. This option is used mostly for debugging.

sleep 235
This command causes the process to go to sleep (wait) for the specified time in seconds.

sort 242
Text file(s) are sorted in different orders.

Option	Operation
-b	Ignores leading blanks.
-d	Uses dictionary order for sorting. Ignores punctuation and control characters.
-f	Ignores the distinction between lowercase and uppercase letters.
-n	Numbers are sorted by their arithmetic values.
-o	Stores the output in a specified file.
-r	Reverses the order of the sort from ascending to descending.

spell 424
This command checks the spelling of a specified document or words entered from the keyboard. It only displays the words not found in the spelling list and does not suggest a correct spelling.

Option	Operation
-b	Checks with British spelling.
-v	Displays the words that are not in the spelling list and their derivation.
-x	Displays plausible stems for each word being checked.

stty 369

This command sets options that control the capabilities of your terminal. There are more than a hundred different settings, and the following table lists only some of the options.

Option	Operation
echo [-echo]	Echoes [does not echo] the typed characters; the default is **echo**.
raw [-raw]	Disables [enables] the special meaning of the metacharacters; the default is **-raw**.
intr	Generates an interrupt signal; usually the [Del] key is used.
erase	[Backspace]. Erases the preceding character; usually the # key is used.
kill	Deletes the entire line; usually @ or [Ctrl=u] is used.
eof	Generates the (end-of-file) signal from the terminal; usually [Ctrl-d] is used.
ek	Resets the **erase** and **kill** keys to # and @ respectively.
sane	Sets the terminal characteristics to sensible default values.

tail 195

This command displays the last part (tail end) of a specified file. This is a quick way to check the contents of a file. The options give the flexibility to specify a desired part of the file.

Option		Operation
UNIX	Linux Alternative	
-l	**--lines**	Counts by lines. This is the default option.
-c	**--chars=num**	Counts by characters.
	--help	Displays the help page and exits.
	--version	Displays the version information and exits.

talk 275

This command is used for terminal-to-teminal communication. The receiving party must be logged on.

tar 420

This command is used to copy a set of files into a single file, called a `tarfile`. A `tarfile` is usually saved on a magnetic tape but it can be on any other media such as a floppy disk. It packs multiple files into a single file (in `tar-format`) that can be unpacked later by **tar**.

Option		Operation
UNIX	Linux Alternative	
-c	**--create**	(**create**) Creates a new `tarfile`. Writing begins at the beginning of the `tarfile`.
-f	**--file**	(**file**) Uses the next argument as a place where the archive is to be placed.
-r	**--concatenate**	(**replace**) Writes a new archive at the end of the `tarfile`.
-t	**-- list**	(**table of contents**) Lists the name of the files in the `tarfile`.
-x	**--extract** or **--get**	(**extract**) Extracts files from the `tarfile`.
-v	**--verbose**	(**verbose**) Provides additional information about the `tarfile` entries.
	--help	Displays a usage message.
	--version	Displays the version information.

tee 238

The output is split. One copy is displayed on your terminal, the output device, and another copy is saved in a file.

Option	Operation
-a	Appends output to a file without overwriting an existing file.
-i	Ignores interrupts; does not respond to the interrupt signals.

test 336

This command tests the condition of an expression given to it as an argument, and returns true or false depending on the status of the expression. It gives you the capability of testing different types of expressions.

time 417

This command provides information about the computer time your command uses. It reports the real time, user time, and system time required by a specified command.

tput 371

This command is used with the terminfo database, which contains codes for terminal characteristics and facilitates the manipulation of your terminal characteristics such as boldface text, clear screen, and so on.

Option	Operation
bel	Echoes the terminal's bell character.
blink	Makes a blinking display.
bold	Makes a boldface display.
clear	Clears the screen.
cup *r c*	Moves cursor to row *r* and column *c*.
dim	Dims the display.
ed	Clears from the cursor position to the end of the screen.
el	Clears from the cursor position to the end of the line.
smso	Starts stand out mode.
rmso	Ends stand out mode.
smul	Starts underline mode.
rmul	Ends underline mode.
rev	Shows reverse video, black on white display.
sgr0	Turns off all the attributes.

trap 368

This command sets and resets the interrupt signals. The following table shows some of the signals you can use to control the termination of your program.

Signal Number	Name	Meaning
1	**hang up**	Terminal connection is lost.
2	**interrupt**	One of the **interrupt** keys has been pressed.
3	**quit**	One of the **quit** keys has been pressed.
9	**kill**	The **kill -9** command has been issued.
15	**terminator**	The **kill** command has been issued.

type. 417

This command gives more information about another command, such as whether the specified cmmand is a shell command or a shell built-in command.

wall 274

This command is used mostly by the system administrator to warn users of some imminent events.

wc 184

This command counts number of characters, words, or lines in the specified file.

Option		Operation
UNIX	Linux Alternative	
-l	**--lines**	Reports the number of lines.
-w	**--words**	Reports the number of words.
-c	**--chars**	Reports the number of characters.
	--help	Displays the help page and exits.
	--version	Displays version information and exits.

who 38

Lists the login name, terminal lines, and login times of the users who are on the system.

Option	Linux	Operation
-q	**--count**	The quick **who**; just displays the name and number of users.
-H	**--heading**	Displays heading above each column.
-b		Displays the time and date of the last reboot.
	--help	Displays a usage message.

write 271

This command is used for terminal-to-terminal communication. The receiving party must be logged on.

APPENDIX D

Summary of vi Editor Commands

This appendix contains a summary of all the vi editor commands covered in this book. For more information, refer to Chapters 4 and 6. Figure D.1 will refresh your memory of the vi editor modes of operation.

Figure D.1
The vi Editor Modes of Operation

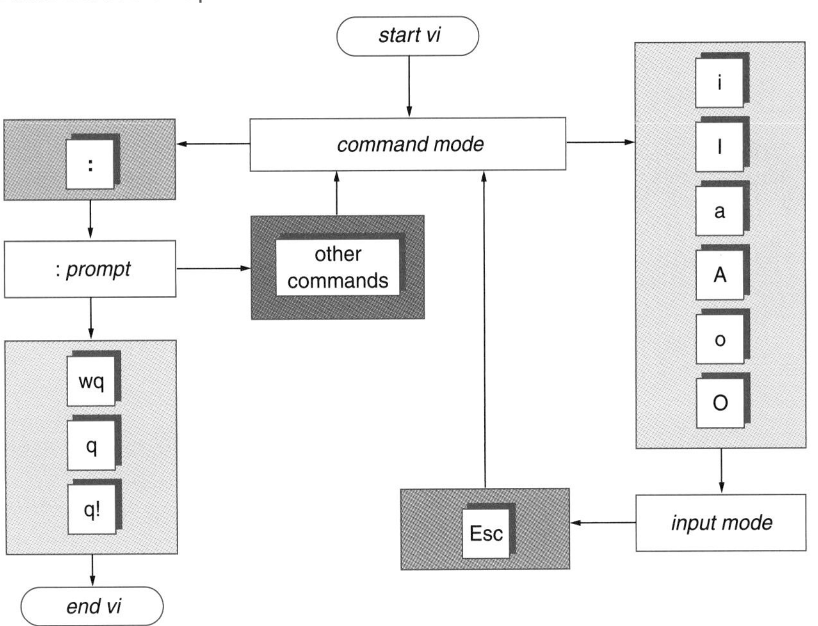

The vi editor

vi is a screen editor you can use to create files. vi has two modes: the command mode and the text input mode. To start vi, type **vi**, press [Spacebar], and type the name of the file. Several keys place vi in the text input mode, and [Esc] always returns vi to the command mode.

The change mode keys

These keys change vi from the command mode to the text input mode. Each key places vi in the text input mode in a different manner. [Esc] places vi back in the command mode.

Key	Operation
i	Places the text you enter before the character that the cursor is on.
I	Places the text you enter at the beginning of the current line.
a	Places the text you enter after the character that the cursor is on.
A	Places the text you enter after the last character of the current line.
o	Opens a blank line below the current line and places the cursor at the beginning of the new line.
O	Opens a blank line above the current line and places the cursor at the beginning of the new line.

Correcting text keys

These keys are applicable in the command mode only.

Key	Operation
x	Deletes the character specified by the cursor position.
dd	Deletes the line specified by the cursor position.
u	Undoes the most recent change.
U	Undoes all the changes on the current line.
r	Replaces a character that the cursor is on.
R	Replaces characters starting from the cursor position, and changes vi to the text mode.
.	Repeats the last text changes.

Cursor movement keys

These keys allow you to move around in your document in command mode.

Key	Operation
h or **[Left Arrow]**	Moves the cursor position one space to the left.
j or **[Down Arrow]**	Moves the cursor position one line down.
k or **[Up Arrow]**	Moves the cursor position one line up.
l or **[Right Arrow]**	Moves the cursor position one space to the right.
$	Moves the cursor position to the end of the current line.
w	Moves the cursor position forward one word.
b	Moves the cursor position back one word.
e	Moves the cursor position to the end of the word.
0 (zero)	Moves the cursor position to the beginning of the current line.
[Return]	Moves the cursor position to the beginning of the next line.
[Spacebar]	Moves the cursor position one space to the right.
[Backspace]	Moves the cursor position one space to the left.

The quit commands

With the exception of the **ZZ** command, these commands start with **:**, and you must end a command line with [Return].

Key	Operation
wq	Writes (saves) the contents of the buffer and quits the vi editor.
w	Writes (saves) the contents of the buffer but stays in the editor.
q	Quits the editor.
q!	Quits the editor and abandons the contents of the buffer.
ZZ	Writes (saves) the contents of the buffer and quits the vi editor.

The search commands

These keys allow you to search forward or backward in your file for a pattern.

Key	Operation
/	Searches forward for a specified pattern.
?	Searches backward for a specified pattern.

Cut-and-paste keys

These keys are used to rearrange text in your file.
They are applicable in vi's command mode.

Key	Operation
d	Deletes a specified portion of the text and stores it in a temporary buffer; this buffer can be accessed by using the put operator.
y	Copies a specified portion of the text into a temporary buffer; this buffer can be accessed by using the put operator.
P	Places the contents of a specified buffer above the cursor position.
p	Places the contents of a specified buffer after the cursor position

Scope keys

Using the vi commands in combination with the scope keys gives you more control in your editing tasks.

Key	Operation
$	The scope is from the cursor position to the end of the current line.
0 (zero)	The scope is from just before the cursor position to the beginning of the current line.
e or **w**	The scope is from the cursor position to the end of the current word.
b	The scope is from the letter before the cursor backward to the beginning of the current word.

Paging keys

The paging keys are used to scroll a larger portion of your file.

Key	Operation
[Ctrl-d]	Scrolls the cursor down toward the end of the file, usually 12 lines at a time.
[Ctrl-u]	Scrolls the cursor up toward the beginning of the file, usually 12 lines at a time.
[Ctrl-f]	Scrolls the cursor down (forward) toward the end of the file, usually 24 lines at a time.
[Ctrl-b]	Scrolls the cursor up (backward) toward the beginning of the file, usually 24 lines at a time.

Setting the vi environment

You can customize the behavior of the vi editor by setting the vi environment options. You use the **set** command to change the options' values.

Option	Abbreviation	Operation
autoindent	**ai**	Aligns the new lines with the beginning of the previous ones.
ignorecase	**ic**	Ignores the uppercase/lowercase difference in search operations.
magic		Allows the use of the special characters in a search.
number	**nu**	Displays line numbers.
report		Informs you of the number of lines affected by the last command.
scroll		Sets the number of lines to scroll when [Ctrl-d] command is given.
shiftwidth	**sw**	Sets the number of spaces to indent; used with the autoindent option.
showmode	**smd**	Displays the vi editor modes in the right corner of the screen.
terse		Shortens the error messages.
wrapmargin	**wm**	Sets the right margin to a specified number of characters.

APPENDIX E

The ASCII Table

Character/Key	Decimal	Hex	Octal	Binary
CTRL-1 (NUL)	0	00	000	0000 0000
CTRL-A	1	01	001	0000 0001
CTRL-B	2	02	002	0000 0010
CTRL-C	3	03	003	0000 0011
CTRL-D	4	04	004	0000 0100
CTRL-E	5	05	005	0000 0101
CTRL-F	6	06	006	0000 0110
CTRL-G (BEEP)	7	07	007	0000 0110
CTRL-H (BACKSPACE)	8	08	010	0000 1000
CTRL-I (TAB)	9	09	011	0000 1001
CTRL-J (NEWLINE)	10	0A	012	0000 1010
CTRL-K	11	0B	013	0000 1011
CTRL-L	12	0C	014	0000 1100
CTRL-M (RETURN)	13	0D	015	0000 1101
CTRL-N	14	0E	016	0000 1110
CTRL-O	15	0F	017	0000 1111
CTRL-P	16	10	020	0001 0000
CTRL-Q	17	11	021	0001 0001
CTRL-R	18	12	022	0001 0010
CTRL-S	19	13	023	0001 0011

Character/Key	Decimal	Hex	Octal	Binary
CTRL-T	20	14	024	0001 0100
CTRL-U	21	15	025	0001 0101
CTRL-V	22	16	026	0001 0110
CTRL-W	23	17	027	0001 0111
CTRL-X	24	18	030	0001 1000
CTRL-Y	25	19	031	0001 1001
CTRL-Z	26	1A	032	0001 1010
CTRL-[(ESCAPE)	27	1B	033	0001 1011
CTRL-\	28	1C	034	0001 1100
CTRL-]	29	1D	035	0001 1101
CTRL-^	30	1E	036	0001 1110
CTRL-_	31	1F	037	0001 1111
SP (SPACEBAR)	32	20	040	0010 0000
! (EXCLAMATION MARK)	33	21	041	0010 0001
" (DOUBLE QUOTATION MARK)	34	22	042	0010 0010
# (NUMBER SIGN)	35	23	043	0010 0011
$ (DOLLAR SIGN)	36	24	044	0010 0100
% (PERCENT SIGN)	37	25	045	0010 0101
& (AMPERSAND)	38	26	046	0010 0110
' (SINGLE QUOTATION MARK)	39	27	047	0010 0111
((LEFT PARENTHESIS)	40	28	050	0010 1000
) (RIGHT PARENTHESIS)	41	29	051	0010 1001
* (ASTERISK)	42	2A	052	0010 1010
+ (PLUS SIGN)	43	2B	053	0010 1001
, (COMMA)	44	2C	054	0010 1100
- (HYPHEN/MINUS SIGN)	45	2D	055	0010 1101
. (PERIOD)	46	2E	056	0010 1110
/ (SLASH)	47	2F	057	0010 1111
0	48	30	060	0011 0000

Character/Key	Decimal	Hex	Octal	Binary
1	49	31	061	0011 0001
2	50	32	062	0011 0010
3	51	33	063	0011 0011
4	52	34	064	0011 0100
5	53	35	065	0011 0101
6	54	36	066	0011 0110
7	55	37	067	0011 0111
8	56	38	070	0011 1000
9	57	39	071	0011 1001
: (COLON)	58	3A	072	0011 1010
; (SEMICOLON)	59	3B	073	0011 1011
< (LESS THAN SIGN)	60	3C	074	0011 1100
= (EQUAL SIGN)	61	3D	075	0011 1101
> (GREATER THAN SIGN)	62	3E	076	0011 1110
? (QUESTION MARK)	63	3F	077	0011 1111
@ (AT SIGN)	64	40	100	0100 0000
A	65	41	101	0100 0001
B	66	42	102	0100 0010
C	67	43	103	0100 0011
D	68	44	104	0100 0100
E	69	45	105	0100 0101
F	70	46	106	0100 0110
G	71	47	107	0100 0111
H	72	48	110	0100 1000
I	73	49	111	0100 1001
J	74	4A	112	0100 1010
K	75	4B	113	0100 1011
L	76	4C	114	0100 1100
M	77	4D	115	0100 1101
N	78	4E	116	0100 1110
O	79	4F	117	0100 1111

Character/Key	Decimal	Hex	Octal	Binary
P	80	50	120	0101 0000
Q	81	51	121	0101 0001
R	82	52	122	0101 0010
S	83	53	123	0101 0011
T	84	54	124	0101 0100
U	85	55	125	0101 0101
V	86	56	126	0101 0110
W	87	57	127	0101 0111
X	88	58	130	0101 1000
Y	89	59	131	0101 1001
Z	90	5A	132	0101 1010
[(LEFT SQUARE BRACKET)	91	5B	133	0101 1011
\ (BACKSLASH)	92	5C	134	0101 1100
] (RIGHT SQUARE BRACKET)	93	5D	135	0101 1101
^ (CIRCUMFLEX)	94	5E	136	0101 1110
_ (UNDERSCORE)	95	5F	137	0101 1111
` (GRAVE ACCENT MARK)	96	60	140	0110 0000
a	97	61	141	0110 0001
b	98	62	142	0110 0010
c	99	63	143	0110 0011
d	100	64	144	0110 0100
e	101	65	145	0110 0101
f	102	66	146	0110 0110
g	103	67	147	0110 0111
h	104	68	150	0110 1000
i	105	69	151	0110 1001
j	106	6A	152	0110 1010
k	107	6B	153	0110 1011
l	108	6C	154	0110 1100
m	109	6D	155	0110 1101
n	110	6E	156	0110 1110
o	111	6F	157	0110 1111

Character/Key	Decimal	Hex	Octal	Binary
p	112	70	160	0111 0000
q	113	71	161	0111 0001
r	114	72	162	0111 0010
s	115	73	163	0111 0011
t	116	74	164	0111 0100
u	117	75	165	0111 0101
v	118	76	166	0111 0110
w	119	77	167	0111 0111
x	120	78	170	0111 1000
y	121	79	171	0111 1001
z	122	7A	172	0111 1010
{ (LEFT BRACE BRACKET)	123	7B	173	0111 1011
¦ (VERTICAL LINE)	124	7C	174	0111 1100
} (RIGHT BRACE BRACKET)	125	7D	175	0111 1101
~ (TILDE)	126	7E	176	0111 1110
DEL (DELETE KEY)	127	7F	177	0111 1111

Index